ARCHAEOMETALLURGY
IN INDIA

ARCHAEOMETALLURGY IN INDIA

Editor

Vibha Tripathi

1998

Sharada Publishing House

Delhi - 110035

First Published 1998

ISBN : 81-85616-46-9

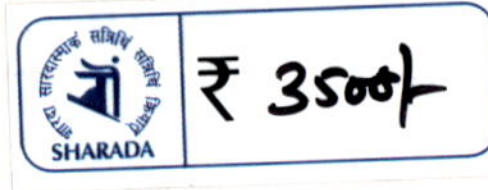

Published by B.L. Bansal for
SHARADA PUBLISHING HOUSE
40, Anand Nagar, Inderlok, Delhi - 110035
Phones: 5117390, 5116763
Typeset by Neographics
1811 Gyani Bazar, Kotlamubarakpur, New Delhi - 110003
and Printed at Santosh Offset,
Delhi - 110035

Dedicated to
MAHAMANA PT. MADAN MOHAN MALAVIYA
The Founder of
Banaras Hindu University
whose vision of bringing together
Pratichi-Prachi—The Modern and Ancient
Inspired this venture

Contents

Foreword xi
Introduction xiii
Prologue xv
Inaugural Address xxv
List of Contributors / Participants xxxi

SECTION I
MINING & METALLURGY

1. Archaeo-Materials in India: Some Residual Problems and Future Prospects
ARUN KUMAR BISWAS 3

2. The Metallurgical Tradition of the Harappans
D.P. AGRAWAL AND RAJAM SESHADRI 10

3. Transition from Neolithic to Copper Age at Kushadwip, District Bankura, West Bengal Archaeometallurgical Studies
SUDHIR CHANDRA DUARI AND PRANAB K. CHATTOPADHYAY 17

4. Metallographic and Chemical Study of Copper Artifacts from Senuwar
PRANAB KUMAR CHATTOPADHYAY AND BIRENDRA PRATAP SINGH 20

5. Metallographic Studies of Copper Antiquities from Manjhi (District Saran, Bihar), 400 B.C. - 300 A.D.
HARI NARAIN, JAI PRAKASH, M.V. NAIR AND O.P. AGRAWAL 27

6. Iron Technology at the Chalcolithic Phase of West Bengal
ASOK DATTA 36

7. Metallographic Studies of Iron Artifacts of Domestic Use, Tools and Weapons from Sringverpura (250 B.C. - 600 A.D.) Uttar Pradesh
HARI NARAIN, JAI PRAKASH, O.P. AGRAWAL AND M.V. NAIR 44

8. Iron Technology: Its Role in the Historical Development in the Morar Valley, M.P
R.P. PANDEY 54

9. Metallurgical Study of Ancient Swords
B. PRAKASH 64

10. Glory of Indian Iron and Steel Technology in Archaeology
SYED JAFAR MAHMUD 81

11. Studies and Archaeometallurgical Significance of Some Gold, Silver and Copper Coins from Rajghat (Varanasi) Excavations — A Brief Review
SOMNATH MISRA AND T.N. TIWARI 93

12. The Development of Zinc Technology in Ancient Gujarat
SWARNA KAMAL BHOWMIK 104

13. Ancient Industry of Lead, Silver & Zinc in Rajasthan
P.T. CRADDOCK, I.C. FREESTONE, LYNN WILLIES, H.V. PALIWAL, L.K. GURJAR AND K.T.M. HEGDE 108

14. Metals in Indian Medicine
K.N.P. RAO 134

15. A Scientific Analysis of Iatro-Metallurgical Practices of Ancient Indians
K. NAGESH RAO AND P.R. RANGESH 141

16. An Historical Review on Ancient Mining with Special Reference to India
R.D. SINGH 148

17. The Ecology and Archaeology of the Seruwila Copper-Magnetite Deposit, North East Sri Lanka
SUDHARSHAN SENEVIRATNE 156

18. Bronze Tools and Technology in Protohistoric Afghanistan: A Review
V.C. SRIVASTAVA 176

19. Probable Sources of Silver in Ancient India: A Historico-Scientific Approach
BHAGIRATHI PRASAD AND NISAR AHMAD 184

20. Mineral Resource: A Possible Archaeometallurgical Correlation
AMIT TRIPATHI AND VIBHA TRIPATHI 191

SECTION II

ETHNO-ARCHAEOLOGICAL EVIDENCE ON ARCHAEOMETALLURGY

21. Iron Working in Ancient India - A Case Study
VIBHA TRIPATHI 209

22. Evaluation of Earlier Ironmaking Processes and their Relevance in Indian Context
B.B. AGRAWAL, K.K. PRASAD, G.I.S. CHAUHAN
P.K. CHAUDHURI AND S.M. AERON 218

23. Revival of Rural Ironmaking in Orissa
N.R. SRINIVASAN 228

24. High Tin Bronze Working in India : The Bowl Makers of Kerala
SHARADA SRINIVASAN 241

25. Metal Working in Medieval Assam
R. DASGUPTA 251

26. Is Iron Appearing in Archaeological Context in Eastern India A Product of Tribal Expertise? An Ethno-Archaeological Enquiry
BHUPENDRA PAL SINGH 263

27. Iron Metallurgy in Ancient Madhya Pradesh
K.D. BAJPAI 274

28. Metal Implements of Northern India (C. 1000 B.C. to 500 B.C.)
J.P. UPADHYAYA 278

29. Newly Discovered Anthropomorphic Figures from Nurpur, U.P.
MADHURI SHARMA AND D.P. SHARMA 286

30. Two Unique Lugged Shouldered Copper Axes from the National Museum of India Collection
D.P. SHARMA 292

SECTION III

LITERARY EVIDENCE ON ARCHAEOMETALLURGY

31. A Literary Perspective on Archaeomaterials and Scientific Concepts in Ancient India
SULEKHA BISWAS AND ARUN KUMAR BISWAS 299

32. Early Indian Literature as a Source of Archaeometallurgy
P.K. AGRAWALA 307

33. Antecedent Stages in the Evolution of Metal Technology
V.S. PATHAK AND PREM SAGAR CHATURVEDI 312

34. Literary References on Metals, Metallic Objects of Art and Metal Technology
SWARNA KAMAL BHOWMIK AND MUDRIKA JANI 319

35. Iron in Early Buddhist Text
PRIYA SRIVASTAVA AND VIBHA TRIPATHI 339

36. 'Ayas' in the Rigveda - A Note
D.N. TRIPATHI 346

37. The Search of Metals and Techniques for Minting Earliest Coins in India
B.R. MANI 350

SECTION IV

SOCIO - ECONOMIC IMPLICATIONS OF METALS

38. Social Implications of Technology : A Study of Iron in pre- C. 200 B.C. India
VIJAY KUMAR THAKUR 359

List of Illustrations 387

Illustrations

Foreword

India had been at the fore fronts of metallurgical skills since time immemorial. The expertise is manifest in some of the famous reminiscences like the Delhi Iron Pillar, the Dhar Iron Pillar, the beams a Konark. Their mastery in the craft is well recognised now with the evidence of Zinc production as early as 3rd century B.C. while the rest of the world learnt the technique only a couple of centuries back. Our knowledge about these skills is still far from satisfactory. To understand the enigmatic issues involved - both Technological and Archaeological - in ancient Indian metallurgy, a National Seminar on Indian Archaeometallurgy was organised in the Deptt. of A.I.H.C. & Arch. in collaboration with the Deptt. of Metallurgical Engineering, I.T., BHU.

Banaras Hindu University has a pioneering tradition in multi-disciplinary studies. This spirit is imparted to it by its Founder Mahamana Pt. Madan Mohan Malaviya Ji who believed in combining the wisdom of the ancient Indian Learning with the modern scientific advancements. This was the first such Seminar in the country on this subject. It aimed at bringing together experts from diverse fields like Indology, Archaeology, History, Ethnology and Metallurgy. For a worthwhile reconstruction of history of metals and metallurgy in India, a free flowing dialogue amongst the experts of these diverse disciplines is a pre-requisite. It was with this view that this seminar was organised.

We, at Banaras Hindu University are fortunate to have two well established departments both being leading institutes of its kinds in the country. If we join our expertise and resources, we are likely to be at the fore fronts of Archaeometallurgical researches which has yet to come of age in India.

I compliment and congratulate Dr. V. Tripathi, Reader in Ancient Indian History, Culture and Archaeology who has taken a lead in this direction and for having produced this outstanding volume.

HARI GAUTAM
MS FRCS (Edin) FRCS (Eng)
FAMS FACS FICS FIACS
Vice-Chancellor

Introduction

The archaeological community has to be beholden to the BHU, for organising this very wonderful seminar where ancient metal technology of India could be discussed from different angles - in a multidisciplinary perspective. Lets admit that such seminars are few and far between in India and therefore this effort is all the more laudable.

The papers presented were not only on copper and iron but also on other metals. But technological aspects and archaeological significance of the findings received equal emphasis. The fact that ethnology may have much to explain and elucidate the ancient metallurgical practices was fully realised as is evident from the significant contributions on this aspect. India has a hoary literary tradition. We have to glean facts about archaeometallurgy from ancient literary sources too. I am happy to see important papers covering this much neglected aspect.

What is more encouraging is that younger scholars are taking equal, if not more, interest in these technical studies. There are a number of outstanding problems in this field and one can reconstruct the story of ancient metal technology of India only with the active interest and contributions of the younger generation.

In the field of copper technology we do not know which mineral sources were actually used. Some work has been done, including mine, on trace impurity patterns of ores and artefacts, but I must admit that its not enough. We need to carry out comparison of lead isotope ratios between artefacts and minerals to be able to locate actual mineral belts used by the different protohistoric cultures. We have the required mass spectrometers in India but there is still a gulf between the two cultures which does not allow the use of such technologies for archaeometallurgy. It needs to be emphasised that India today can boast to have the latest technologies in a number of fields which can be very profitably used for archaeological research. There is enough interest among the scientists but we have yet to evolve a common language between scientists and archaeologists so that all the latest technologies are harnessed for the sake of archaeology.

One much neglected area of study is the study of ancient mining technology. In Rajasthan there are a large number of ancient workings but they are still awaiting a systematic study. In contrast, for example, the Chinese have worked out elaborate details about the ancient mining technology in China.

The situation is worse as far as the iron technology is concerned. We know very little about the ores used, about the history of casting in India, history of carburisation and so forth. We do need a centre where such studies could be carried out in an interdisciplinary environment. At BHU we have both—an excellent department of metallurgy and a reputed group of archaeologists. BHU could thus prove an excellent choice for such a centre to be nucleated. Unless we have such a centre we would not be able to unravel the history of ancient metal technology in India in all its dimensions. I was much impressed by the effort of the Chinese in this field. In a meeting I attended, there were 40 metallurgical groups represented. Naturally the amount of work they have been able to carry out in this field is simply staggering.

I wish such seminars are held more often and lead to closer collaboration between the scientists and the archaeologists which alone will usher into an era of an integrated yet multidisciplinary research in India. We have great hopes from the BHU group to lead us all in this much neglected field.

10, Jayanti Lal Park,
Amli-Bopal Road,
Ahmedabad - 380 058.

D.P. AGRAWAL
ICHR National Fellow

Prologue

The high level of Indian metallurgical skill has been acclaimed right since the ancient times. Some of the famous examples of advanced indigenous metallurgical technique are still available to us in form of 7.26 m. high, 6 ton Delhi iron pillar of 5th-6th century A.D. standing unrusted in open air for centuries, the impressive beams used in temple structure at Konark as early as 10th century A.D. or the famous Dhar pillar of 10th-11th century A.D. The detailed description of metallurgical processes and their intricate properties in the 10th-11th century alchemy-text, the *Rasaratna Samuchchaya* is yet another proof of a high level of knowledge of metallurgy. This tradition continued in varied forms right upto the 18th century. The glorious past of history of ancient metallurgy still remains enigmatic. The significance of metals in rise of civilizations cannot be minimised. It is indeed in recognition to the role played by metals in the rise of human civilizations that the labels attributed to different stages of cultural growth are after dominant metals being used at that particular time, viz. the Copper-Bronze Age or Iron Age etc. During recent years with increasing application of scientific knowledge and new techniques of investigation and analysis in the field of archaeology, archaeometallurgy has become an independent branch of study. In India, however, the archaeometallurgical studies are still in infancy. We have yet to reconstruct a systematic history of ancient metallurgy in India.

Metallurgy in the Indian subcontinent dates back to circa 6000 B.C. with occurrence of a few of copper objects, from the early phases of Mehargarh (Pakistan). Gold had been used in Neolithic context at Tekkalkota along with a few objects of copper in peninsular India. Chronologically the occurrence of gold from the Harappan levels is earlier but its use by neolithic community is significant indeed. The proximity of the sites with rich gold mines lead to recognition and exploitation of the shining precious metal which could easily be shaped into desireable form with elementary technological know-how which they seem to have possessed. Coming back to the Indus valley, it is well known that they had mastered complex metallurgical skill of forging, casting and alloying minor metals like lead, tin and arsenic. These metals were used as alloy with copper, silver and gold or independently. In a different ecological zone though in same time frame as the

Indus valley, we came across the Copper Hoards of the Gangetic Plains. The heavy copper objects were hoarded in large sachets mostly in the Upper Gangetic Plains, also in plateau of Bihar, Bengal, Orissa, a region which understandably was also the source of copper used in the manufacture of these artifacts. Metallurgically, the Copper Hoard objects do not exhibit a high level of technological skill. Mostly these are bulky objects of pure copper with a few specimens displaying arsenic alloying. The fact that these objects were valuable commodities rather than utilitarian objects of day-to-day life is borne out by 1. the way they were hoarded, 2. virtual absence of wear marks on the edges of these well preserved objects, 3. the massiveness of objects in weight, height and length. In his 'Metalwork of the Bronze Age in India', an excellent study of copper bronze objects, Paul Yule has classified several of the axes and celts as ingots deducts and "the appearance of harpoons, exaggerated weight and size, all present grounds to doubt their primary utilitarian function". A later contemporary of the Indus valley, and the Copper Hoards are the chalcolithic cultures scattered in plethora of tiny settlements in Gujarat, Rajasthan, Madhya Pradesh and Deccan. Most of these cultural groups appear to be independent cultural pockets secluded by natural barriers having geographical interactions within their ecological limits. Inter-regional relations are hardly discernible in material remains. The technological skill is also largely area-exclusive. The sites closer to the ore deposits fully exploited them and supplied them to the neighbouring settlers. Ahar and Ganeswar in Rajasthan may be good examples of the case. Similarly, the Central Indian and Deccan folks exploited their own local resources which were rather scanty. Technologically, they do not stand a comparison with the Indus coppersmithy. The blow holes or the visible joints in the cast objects bear testimony to their elementary level of the metallurgical skill. Alloying was known but the total percentage of alloyed objects is very low.

It is in this techno-cultural milieu that iron makes its earliest emergence in India. Whether iron came as a result of diffusion from west or grew independently is still an issue not resolved to the satisfaction of all. But the cultural milieu of the appearance of iron is hardly conducive of western contact i.e., from Iran-Afghanistan borders. Chronologically, the bordering lands do not show an earlier beginning of iron-having a date around 1100-1000 B.C., Culturally, there is little by way of pottery forms or designs, habitational remains or settlement system to suggest interaction of any kind at this point of time. Even the tool-types at the earliest stage found in India are too elementary compared to the Iranian cemeteries where the earliest iron occurs. The scale tilts heavily in favour of an independent origin of iron in India, in every likelihood as an off-shoot of copper technology. The elementary level of technological skill of pre-iron cultures as well as the scarcity of copper available to these cultural groups must have constantly goaded them to search for newer sources of metal. Experiments with alternative ores could have led to discovery of iron. There exists a greater possibility that "iron smelting was not invented by advanced copper smelters" as comments Tylecote. The Harappan artisans were too well adept in copper technology to commit mistakes. They successfully fluxed out iron oxide from the sulphide ores that they used (chalcopyrite). We have come across such discarded lumps of iron on several Afghanistan sites of Indus valley period as described by Jim Shaffer in 'Bronze Age iron in Afghanistan' etc. Wherever the coppersmiths were not careful enough some low percentage of iron has got into the copper objects as evidenced 1.22 and 6.48 percent iron in Ahar copper objects and also in object from Rangpur (with 1.88 percent Fe.). The chalcolithic coppersmiths of Rajasthan, Madhya Pradesh, Bihar and Bengal might have been the likely groups to have experimented with variety of locally available ores. Chalcolithic cultures like Mangalkot, Hathigra, Pandurajar Dhibi in Bengal, Chirand in Bihar or the chalcolithic societies of Eran and Prakash in Madhya Pradesh and Maharastra etc., manifest an early use of iron, some of them coming from the earliest horizons of chalcolithic culture. The early evidence of iron at Ahar in Dist. Udaipur has been raised by Sahi. No serious effort has been made to controvert the issue by the excavators so far leading to an acceptance of iron in 1200-1300 B.C. by archaeologists like Allchin.

Turning to the growth of technology, it remains at an elementary level for a long time, confined to production of wrought iron of high purity. It is at a much later stage, nearly after four five centuries that the metallurgy shows signs of improvement. Carburization followed by tempering, quenching appear around 500-400 B.C. This is synchronised by a social adaptation of iron technology. There is a diversification in tool typology at this juncture. Could it be because of an improved technological skill capable of production of better tools and implements? The social adaptation of technology must have led to technological - innovations the two being inter-linked closely which in turn must have accelerated economic growth.

The noble metals, both gold and silver trace their antiquity to the Indus valley civilization. But more interesting is the use of gold ornaments by the Neolithic settlers of South India though chronologically they may not be earlier than the Harappans but culturally and contextually it is significant, indeed. It is also significant for the reason that the settlements lie in close proximity of gold mines. It is this very region which is said to have supplied gold to the chalcolithic cultures of Deccan. The time frame of the two is not far apart.

Zinc was first produced in its metallic form in India. The very complex procedure involved in smelting of zinc may be responsible for its late introduction and utilization among the metals used in the antiquity. Traces of ancient mining and metallurgical processes have been located in abundance near Zawar mines in Rajasthan. Extensive field work and laboratory investigations in ancient zinc metallurgy have been undertaken by the British Museum and Hindustan Zinc Ltd. leading to proper understanding of its technology. It is high time that a proper appraisal is made of growth of metallurgical studies upto now and a direction is given for further studies in archaeometallurgy.

A national seminar was organised on Archaeometallurgy in the Dept. of Ancient Indian History, Culture and Archaeology at Banaras Hindu University in Oct. 1992 with an aim to take stock of existing knowledge in the field of study of metals. It was an endeavour to bring together specialists of related areas to start a healthy trend of interaction for an interdisciplinary research on the subject.

Important contributions have been made on various aspects of metals by eminent scholars. But so far, the nature of such investigations are generally confined either to the so-called important cultural periods like the Indus Valley or the Iron Age. Studies are limited to individual metals like copper-bronze, iron etc. and are more descriptive and less analytical. It is high time that the efforts are intensified to analyse metal objects from different cultural horizons and studied together jointly by metallurgists and archaeologists. This kind of study need to be substantiated by literary data. Only through such a concerted effort, a complete picture of growth of metallurgy and a concise history of metals will emerge. In the present volume which is a result of the deliberations of the National Seminar on Indian Archaeometallurgy, the views of historians, archaeologists, indologists, ethnologists and metallurgists find place. The discussions focused on a wide cross-section of issues expanding over a period of thousands of years—right from the earliest stages of advent of metallurgy to the prevalance of pre-industrial metal working till recent past in several parts of India. The Seminar also touches on diverse metals—from the utilitarian copper, iron and their alloys to the noble metals. All possible sources viz. archaeology, literature, ethnology or historical records have been tapped to retrieve data on the subject of metals and metallurgy. A dialogue has thus been initiated amongst the scholars engaged is research in these diverse areas.

The deliberations cover all the metals which were being used in the antiquity, covering a large span of time and place. There are papers on Harappan metallurgy (Agrawal & Seshadri) and Copper-Hoards (Sharma). The paper by S.C. Duari, an enthusiastic amateur in the field of archaeology reflects a transition from neolithism to chalcolithism. The chalcolithic culture had so far been a neglected period, the paper by P.K. Chattopadhyay and Birendra Pratap Singh provides an insight into the intricacies of Chalcolithic Copper metallurgy at Senuwar in District Rohtas in Bihar. A development in metallurgical skill from Chalcolithic Senuwar to early-historic Manjhi can easily be made out by the paper by Hari Narain and Jai Prakash etc. Iron has naturally drawn greater attention compared to other metals. Iron technology has been discussed at its beginning at the Chalcolithic phase of West

Bengal by Asok Datta and in the Morar Valley of Madhya Pradesh by R.P. Pandey. The period extends further in the study of ancient swords of recent centuries by Bhanu Prakash. The uniform structure as visible in the micrograph indicates the mastery of the art of steel making of indigenous craftsmen. It has aptly been eulogised in Jafar Mahmud's article 'Glory of Indian Iron & Steel' tracing back the metal technology from the Ṛgvedic times. Coinage is a comparatively late advent in the history of Indian civilization. Special metallurgical techniques were involved in minting of coinage. Results of analysis of gold, silver and copper coins from Rajghat excavations have been discussed by Somnath Misra & T.N. Tiwari. It is now commonly believed that India is the earliest user of zinc in its metallic form. The paper covers literary as well as other sources for highlighting the evidence of zinc. Equally exhaustive is the paper by Craddock, Freestone, Paliwal, Gurjar et. al. on the evidence of lead, silver and zinc in Rajasthan, one of the ancient most zinc mines ever exploited by man.

With the growth of metallurgy, the nature and properties of different metals and minerals were understood and exploited in ancient India. A full-fledged system of alchemy grew to cure diseases, inventing intricate details for curing and maturing metallic properties for specific needs. The subject has been expertly dealt with by K.N.P. Rao in 'Metals in Indian Medicine'. Whether the metals have any somatic appeal becomes clear from the paper by Nagesh Rao and Rangesh in 'A Scientific analysis of Iatro Metallurgical Practice of Ancient India'.

The evidence on mining in ancient time is not very exhaustive, perhaps because with abundance of ores, they could be prospected and collected easily to meet the limited requirements of early Indian societies. However, some evidence is forthcoming with growing interest in the field and also with involvement of mining engineers in history of sciences and technology. Mining shafts and pits are gradually coming to light. Sudharshan Seneviratne has dealt with the evidence of coppermagnetite deposits in north-east Sri Lanka and correlated it with the ecological and archaeological factors. The paper is significant from the view point of resource zone study in the field of archaeometallurgy. Like Sri Lanka in the South, Afghanistan is another neighbouring country in the North having a vital cultural link with India from the earliest times. V.C. Srivastava's paper reviews the 'Bronze Tools & Technology' in that region. Silver is amongst the scarce metals in India, though was being used from an early context. The probable source of silver have been studied by Bhagirathi Prasad & Nisar Ahmad. The subject of mineral resources and its correlations with archaeological evidence has been brought out by Amit Tripathi and Vibha Tripathi in an attempt to highlight the significance of this kind of approach in the present archaeological researches in India.

The section on Ethno-archaeological evidence on archaeometallurgy focuses on ethnological and archaeological data. The significance of ethnological evidence to understand ancient technology can not be over emphasised. Seeing the large size of India and number of metal-working societies that inhabit it, the number of papers on this subject is dismally small, but as a first effort, it may be worthwhile. My own paper 'Iron working is Ancient India : A case study' is an attempt to reconstruct ancient iron working on the basis of archaeological and ethnological evidence. The metallurgists at Steel Authority of India have taken a keen interest in the pre-industrial iron working which inhabitated the area covered by rich ore deposits. They have even promoted the traditional working. It coexists side by side with a modern steel plant. The paper by B.B. Agrawal et. al. on 'Evaluation of earlier iron making processes and their relevance in Indian context' is an account of traditional iron working surviving in Bihar and adjacent borders of Madhya Pradesh.

N.R. Srinivasan has reviewed the evidence which survives in Orissa and has studied the details of their technology and tried to work successfully on their model. The results are presented here in 'Revival of rural Ironmaking in Orissa'. Bhupendra Pal Singh has raised relevant problems in ethno-archaeological evidence in iron working in eastern India. In eastern India itself but around Assam, R. Dasgupta reviews the metal art as it existed during medieval period. The evidence of bronze making in South India, specially Kerala has been highlighted by Sharada Srinivasan in her paper 'High Tin Bronze working in India—the bowl makers of Kerala'.

Researches in Indian Archaeology have not yet given sufficient emphasis to study of metals. It is also borne out by the number of papers based on purely archaeological evidence in the present volume. The excavators have yet to realise the significance of metal objects as an independent commodity to be correlated with life of common man at a given period of time. A very encouraging response came from (Late) Prof. K.D. Bajpai of Sagar who contributed his last paper on the evidence of 'Iron metallurgy in Ancient Madhya Pradesh'. He was large-hearted enough to have agreed to the suggestion on incorporating some more data from the famous site of Eran but as a great misfortune of Indian archaeology, he passed away without fulfiiling his promise to us. With great reverence and gratitude, that paper is being included here without any alterations.

Metal objects from excavations of North Indian sites have been studied by J.P. Upadhyaya. Objects of Copper-Hoard family have been dealt with in detail by Madhuri Sharma and D.P. Sharma in separate papers. Their identifications and correlations generated heated discussions in course of discussions during the Seminar.

The third section is devoted to literary evidence on Archaeometallurgy. The rich literary tradition of India has great potentials of providing invaluable insight into the subject of metal technology and amount and types of objects being used in day-to-day life or on special occasions. Interesting discussions followed the papers opening new horizons to further researches in the area. Sulekha Biswas and A.K. Biswas conceptualised in 'A Literary perspective in archaeomaterials and scientific concepts in Ancient India'. P.K. Agrawala threw light on significance of literary data as source of understanding metallurgy and the appropriate methodology to be adopted. The paper by V.S. Pathak & Prem Sagar Chaturvedi traces 'Antecedent stages in the evolution of metal technology'. How the pyrotechnology evolved from pottery kilns to metal furnaces has been traced affectively through literary evidence. Silver, one of the rare metals and with fewer sources in India surprisingly has its beginnings in Indo-European languages tracable through Rigveda to subsequent stage in common words and sources in Iran and Afghanistan by V.S. Pathak and J.N. Tiwari. The other important paper is by V.S. Pathak and Maheshwari Prasad on sources of gold and the origin of Gupta gold coinage. These two papers, however could not be included in the present volume. Various literary references are available in ancient literature on metals, metallic objects of art and metal technology which find place in the paper of Suwarna Kamal Bhowmik and Mudrika Jani. The paper by Priya Srivastava & Vihba Tripathi focuses on evidence of metals in early Buddhist Literature. While dealing with the types of objects used in the contemporary society, the paper attempts of scoop up data on metallurgical processes too. The earliest Indian text, Rigveda uses the term Ayas which has assumed controversial denominations. It has been taken up in the note 'Ayas in Rigveda' by D.N. Tripathi.

Last, but not least is the section —The socio-economic implication of adaptation of metals. Unfortunately despite assurances and presentation of abstracts of papers on this theme, some esteemed scholars on the subject disappointed me by not contributing their final papers on the much debated subject. We have just one paper on this theme by Vijay Kumar Thakur. It compensates for lack of papers on the theme, as it is an exhaustive review paper and tackles the issue with thoroughness.

A panel discussion was organised at the concluding session of the Seminar. Most of the scholars participated in the discussion. Resolutions unanimously adopted are :

Centre of Archacometallurgy at Banaras Hindu University

It was resolved at the conclusion of the Seminar during panel discussion that a centre for archaeometallurgy be established at Banaras Hindu University in the Dept. of Ancient Indian History, Culture and Archaeology.

All over the world researchers in archaeometallurgy are being actively pursued. In India, however, such study is still in infancy. There are a few National Laboratories or individual scientists who are involved in conducting some researches in this field. It is high time that a subject of such vital significance to the cultural reconstruction be studied in India in a planned and systematic way. A centre of archaeometallurgy will go a long way in initiating and consolidating studies on ancient metals. Cultural labels like Copper-Bronze Age, Chalcolithic or Iron Age have seen assigned because of presence of a particular metal at a given time. Therefore, metals need to be studied in their cultural perspective, from technological view point and the socio-economic angles. It can be achieved by involving metallurgists, archaeologists, historians and experts on ancient Indian literary sources. The centre will fulfil this long overdue objectives of study. Following are its chief aims :

(1) To build a data base for researches on metal analysis. Excavators may be requested through the good offices of Archaeological Survey of India to provide samples for study to the centre.

(2) To conduct experimental metallurgy — by laboratory simulated programme of metal production on past pattern. A fuller understanding of ancient technology may thus be achieved.

(3) To undertake ethnological survey and invite ethnic metal working groups to demonstrate their traditional skill and make efforts to bring such societies to light and make a proper recording of their methods lest they are last to us.

(4) To make an indepth survey of ancient Indian texts to bring out data on the subject. It is only through such a co-ordinated effort that a total picture of ancient metal crafts can be redrawn.

(5) To form interaction groups, if possible in various parts of the country who conduct areawise research and then interact with each other to exchange information.

(6) Seminar/Symposia and Publications

Annual seminar on archaeometallurgical studies should be held and a regular publication regimen should be followed. Since expertise and basic infrastructure data are available in the fields of archaeology, metallurgy, historians, Sanskrit scholars all in one campus at Banaras Hindu University, it was unanimously agreed upon that this is the most appropriate place for such a centre. We propose to follow these measures to the best of our capabilities in future.

I am indeed grateful to all the scholars who participated in the seminar and contributed their valuable research papers. I am thankful to those also who initially submitted the abstracts of their papers and intended to participate but because of the disturbances that followed due to the report on Mandal Commission, had to cancel their trip at the last moment. Many of these papers are being included here.

I would like to acknowledge the blessings of very senior scholars like Prof. B.P. Sinha, Prof. N.R. Banerjee, Prof. R.S. Sharma, (now late) Prof. S.B. Deo with deep gratitude. I am deeply beholden to Prof. A.K. Narain, retired professor, University of Wisconsin, the versatile scholar of the senior generation for inaugurating the seminar with his characteristic inspiring note. I also take this opportunity to thank my colleagues in the department of Ancient Indian History, Culture & Archaeology of Banaras Hindu University for their co-operation, suggestions and guidance in organizing this Seminar. I am specially thankful to Sh. O.P. Tandon who allowed me to use the facilities of Bharat Kala Bhawan and helped me in every possible way. The then head of the Department Prof. P. Singh was a source of constant encouragement and help. I am deeply indebted to him. The Dept. of Metallurgical Engineering actively supported and participated in the Seminar in the spirit of collaborative inter-disciplinary research. I am thankful to Prof. H.C. Bharadwaj who had joined Centre of Advanced Study at Simla but actively associated with organisation of this Seminar. I am indeed very thankful to Prof. D.P. Agrawal for his guidance and inspiration at every stage. Prof. Bhanu Prakash of Deptt. of Metallurgical Engineering who, if I may be permitted to comment lightly, has changed his sides and loyalty and in now almost full time archaeometallurgist. He was constantly associated in organizing this Seminar, I am deeply indebted to him. I am also thankful to Sunil Kumar, who offered help whenever it was needed most. Last but not least are my thanks to my dynamic young colleague Dr. R.N. Singh who had taken upon himself all the responsibilities of organizational management sparing me to actively participate in the academic deliberations. I shall be failing in my duty if do not thank my students in the Department of AIHC & Arch. who constantly worked, even at odd hours during the course of the seminar. I express my deep gratitude to all those scholars who took the trouble of participating in discussion and contributing papers in the Seminar, making it a success. I am personally indebted to them. I own responsibility to all the omissions and commissions which have crept into this volume.

In the end I express my thanks to the Indian Council of Historical Research, New Delhi and Indian National Science Academy, New Delhi whose financial support made it possible to organise this National Seminar. I am indeed very grateful to the Chairman and Secretary of these bodies, who saw the significance of this kind of academic exercise and came up with generous financial support. I do hope that they will continue to support such ventures in future also.

In the end I pay my homage to my Guru Prof. K.K. Sinha and Prof. K.D. Bajpai and Prof. S.B. Deo who passed away before the publication of the proceeding of the Seminar on the Indian Archaeometallurgy. At the final stages of preperation of this volume, our colleague Prof. R. Dasgupta whose paper is included here, passed away. We mourn his untimely departure. The loss to the field of history of Art cannot be compensated. Their blessings, I am sure are there with me as a result of which the volume could acquire its present shape.

Department of A.I.H.C. & Arch.
Banaras Hindu University
Varanasi - 221005

VIBHA TRIPATHI

Inaugural Address

S.B. DEO
(Prof. Emeritus Deccan College, Pune)

Distinguished Colleagues & Friends,

I am indeed thankful to you all for the honour you have bestowed on me by asking me to inaugurate this Seminar on Archaeometallurgy and thus providing me an opportunity to have a meaningful dialogue and discussion with the distinguished scholars and scientists that are present here. More than anybody else, I am conscious of the fact that I am not a scientist, though I had the privilege of coming in contact with scientists specialized in different disciplines during the last four decades or more. This contact has been through the medium of archaeological research which in recent years has been depending heavily on allied sciences dealing with human activity and environment. I have had the privilege of knowing most of you and I must say that this has enriched my limited knowledge and expanded my horizons of understanding. It is therefore with a sense of gratitude to you all that I stand before you more with a desire to learn from you than to contribute.

It is but appropriate that a premier university like the BHU and its well-known department of Ancient Indian History, Culture and Archaeology should think of and plan a Seminar on archaeometallurgy. More than that, this geographical area has made the most outstanding contribution to Indian History and Culture as the Gangetic doab has virtually been the heart of Indian culture, history, religion, philosophy, archaeology and the unabated faith of the Indian people have immortalized Varanasi.

More than anything else, I guess, there is a special reason why this seminar has been organised here. It was Hastinapur in this region that first pinpointed the attention of scholars on the problem of the antiquity of iron and the debate still continues. Varanasi thus could have been the only choice for holding a seminar on archaeometallurgy.

It will be agreed that metals have played a very significant part in human life and culture; especially iron which has been termed as 'democratic metal' and a metal of war and peace. However, the role of metals in archaeological context did not go beyond a descriptive level for a very long time as our excavation reports would show. Several of the publications give more or less an inventory of metal objects. However, especially after independence, the emphasis and approach have changed. No more do we find the appendix on metal analysis as a mere ornamental padding to the report. In recent years, there has been a growing awareness of going beyond mere description and to lay emphasis on technology and the role it played in the then contemporary society.

This work has gained momentum because of universities and research institutions and their laboratories entering the field. A few universities and some research institutions have set up laboratories at which metallurgical and allied analysis is being carried out and younger scholars like D.P. Agrawal, O.P. Agrawal, H.C. Bhardwaj, K.T.M. Hegde, V.D. Gogte, V.T. Athavale, to mention a few have made a significant contribution to our understanding of archaeometallurgy.

However, it may be mentioned that most of the work done so far is site-oriented rather than having a larger framework with the result that most of such studies are in the form of sporadic articles or appendices in an excavation report. What one expects is an All India picture through the ages in respect of metal technology. Secondly, out of such a presentation, it would be worthwhile to attempt to understand the innovations in techniques and their usefulness to the society at large. It must be realised that technological innovations have the potential to initiate changes in socio-economic structure as also bring about a change in the value-system of a society.

It is indeed not an easy task to work out these aspects for it requires acquaintance with source material besides that from archaeological excavations. It is well-known that India has immense literary data which is distinguished by topical variety as also a wide chronological range. Though it has its limitations which are well-known to those who have studied it, yet there is indeed a mine of information. And the data begins with the Vedic literature and continues even in the late medieval period. The Rigveda, the Yajurveda and the Atharvaveda provide evidence of the use of various metals which implies the knowledge of metal technology. For instance, the Vajasaneyi Samhita (XVIII. 13) refers to gold, iron (ayas), dark metal (syamam), reddish metal (loham) and tin (trapu). There is no mention of 'Loha' in the Rigveda, whereas there seems to be some difference of opinion regarding the exact meaning of the term 'Ayas'. We are not concerned here with the controversy. Suffice is to say that metals are mentioned and a number of objects of gold which attest to the expertise in metal technology.

In the texts subsequent to the Vedic, there is lot of information regarding metal technology as for instance in the Upanishads, the Arthasastra of Kautilya, Panini, Susruta, Kamasutra, Varahamihira and several others. However, the archaeological data for this period has also been available. What is needed, if possible, is to compare and to integrate or otherwise these two categories of data. The scientists and the scholars of literary data could come together. The tendency to isolate each other will not be helpful for objective inquiry.

Yet another aspect needs to be emphasised. There are still itinerant smiths in various parts of rural India as also tribal pockets where metallurgy is still practised on the basis of traditional know-how. For instance, such practices are still in vogue in Rajasthan, Kutch, Maharashtra, Andhra-Karnataka, Bastar and Tamilnadu. For instance, Gururaja Rao states that "In south India ...... steel is produced by carburization of wrought iron which is filled into small clay crucibles along with a small quantity of fire wood or charcoal, then sealed with clay, placed into the furnace and heated. The molten liquid is allowed to become solid slowly, after absorbing sufficient quantity of carbon". It would be worthwhile to try this in a laboratory attached to an institution to test the veracity of this process as also to understand the economic viability of this method.

That metallurgy appears to have been practised as 'a rural craft' rather than an industry appears to have been the case in ancient India. The Sanskrit texts also refer to various technologies, different moulds, techniques of casting, the use of flux etc. The high quality of metal execution has also been referred to by foreign travellers visiting India in different periods. As early as the times of Alexander, Indian steel was highly valued because of its purity and sturdiness. Expertise passed on from generation to generation must have contributed to this excellence.

Against this background, the archaeological data needs to be assessed. As is well-known right from the Early Harappan times, we have a fairly good data in terms of the artefacts of copper, bronze, gold, silver, electrum and subsequently of iron. Not only that, at some sites we have evidence of the remains of smelting hearths as well. Some work in respect of the metal analysis, ore-source and the technique of manufacture has also been done. However, it appears that this analytical work has to be undertaken on a much larger scale than hitherto done. And there is justification for this as well. The regional extent of the Indus Civilization has been found to have been much more than that it was thought to be prior to Independence. Several regional chalcolithic cultures have come to light during the last few decades. These are post-Harappan in terms of chronology, though a few like Ahar and Kayatha as also possibly Sawalda seem to have been contemporary with late Harappan. Harappan impact seems to have been felt as far south as Maharashtra. Even in the Neolithic of Karnatak, metal artefacts like those in gold have been evidenced. The wide and varied occurrence of megalithic monuments in various parts of India, especially peninsular India, has been characterised by scores of objects of iron, copper, bronze, gold and electrum. This explosion of archaeological data and its potential has opened up new venues of researches in archaeometallurgy. It is also qualified by the fact that most of these data have been put in a more or less a precise chronological bracket so that studies in archaeometallurgy can now be carried out with reference to time and space.

Just as new discoveries have brought out immense data, it has also posed new problems and new perspectives. A case in instance may be placed before this august gathering which as a non-scientist, I would like to understand. Take the case of three Jorwe Culture sites like Jorwe, Nevasa and Chandoli. Here, whereas the bangles and the beads are made of almost pure copper, the axes are of lowgrade bronze, that from Jorwe having 1.73% tin and that from Nevasa showing 2.72% tin. It might appear that for sophisticated objects of ornamentation, pure copper was used whereas for axes it was not so, for obvious reasons related to their use. However, the technology adopted does not seem to have been of a uniform expertise as the axe from Chandoli shows poor casting resulting in porous surfaces. Does it mean that within the same cultural unit, technology differed ? Navdatoli in central India has given axes with 3.1% tin. Does it relate to ore, technology or regional/traditional expertise ?

Another case which is interesting is that of the iron artefacts from the Megaliths in peninsular India. As is well known the Vidarbha megaliths fall within the time-bracket of 7th to 5th cent. BC in C 14 showing. The beginning may go back to a still earlier period. The megalithic burials in this region have given literally hundreds of iron objects remarkable not only in their range but also in the purity of iron, entitling it to be designated possibly as steel. Naikund near Nagpur has also given the evidence of iron-smelting furnace and also confirmed the utilization of iron ore from the same region as confirmed by the work of Gogte of the Deccan College. However, further South, in Andhra, sites like Raigir, though having the same cultural context, tell a different story. Similar is the case with copper/bronze, the megalithic people of Vidarbha used a purer form of copper as compared to those further south as at Raigir in Andhra Pradesh where bronze crops show 21% tin. How does one explain it ? Is it because of the nature of ore ? Is it because of levels of technology ? Is it because of intentional production of certain categories of objects which were meant for ritualistic use rather than for everyday domestic use? Is it because certain objects had to be so because of funerary usage ? Is it because the craftsmen had categories within their group, some turning out low grade bronze for funerary use and some making high quality sophisticated objects ? Or finally, is it because of the social and economic structure which imposed limitations on certain categories of craftsmen ?

I would also like to draw your attention to one more aspect which might be rewarding in the study of archaeometallurgy. At Naikund, as mentioned earlier, the remains of an iron-smelting furnace were found and it has been stated that 'iron ore was located in the form of rubble in a nala which is about 1 km south-east of the smelting site .... the iron ore has about 80% iron oxide chiefly as Fe2O3 ..... the megalithic smelters used 10 to 12 kg. of iron ore in single smelting operation producing 3 to 3.2 kg. of pure iron'. Now, this information has to be related to firstly to the fact that Naikund is both a habitation and burial site; secondly, it has nearly a hundred burials; thirdly, the out-turn of iron objects both in habitation and burials has been very remarkable and finally whether the craftsmen could cope up with the demand by using such furnaces or whether they were working full time till the year round so as to keep stock of artefacts readily available which implies surplus agricultural production as the plant remains indicate.

I hope this approach would appeal to you. If so, then the combined efforts of archaeologists and scientists would go a long way in answering the process of cultural dynamics associated with technology. Ore deposits, their quality, their proximity or otherwise, transport, technology, demand, social structure and economy and finally contacts—all these have a role to play in the study of archaeometallurgy. It should not restrict itself to stratigraphic position and sequence, nor simply to spectographic analysis. It is the men of different periods that have left their material records for us. They were what they were because of the environment in which they prospered. The craftsmen among them were interwoven into the contemporary economy of which, along with the agriculturists they became an inseparable part. We are fortunate to inherit their technology. We can pay our debt to them by trying to understand them as comprehensively as possible which includes their success and failures within the framework of their social organization. I do hope that this Seminar takes cognizance of this aspect of total study.

I would like to appeal to the archaeologists and scientists gathered here to pool their intellectual mite in an academic endeavour which, I hope, you all will appreciate. Now, that a lot of metallurgical data is available, a number of research laboratories have come up and quite a cadre of young as well as experienced scientists is available, why not plan to prepare a comprehensive History of Indian Technology on the lines of Singer and Holymyard or Forbes ? What we have is a series of sporadic articles or site-related stories or small monographs like the one by Sankalia or the Indian Science Academy. In this work we have also to seek the expertise of the scholars of ancient Indian literature and also to make use of works by scholars like O.P. Jaggi and others like him. Varanasi has been our traditional centre of learning. Let this university take a lead in this matter and bring together scholars and institutions in the planning and execution of this project.

Thanking you.

Oct. 2, 1991 S.B. DEO

Dept. of A.I.H.C. & Arch
Banaras Hindu University, Varanasi

List of Contributors/Participants

1. Mr. S.M. Aeron
   Steel Authority of India Ltd.,
   Research and Development
   Centre for Iron and Steel
   P.O. Hindoo, Doranda
   Ranchi - 834002.

2. Mr. B.B. Agrawal
   Steel Authority of India Ltd.,
   Research and Development
   Centre for Iron and Steel,
   P.O. Hindoo, Doranda
   Ranchi - 834002.

3. Prof. D.P. Agrawal
   Physical Research Laboratory
   Navrangpura,
   Ahmedabad - 380 009.

4. O.P. Agrawal
   Director General
   Indian Conservation Institute
   INTACH
   Lucknow

5. Dr. P.K. Agrawal,
   Dept. of AIHC & Archaeology,
   Banaras Hindu University,
   Varanasi - 221 005.

6. Dr. Nisar Ahmad,
   Dept. of AIHC & Archaeology,
   Banaras Hindu University,
   Varanasi - 221 005.

7. (Late) Dr. K.D. Bajpai

8. Dr. R.K. Baweja,
   Kurukshetra University,
   Kurukshetra.

9. Prof. H.C. Bhardwaj,
   Dept. of AIHC & Archaeology,
   Banaras Hindu University,
   Varanasi - 221 005.

10. Prof. I.K. Bhatnagar,
    National Museum,
    Janpath,
    New Delhi - 110 011.

11. Dr. S.K. Bhowmik,
    Director,
    State Museum, Gujarat,
    Gandhinagar.

12. Prof. Arun Kumar Biswas,
    Dept. of Metallurgical
    Engineering
    Indian Institute of Technology,
    Kanpur - 208 016.

13. Ms. Sulekha Biswas,
    Dept. of Metallurgical
    Engineering
    Indian Institute of Technology,
    Kanpur - 208 016.

14. Dr. Pranab K. Chattopadhyay,
    Training Centre,
    Alloy Steels Plant,
    SAIL,
    Durgapur - 713 208.

15. Dr. P.S. Chaturvedi,
Dept. of AIHC & Archaeology,
University of Gorakhpur,
Gorakhpur - 273 009.

16. Mr. G.I.S. Chauhan
Steel Authority of India Ltd.,
Research and Development
Centre for Iron and Steel
P.O. Hindoo, Doranda
Ranchi - 834002.

17. Mr. P.K. Chaudhuri
Steel Authority of India Ltd.,
Research and Development
Centre for Iron and Steel
P.O. Hindoo, Doranda
Ranchi - 834002.

18. Dr. P.T. Craddock
British Museum
London U.K.

19. (Late) Prof. R. Dasgupta

20. Dr. Asok Datta,
Lecturer,
Dept. of Archaeology,
University of Calcutta,
Calcutta - 700 019.

21. (Late) Prof. S.B. Deo,

22. Mr. Sudhir Chandra Duari,
Head Master,
Kushadwip High School,
P.O. Kushadwip,
Distt. Bankura - 722 207.

23. Dr. I.C. Freestone
British Museum
London, U.K.

24. Dr. L.K. Gurjar,
Hindustan Zinc Ltd.,
Udaipur.

25. Dr. K.T.M. Hegde
M.S. University of Baroda
Vadodara.

26. Mudrika Jani
Gujarat State Museum
Gandhinagar.

27. Dr. S. Jafar Mahmud,
I.N.S.A.,
Bahadur Shah Zafar Road,
New Delhi - 110 002.

28. Dr. B.R. Mani,
Superintending Archaeologist,
Archaeological Survey of India,
New Delhi.

29. Dr. R.S. Misra,
Dept. of History and Indian Culture,
University of Rajasthan,
Jaipur.

30. Prof. Som Nath Misra,
Principal,
Regional Engineering College,
Rourkela - 769 008.

31. Prof. K. Nageshrao,
Visiting Professor,
U.V.C.E.,
Bangalore.

32. Dr. M.V. Nair,
Director,
National Research Laboratory for Conservation of Cultural Property,
Sector E/3, Aliganj Scheme,
Lucknow - 226 020.

33. Prof. A.K. Narain,
University of Wisconsin,
Wisconsin, U.S.A.

34. Dr. Hari Narain,
National Research Laboratory for Conservation of Cultural Property,
Sector E/3, Aliganj Scheme,
Lucknow - 226 020.

35. Dr. H.V. Paliwal,
Hindustan Zinc Ltd.,
Udaipur, Rajasthan.

36. Dr. Rakesh P. Pandey,
Reader,
Dept. of AIHC & Archaeology,
Jiwajee University,
Gwalior - 474 011.

37. Mr. Jai Prakash
Research Officer
Pharmacoporial Laboratory for Indian Medicine,
Ghaziabad (U.P.)

38. Prof. V.S. Pathak,
Ex-Vice-Chancellor,
University of Gorakhpur,
Gorakhpur - 273 009.

39. Dr. Anand Prakash,
Dept. of History of Arts,
Banaras Hindu University,
Varanasi - 221 005.

40. Prof. Bhanu Prakash,
Dept. of Metallurgical Engineering
Institute of Technology,
Banaras Hindu University,
Varanasi - 221 005.

41. Mr. Bhagirathi Prasad,
Research Scholar,
Dept. of AIHC & Archaeology,
Banaras Hindu Univesity,
Varanasi - 221 005.

42. Mr. K.K. Prasad,
S.A.I.L.,
Research and Development Centre for Iron and Steel,
P.O. Hindoo, Doranda,
Ranchi - 834 002.

43. Prof. Maheshwari Prasad,
Dept. of AIHC & Archaeology,
Banaras Hindu University,
Varanasi - 221 005.

44. Dr. P.R. Rangesh
Lecturer in Rasa-Sastra
Govt. College of Indian Medicine
Bangalore.

45. Prof. K.N.P. Rao,
601, XI, Cross,
3rd Phase, J.P. Nagar,
Bangalore - 78.

46. Prof. Sudharshan Seneviratne,
Dept. of Archaeology,
University of Peradeniya,
Peradeniya, Sri Lanka.

47. Rajam Seshadri
M.S. University
Baroda - 390002.

48. Dr. D.P. Sharma,
National Museum,
11, Janpath,
New Delhi - 110 011.

49. Mrs. Madhuri Sharma,
National Museum,
11, Janpath,
New Delhi - 110 011.

50. Dr. M.S. Shukla,
Dept. of AIHC & Archaeology,
Banaras Hindu University,
Varanasi - 221 005.

51. Dr. Arvind Kumar Singh,
Dept. of AIHC & Archaeology,
Jiwajee University,
Gwalior - 474 001.

52. Dr. Bhupendra Pal Singh,
Dept. of AIHC & Archaeology,
Banaras Hindu University,
Varanasi - 221 005.

53. Dr. Birendra Pratap Singh,
Dept. of AIHC & Archaeology,
Banaras Hindu University,
Varanasi - 221 005.

54. Prof. R.D. Singh,
Ex-Head, Dept. of Metallurgy,
Institute of Technology,
Banaras Hindu University,
Varanasi - 221 005.

55. Dr. Ram Shekhar Singh,
A/404, Officers' Hostel,
Bailey Road,
Patna - 800 001.

56. Prof. B.P. Sinha,
Ex-Head,
Dept. of AIHC & Archaeology,
Patna University,
Patna - 800 005.

57. Dr. N.K. Srinivasan,
Research Fellow,
S.A.I.L., Rourkela Steel Plant,
Rourkela.

58. Ms. Sharada Srinivasan,
Institute of Archaeology,
University College, London.

59. Dr. Priya Srivastava,
Dept. of AIHC & Archaeology,
Banaras Hindu University,
Varanasi - 221 005.

60. Prof. V.C. Srivastava
Head, Dept. of AIHC & Arch.
Banaras Hindu University
Varanasi - 221 005.

61. Prof. T.N. Tewari,
Regional Engineering College,
Rourkela - 769 008.

62. Dr. Vijai K. Thakur,
Dept. of History,
University of Patna,
Patna - 800 005.

63. Prof. J.N. Tiwari,
Dept. of AIHC & Archaeology,
Banaras Hindu University,
Varanasi - 221 005.

64. Mr. Amit Tripathi,
Dept. of Geology,
Banaras Hindu University,
Varanasi - 221 005.

65. Dr. Vibha Tripathi
Dept. of AIHC & Archaeology,
Banaras Hindu University
Varanasi - 221005.

66. Prof. D.N. Tripathi,
Dept. of AIHC & Archaeology,
University of Gorakhpur,
Gorakhpur - 273 009.

67. Dr. J.P. Upadhyaya,
Deptt. of Ancient History and Arch.
Christian Ewing College,
Allahabad - 211 003.

68. Dr. Lynn Willies
Peak District Mining Museum,
Matalockbath, U.K.

Section I
Mining & Metallurgy

Archaeo-Materials in India : Some Residual Problems and Future Prospects

ARUN KUMAR BISWAS

Introduction

Archaeomaterial science is wider in scope than archaeometallurgy, just as metallurgy can not be conceived outside the domain of material science. Metallurgy is usually preceded and sometimes followed by the handling of non-metallic phases. Metals have always been made from ores coming from mines. Apart from native gold, silver and copper, other ores are of non-metallic phase, reduction of which result in metals and slag, another non-metallic phase. Raw materials have also been used to produce other end-products such as gems, metallic oxides, refractories, cements, pigments, bricks, potteries etc., none of which have metallic characteristics. Thus, archaeomaterial studies must be viewed as a broader subject with archaeometallurgy as one of its components.

Certain Issues Related to Indian Archaeomaterials

We present certain highlights of one of our recent reviews on minerals and metals in ancient India (Biswas, A.K., 1991). Specially chosen remarks pertain to copper, gold, silver, lead, tin, iron, zinc and diamond.

D.P. Agrawal (1971 : 239) made an equivocal statement "The (chalcolithic) metallurgical know-how had already diffused from Iran, hence we get (in the Harappan Culture) a full blown technology right from the beginning" — which must be reconsidered in the light of fifth millennium B.C. copper artefacts from Mehargarh (Jarrige, J.F., 1982) and pre-Harrappan copper ore mining and smelting at Ganeshwar - Jodhpura (Agrawala, R.C., 1982).

Gold and silver wares have been found in context of the Harappan civilization. It is hard to accept that gold and electrum (natural alloy of gold and silver) came from the underground mines of South India which have not shown C-14 date earlier than 760+150 B.C. (Hatti, Karnataka) for any gold field. It is probable that gold was derived mostly from the alluvial washings. Though Harappan gold contained some silver (electrum), the silver samples, more prolific than gold, contained no gold but some lead and copper. This means that silver was probably obtained from argentiferous galena rather than electrum. The earliest mining archaeological evidence for extraction of argentiferous galena is from Rajpura-Dariba in Rajasthan (100 m. depth, C-14 date 1260-1130 B.C.). Silver could have been obtained from the Afghanistan mines near Herat during the Harappan era.

At Mohenjo-daro, six percent of the tools contained lead. The copper tools were also alloyed with arsenic (pieces of löllingite $FeAs_2$ were found in the Harappan sites) and tin. The problem regarding the possible sources of tin ore in ancient India has been reviewed by Chakraborti (1979). We tend to agree with Tylecote (1962: 63) that the ancient metallurgists might have exploited (and exhausted) placer deposits of 'stream tin' rather than vein deposits which are hard to mine.

We question D.P. Agrawal's 'equation of the early Aryans with the chalcolithic Banasians' (1971: 240); probably he was much too anxious to adopt similar view earlier expressed by Sankalia, Deo and Ansari (1969: 216 and 223) who visualised Western Asians reaching South-eastern Rajasthan and Navdatoli somehow by the sea or any other route.

Similarly, we question N.R. Banerjee (1965: 239) who concludes his pioneering work on Iron Age in India with the statement that 'the prime responsibility for introducing iron in India and spreading it far and wide within the sub-continent can be squarely fixed on the Aryans endeavour without identifying the Aryans in cranial and archaeological terms or explaining the almost simultaneous emergence (around 1200 B.C.) of the indigenous iron age in Uttar Pradesh, Karnataka and Bihar-Bengal nuclear zones.

It is now well-established that the Indian subcontinent was one of the earliest sites for the discovery of the lamination technique viz. forge-welding of alternate layers of uncarburized and carburized sheets of iron (Agrawal, O.P. et. al., 1990). The earliest evidence of steel in ancient India is obtained in a third century B.C. sickle from Pandu Rajar Dhibi in West Bengal. Chattopadhyaya and De (1989) have shown from the electron micrograph that the 255 mm. sickle was carburized (0.22 p.c. carbon) and then quench-hardened and tempered to produce martensitic structure showing VHN 549 microhardness in the martensitic region and 391 in the ferritic region.

Bronson (1986) has questioned whether 'Wootz' was a supersteel or mere carburized iron, and whether wootz or steel used to be commercially made during the early Christian era. Although literary evidence provide affirmative answers, confirmatory archaeological evidence are still awaited.

Cementation brass was produced in Lothal and Atranjikhera. Taxila brass vase of 4th century B.C. assays 34.34 p.c. zinc for which zinc must have been separately produced by the reduction-distillation route. Werner (1970) demonstrated experimentally that brass produced by the cementation process could not contain more than 28 per cent zinc. The literary references to zinc distillation appeared several centuries later. The 7th century A.D. *Rasaratnakara* text and Nagarjuna, the alchemist, flourishing in the 2nd century A.D. referred to zinc distillation.

It is noteworthy that literary references to the metallurgical practices of copper, iron and zinc appeared many centuries after the discovery or introduction of the said metal. On the other hand, we have early literary references to Indian steel regarding which the contemporary archaeological evidence are yet to be obtained.

Kautilya's *Arthasastra* of the Pre-Christian era mentions *vajra* or diamond. The fifth century A.D. text *Brihatsamhita* describes the geographical locations of the diamond mines in India. However we have yet to obtain C-14 dates or some other proof regarding the antiquity of diamond mines in ancient India. Bauer (1968:141) commented that the Indian diamond mine sites known at present do not date back to a period earlier than 1000 A.D. We disagree on the basis of literary records, but we do not have specific C-14 dates or archaeological evidence to counter Bauer's statement.

The Issue of Trace Element Correlation

The issue of ore-artefact correlation through the ascertainment of trace element compositions is highly controversial. We have pointed out (Biswas, A.K. et. al., 1990: 50-51) that the correlation depends upon; (a) the type of ore (b) the nature of impurities and (c) the precise conditions of smelting. Neither the positive nor negative correlations should be accepted as conclusive evidence. Unfortunately many scholars have ignored the cautionary remarks of Thomson (1958).

D.P. Agrawal and K.T.M. Hegde correlated trace elements of copper ore from Khetri with those in the copper artefacts in far-away Mohenjo-daro and Ahar. The raw materials could have been obtained from other mines much closer (than Khetri) to Mohenjo-daro and Ahar.

Hegde (1968: 518) correlated argentiferous lead ore of Zawar with Nagara lead coin but made the hazardous guess that 'the silver metal of Devnimori coin was probably imported' since the elements of Cr, Co, Ni, Bi and Mo, present in the Devnimori coin were not found in the Zawar ore. Eventually Shekhar (1983) found the above trace elements in the Ambamata deposit not very far away from Zawar or Devnimori.

Recently attempts are being made to correlate copper ore with chalcolithic artefact through lead isotope ratio. But Craddock (1985: 59) has rightly warned that lead in the artefact could come not only from the copper ore (as presumed) but also from the flux and deliberate addition of lead during smelting.

The Utility of Slag Analysis

Attempts for ore-artefact correlation should be supplemented with additional evidences for (a) mines of antiquity (C-14 dating) (b) outfits for smelting, furnaces and crucibles, (c) heaps of slag and slag-ore correlation regarding the non-metallic phases and (d) a reasonably large number of artefacts of different stages of manufacture and of composition and properties varying within narrow limits (Biswas, A.K. et., al., 1990: 51).

There are heaps of ancient slag in the excavated sites which conclusively prove indigenous smelting in India. A detailed examination of the slags and slag-ore correlation as suggested above (c) should help us in ascertaining the kinds of ores and fluxes which might have been used during the process of smelting.

It is regrettable that phase studies on Ahar copper slag and Atranjikhera iron slag have not yet been done. On the other hand, our X-ray, electron diffraction and EDAX or Energy Dispersive X-ray Micro-Analysis (in the Scanning Electron Microscope) of the Zawar ancient slag and retort residue have thrown considerable light on the ancient zinc metallurgy in India (Biswas, A.K. et., al., 1984).

Minor Elements and Replication Experiments

The occurrence of minor elements in slags and artefacts can be better explained through replication experiments, which elevate archaeology from arts to science. This may be briefly illustrated in the context of the occurrence of lead and iron in Indian coins and icons.

Lead has often been deliberately added to improve the fluidity of the chalcolithic melt. It is essential to estimate the size distribution of insoluble lead globules in the matrix of the artefact to explain the mechanical properties and the possible method of preparation. High lead copper alloy was known as 'caldarium' in Europe and *Śulvanāga* in India. The *kākatuṇḍī* was possibly leaded brass which when dipped in acidic sour gruel turned black.

The earliest copper artefacts contained little iron. With improvement of furnace technology, the temperature of the copper-reducing assembly could reach or even exceed 1400° C when iron was reduced. On cooling, the iron phase precipitated as alpha-iron dendrites in the matrix of copper-rich epsilon phase (Merkel, 1982 and 1983). Scientific investigations are necessary to establish whether the above phenomenon adequately explains the presence of 4-6 p.c. iron in Nalanda bronzes (containing 8-10 p.c. Sn, - 5 p.c. Pb, a few p.c. Zn) and 21-23 p.c. iron in 7th century A.D. Paunar coins (75 p.c. Cu, 3 p.c. Sn, 0.3 Pb).

Existence of lead or gold in ancient silverware suggests whether the raw material was argentiferous galena or electrum. Arsenic in a chalcolithic object could have appeared from arsenious copper ore or due to deliberate alloying. The 12th century A.D. West Tibetan 'golden' brass statue of *Manjuśrī* on a white 'silvery' lion and base had been analysed in detail. The lion and base are low-zinc brass containing 6.8 and 7.4 p.c. arsenic respectively. The surface contains higher proportion (14 p.c.) of arsenic. X-ray fluorescence and scanning electron microscopy with EDAX analysis revealed that the arsenic-rich layer was formed due to inverse seggregation which occurs in alloys freezing over a long temperature range and contracting during solidification (La Niece et. al., 1989: 12). Trace detection of mercury in the gilt brass images of Tibet suggests that gilding was done by the amalgam route.

Sophisticated Instrumental Analysis

The recent trend of archaeomaterial research involves the extensive use of sophisticated instruments. Transmission and scanning electron microscopy combined with diffraction study and EDAX facilities permitted us to study heterogeneity, elemental composition, grain size and grain phase composition of ancient zinc slag in a comprehensive manner (Biswas A.K. et. al., 1984: 236-237). Using a fine electron probe (~50 μm diameter) we could scan a grain as small as 0.5 um and quantitatively analyse ten elements therein and report the presence of more than a dozen specific phases. Nieuwenhuysen et. al. (1984: 105) have reported the result of EDAX analysis of 23 Himalayan metel statues. Secondary and back-scattered electrons were employed in the scanning electron microprobe analysis for simultaneous detection of a large number of elements in a narrow area. Craddock (1981: 20) analysed 121 Tibetan and Himalayan icons/metal works by atomic absorption spectrophotometer which permitted him to assay 13 elements in each sample down to 10 ppm level.

Mössbauer spectroscopy and neutron activation analysis have been widely used in the context of archaeological research. The family of photo-electron spectroscopy (PES) techniques such as ESCA or electron spectroscopy for chemical analysis provides special information about a very thin surface layer. Many other sophisticated techniques such as DTA, high-sensitivity X-radiography, quantitative image analysis for phase contents, electron energy loss spectroscopy (EELS) etc. may be employed for unravelling the mysteries of archaeometallurgy and archaeomaterials. Insitu or hot-stage electron microscopic techniques may find major applications.

Sophisticated instrumentation analysis should be extended not only to metallic icons and artefacts but also to ancient gems, slags and non-metallic artefacts of archaeological importance.

Literary Data on Archaeomaterials

Literary informations provide many clues as to the areas in which fresh archaeological discoveries may be attempted. *Arthasastra* tells us about diverse types of alloys and *Brihatsamhita* describes different producers of quenching of steel swords. We should aspire to discover archaeological samples which might corroborate the literary references. *Rasasastra* texts refer to different methods for extracting *sattva* or the metallic essence from the metallic ores. A Chinese method for precipitating copper from mine water by scrap iron seems to have been adapted in the Indian condition. Attempts should be made to substantiate the literary references by archaeological discoveries.

Concluding Remarks

The pursuit of our studies on archaeomaterials in India is likely to yield rich dividends. The future prospects are diverse and numerous. Scientific analysis should give us not only accurate measure of the glorious past but also possible applications for the future. The special fibre-reinforced refractory crucible in which wootz used to be made can be utilised in the modern context. Sherby and Wadsworth (1985) have performed replication experiments to prove how wootz used to be converted to steel sword with damask pattern; a very similar technique can produce damask-free ultra high carbon steels which are stronger and yet more ductile than conventional steels. With their superplastic properties, they can be shaped with precision into complicated objects. This is a glorious example how an ancient Indian metallurgical technique can be adapted to the modern and even future needs.

REFERENCES

Agrawal, D.P., 1971. *The Copper Bronze Age in India.* Munshiram Manoharlal, New Delhi.

Agrawal, O.P., Hari Narain, Jai Prakash and Bhatia, S.K., 1990. Lamination Technique in Iron Artefacts in Ancient India. *Historical Metallurgy,* Vol. 24, No. 1, 1990, pp. 12-26.

Agrawala, R.C. and Vijay Kumar, 1982. Ganeshwar-Jodhpura Culture, in *Harappan Civilization,* edited by G.L. Possehl, Oxford and IBH, New Delhi, pp. 125-134.

Banerjee, N.R., 1965. *The Iron Age in India.* Munshiram Manoharlal, Delhi.

Bauer, Max, 1968. *Precious Stones* (edited translation of Edersteinkunde, 1896), two volumes, Dover Publications, Vol. I (pp. 1-260); Vol. II (pp. 261-627).

Biswas, A.K., Gangopadhyay, A. and Ramchandran, T.R., 1984. Phase Studies on Zinc Residues of Ancient Indian Origin. *Transactions of the Indian Institute of Metals,* Vol. 37, No. 3, June 1984, pp. 234-241.

Biswas, A.K., Biswas, S. and Chakravorty, N.A., 1990. Archaeomaterial Studies in India and Literary Evidences, in *Historical Archaeology of India,* edited by Amita Ray and Samir Mukherjee, Books and Books, New Delhi, pp. 49-66.

Biswas, A.K., 1991. *Minerals and Metals in Ancient India,* a two-volume monograph submitted to the Indian National Science Academy, Vol. I Archaeological Evidences, 21 chapters.

Bronson, Bennet, 1986. The Making and Selling of Wootz, A Crucible Steel of India. *Archaeomaterials,* 1986, Vol. I, pp. 13-51; also 1989, reply to Bhanu Prakash, *Ibid.,* Vol. III, pp. 95-96. (Published - 1995 D.K. Printworld, New Delhi)

Chakraborti, D.K., 1979. The Problem of Tin in Early India - A Preliminary Survey. *Man and Environment,* Vol. III, pp. 61-74.

Chattopadhyaya, P.K. and De, S. 1989. Iron objects from Pandurajar Dhibi: Archaeometallurgical Studies, *Steel India,* 12, 1, April 1989, pp. 33-41.

Craddock, P.T., 1981. The Copper Alloys of Tibet and Their Background, in *Aspects of Tibetan Metallurgy* edited by Z.A. Oddy and W. Zwalf, Occasional Paper No. 15 British Museum, pp. 1-31.

Hegde, K.T.M., 1968. Source of the Metal in the Lead Coins of the Kshatrapa Period. *Current Science,* September 20, 37, No. 18, pp. 518-520.

Jarrige, J.F., 1982. Excavations at Mehargarh, in *Harappan Civilization,* edited by G.L. Possehl, Oxford and IBH Publishing Co., New Delhi, pp. 79-84.

La Niece, Susan and Carradice, Iran, 1989. White Copper: The Arsenical Coinage. *The Journal of the Historical Metallurgy Society,* 23, 1, 1989, pp. 9-15.

Merkel, J.F. 1982. Bronze Age Copper Smelting Experiments, Ph. D. Thesis, Institute of Archaeology, London, U.K. ; also *MASCA Journal,* 2, (6), 1983, pp. 173-176.

Nieuwenhuysen, P. and Adams, F., 1984. Elemental Analysis of Himalayan Metal Statues. *Historical Metallurgy,* 18, 2, pp. 105-108.

Sankalia, H.D., Deo, S.B. and Ansari, Z.D. 1969. *Excavations at Ahar,* Poona.

Shekhar, N.C., 1983. Antiquity of Mining and Metallurgical Activities at Ambaji, Kumbari and Deri, Gujarat and Rajasthan. *Indian Journal of History of Science,* 18, 2, pp. 176-183.

Sherby, O.D. and Wadsworth, J., 1985. Damascus Steels. *Scientific American,* February 1985, pp. 112-120.

Thomson, F.C., 1958. *Man,* 58 (i), p. 1.

Tylecote, R.F., 1962. *Metallurgy in Archaeology,* Edward Arnold, London

Werner, O., 1970. Über das Vorkommen von Zink., *Erzmetall,* 23, 1970, pp. 259-269.

The Metallurgical Tradition of the Harappans

D.P. AGRAWAL AND RAJAM SESHADRI

The advent of metallurgy in the Indian sub-continent has been one of the major factors leading the Harappans to the threshold of urbanization. The invention of metal technology, including the physical and chemical processing of the ore, gave rise to a variety of arts and crafts and a sophisticated production technology. The highly refined and sophisticated metal technology prevailing during the Harappan times is amply attested to by the numerous copper based artefacts unearthed during the excavations. The Harappans made prolific use of gold, silver, copper and lead in addition to the alloys of these base metals. However, copper and its alloys were naturally used more extensively in comparison to the noble metals.

The Pre and Early Harappan contexts are generally poor in copper. From the Pre-Harappan levels of Kalibangan, we have only a few copper fragments. But the scarcity of artefacts need not imply that the Pre-Harappan technology was in any way inferior to that of the Harappans (Hegde, 1981). The copper mirror from Mehi representing a stylized female figure and the copper axe from Nal are examples of high metallurgical skill. Contrary to the Early Harappans, the Mature Harappan phase witnessed a sudden upsurge in the metal objects, especially copper, in various forms ranging from utilitarian objects to intricate pieces of art and ornamentation. Although there was no dearth of copper tools, the Harappans continued to use blades of chert and chalcedony. The metal tool-kit of the Harappans comprised flat celts, knives, arrowheads, spearheads, chisels, razors, fishhooks, saws, drill. Besides, the tubular drill, the true saw and the needle, with eye at the pointed end, are contributions of the Harappans to the world of instruments.

The Harappans were the only people in India, who used metal lavishly for vessels, including a variety of vases (Agrawal, 1970; 1971). Most of them seem to have been made in two parts, the bottom and the top, and then joined together. For vessel fabrication they used running-on and rivetting techniques. Evidence of soldering in noble metals has also been reported from the Harappan artefact repertoire. In contrast, there is no evidence of copper being used for vessels in the Chalcolithic or the Copper Hoard Cultures.

Human and animal figurines were cast by the *cire perdue* (lost wax) process. The stark simplicity of the Harappan bronzes is very manifest. They are neither as ornate as the Chinese nor as intricate as the Mesopotamian examples. A number of axes from Chanhudaro and Mohenjodaro show puckered surface, indicating that they did not always use green poling in open moulds. Stirring molten copper with green branches, called green poling in open moulds, releases hydrocarbons which reduce the oxides, thus avoiding blow holes.

Copper is not a very hard metal but it is ductile and has melting point low enough to bring casting within the capabilities of the Harappan furnace technology and it is easily hardenable by cold work. At the same time, it is not ideal for accurate casting. It requires a high degree of surface finish and having strong affinity for oxygen and hydrogen results in a large number of gas holes while casting. Its strength, hardness and wear resistance are limited. This, the ancient smiths learnt by experience and realised that certain combinations of ore minerals and heat conditions produce a more suitable and adaptable metal. These metal combinations of the ancient smiths resulted in the formation of what we now define as alloys. The development of bronze may well have followed such an experimental course, like smelting of minerals, resulting in the production of tin and the experiments in the mining of metals resulting in a metal more suitable than copper in terms of strength, toughness and corrosion resistance. The great achievements of Harappan metallurgy can be studied through : the surviving metal work and metal working debris and by analysing their composition, structure and properties.

Several studies have been carried out on the surviving metal work and debris in order to understand the metallurgical traditions of the Harappans. Representative samples from Rojdi (Bhowmick, Ph. D. thesis), Harappa (Sanaullah, 1931; 1940), Lothal (Rao, 1979), Somnath (Hegde, 1965), Nagwada (Sheshadri, unpublished thesis) and several other sites (Agrawal, 1971; 1982) have been subjected to chemical analysis and metallographic examination.

We have now more than 200 analyses of Harappan artefacts (Agrawal, et. al., 1981) of which only 23% artefacts were alloyed with tin, 12% with arsenic and 80% with lead. In addition to this, we also have a few artefacts whose composition is a mixture of these elements. Here it may be mentioned that the Harappans used for alloying both arsenic and tin; the Copper Hoard Culture used exclusively arsenic alloying; and the Chalcolithic cultures only tin alloying. Two copper artefacts recently analysed from Nagwada, a mature Harappan site in North Gujarat, the axe contains 2.72% of arsenic while the chisel contains 10.60% of tin (Sheshadri, unpublished thesis).

We know from West Asia that the earliest alloying was done with arsenic only but later on tin takes over almost globally. Perhaps the injurious effects of arsenic alloying on the health of the coppersmiths became known and it was given up. In this context, the use of arsenic by the Copper Hoard Culture appears a bit enigmatic; either it is due to the fact that the ores themselves contain significant amounts of arsenic or the Copper Hoards' is the oldest copper technology in India.

Metallurgy includes not only smelting of the ore, separating of the pure metal in the slag and intermixing various metals and non-metals to form alloys, but also fashioning of the metal in order to give it the required shape. The production of metal artefacts can be done in two ways, viz., casting the molten metal into moulds of required shape and building up objects from either solid metal pieces or thin sheets joined by rivets. Among the two, the former method is more advanced, whereas the latter forging the metal to impart a desired shape, is a primitive one. The coppersmiths of the Harappan period employed both casting and forging techniques. But the quality of casting differed from object to object. Some were cast in good well-ventilated moulds while others were done in crude open moulds (Agrawal, 1971). Forging techniques also show variations depending upon the use of the objects: while cutting instruments were cold-worked for a better and enduring working edge, objects of art and jewellary were left in their cast condition. The metallographic studies of the Harappan copper repertoire has shown evidence of cold-work, hot-work and annealing techniques by their smiths (Agrawal, 1971).

Ore Artefact Correlation

Although copper deposits are widely distributed in India they are never found anywhere in great concentrations.

Tin, lead and arsenic deposits in India are found in very meagre quantity and are mostly considered uneconomic sources. Main deposits of tin in India are found in Bihar, Gujarat, Karnataka and Rajasthan and some minor sources are located at Murad Bagh, Kastamuni and the Burma belt.

The main copper deposits in India are in the form of Chalcopyrite ($CuS\ Fe_3$), Chalcocite (Cu_2S), Tetrahedrite ($4Cu_2\ Sb_2O_3$), Covellite (CuS), Malachite ($CuCO_3.\ Cu\ (OH)_2$) and Azurite ($2CuCO_3\ Cu\ (OH)_2$). The Singbhum copper mine in Bihar extends upto 130 x 8 km. and contains about 2.47% of copper. In Rajasthan, the major copper deposits are located in the Khetri and Dariba belt. Guntur in Andhra Pradesh also has a 4.8 km. copper belt (Agrawal, 1971).

Attempts have been made with some success (Agrawal, 1972) to trace the ore from which the metal was extracted for the manufacture of the copper antiquities during the Harappan times. Impurities or inclusions which were not delibrately added, but were obtained into the constitution of metal from the ore, form the basis of such studies.

Various elements like gold, silver, tin, nickel, cobalt bismuth pass on to the composition of the metal without much loss in their quantity. Besides these, elements like iron, manganese, zinc, aluminium, chromium, molybdenum, vanadium, zirconium, tungsten and titanium also form a part of the constitution of the metal, if present in the parent ore.

Among the major copper mining centres, the Aravalli belt in Rajasthan, on the basis of proximity in location and the evidence of ancient mining, is believed to have been exploited during the Harappan times. Besides, the trace impurity pattern shows that gold is absent both in the Harappan artefacts and the Khetri ores. The common presence of Ag, Fe, As, Sb, Pb, Bi, Sn, Ni, Zn, Mn, Co, Al, Cr, Mo, Ti, Mg and V in both the Harappan artefacts and the Khetri ores suggests that the Khetri belt was exploited by the Harappans (Agrawal, 1971; 1972).

Analysis of lead isotopes is generally regarded as a more reliable method for ore-artefact correlation but not much work has been done in this field in India. Some lead isotope assays of copper ore samples collected from the Aravalli Hills have been carried out (Hegde and Ericson, 1985).

In an effort to locate the copper sources used by the Harappans, a large number of ancient copper workings have been radiocarbon dated (Table 1; Agrawal et al., 1976), but none of the dates so far is older than 3120 B.P. With further work, however, older workings are expected to be found.

TABLE 1
^{14}C DATES OF ANCIENT COPPER WORKINGS
(All ^{14}C dates are uncalibrated)

Sample No.	*Site*	*Date in yrs. BP (5730 yrs half-life)*	*Locus*	*Samples submitted by*
TF-1117	Dariba Mines, Dist. Udaipur, Rajasthan	2310±105	Depth 64m, Layer 436	GSI, Udaipur
PRL-684	Junjhunu, Rajasthan	Modern	Ancient Mine	Hindustan Copper Ltd.
PRL-685	-do-	Modern	-do-	-do-
PRL-927	Alwar, Rajasthan	Modern	Old mine shaft, depth 40-50m	Delhi University
PRL-929	Khetri, Rajasthan	250±80	Old mine shaft, depth 20m	-do-
TF-1221	Kumbaria, Dist. Banaskantha, Gujarat	535±90	Dumps behind Jain Temples, Kumbari	GSI, Ahmedabad
TF-1222	-do-	905±85	-do-	-do-

Sample No.	*Site*	*Date in yrs. BP (5730 yrs half-life)*	*Locus*	*Samples submitted by*
PRL-53	Ambamata, Dist. Banaskantha, Gujarat	2110±200	12m distance in No. 3, West Drive, of the underground mine	Mineral Exploration Corp. Ltd., Ambaji
PRL-66	-do-	850±100	Bore hole, No. 100 N.W. Zone of Ambata deposit	GSI, Ahmedabad
TF-805	Bandlomattu Hill, Dist. Guntur, A.P.	900±80	From contact of dolomite-phyllite of Cumbum stage of Nallamallai series	Hindustan Copper Ltd., Bollapalle
TF-806	Bandlomattu Hill, Dist. Guntur, A.P.	655±90	From contact of dolomite-phyllite of Cumbum stage of Nallamallai	Hindustan Copper Ltd., Bollapalle
TF-373	Mailaram, Dist. Khammam, A.P.	535±90	1km west of Mailaram from a slag heap, 1m depth	GSI, Hyderabad
PRL-392	Ranihat, Tehri, Garhwal	1690±140	Layer 10, depth 3.10m	Garhwal University
PRL-394	-do-	1570±140	Layer 8, depth 2.75m	-do-
PRL-208	Dariba Mines, Dist. Udaipur, Rajasthan	2200±100	East Lode old working, depth 4 to 5m	Hindustan Zinc Ltd., Udaipur
PRL-209	-do-	3120±160	East Lode old working, depth 263m	-do-
PRL-210	-do-	1840±130	Main Lode old working, depth 100m	-do-
PRL-252	Ingaldhal, Dist. Chitradurga	1730±110	Shaft of the ancient copper mine	A.S.I., New Delhi

Before we conclude, a few words about the Daimabad bronzes.

Daimabad Bronzes

It may be relevant here to mention the finds of some massive bronzes from Daimabad, from near the Chalcolithic site of the same name from Maharashtra, though in no way stratigraphically connected with the main site. The cache is composed of a rhino, an elephant, a bull and a man driving a chariot. They are very heavy: the heavist is 29 kilograms! This sort of lavish use of metal for non-utility artefacts is unknown in India. Only the Harappans had some animal and human figurines and their weights are in the range of grams only. We carried out chemical analysis of these objects, and found that they contain minor amounts of nickel, iron and lead but no tin. Arsenic, however, is significant: from 0.6 to 1.3%. Neither the iconography, nor the weight, nor even the chemical composition allows one to assign them to the protohistoric period. Such massive use of copper for non-utilitarian or ritualistic use comes into vogue only in the historical period (Agrawal, 1978).

CONCLUSION

The Harappan, in contrast to the Chalcolithic and the Copper Hoard Cultures, show greater abundance of metal artefacts, variety in their tool repertoire, and better mastery in metallurgical skills. Unlike the other protohistoric cultures, the Harappans alone used copper for pots and pans and used both arsenic and tin for alloying. Despite the relative abundance of copper artefacts, the Harappans continued to use chert blades. The role of copper technology, access to rich mines and a semi-arid ecology seem to have contributed a great deal to the formative processes of the Indus Civilization.

REFERENCES

Agrawal, D.P., 1971. *The Copper Bronze Age in India*, Munshiram Manoharlal, Delhi.

Agrawal, D.P., 1972. Problems of Protohistoric Copper Artefacts and their Ore Correlations, *Archaeological Congress and Seminar*, pp. 35-39 (Eds.) U.V. Singh, Kurukshetra University, Kurukshetra, 1976.

Agrawal, D.P., 1982. The Indian Bronze Age Cultures and their Metal Techonology, (Ed.) F. Wendorf and A.E. Close, *Advances in World Archaeology*, Academic Press, New York, pp. 213-264.

Agrawal, D.P.,1984. Metal Technology of the Harappans, in *Frontiers of Indus Civilization*, (Ed.) B.B. Lal and S.P. Gupta, Books and Books, New Delhi, pp. 163-168.

Agrawal, D.P., 1990. Metal Technology of the Harappan Culture and its socio-economic implications, *Indian Journal of History of Science*, 5, No. 2, pp. 238-252.

Agrawal, D.P., R.V. Krishnamurthy and Sheela Kusumgar, 1978. On the Affiliation of the Daimabad Bronzes : Some Fresh Data, *Intl. Symp. on Archaeometry and Archaeological Prospection*, Bonn (Ed.) I. Scollar, pp. 8-13.

Agrawal, D.P., R.V. Krishnamurthy and Sheela Kusumgar, 1981. Arsenical Coppers in the Indian Bronze Age, *Recent Researches in Indian Archaeology and Art History,* (Eds.) M.S. Nagaraja Rao, Agam Kala Prakashan, New Delhi, pp. 9-18.

Agrawal, D.P., C. Margabandhu and N.C. Shekhar, 1976. Ancient Copper Workings. Some new 14C dates, *Indian Journal Hist. of Sci.*, 11, No. 2, pp. 133-136.

Bhowmik, S.K., 1972. *Early Copper and Bronze Technology of Gujarat* - unpublished thesis, M.S. University, Baroda.

Hegde, K.T.M., 1981. Scientific basis and technology of ancient Indian copper and iron metallurgy, *Indian Journal of History of Science,* Vol. 16, No. 2, pp. 189-201, Nov., 1981.

Hegde, K.T.M., and Ericson, J., 1985. Ancient Indian copper smelting furnaces, *British Museum Occasional Paper no.* 48, edited by P.T. Craddock and M.J. Hughes.

Rajam, S., 1990. *Copper Artefacts from Nagwada - A Technical Study.* Unpublished M.A. Dissertation, M.S. University, Baroda.

Rao, S.R., 1979. *Lothal Excavations Report,* ASI, Delhi.

Sanahullah Khan, 1931. In Marshall, J., *Mohenjodaro and the Indus Civilization,* London.

Sanahullah Khan, 1940. In Vats, M., *Excavations at Harappa,* Vol. I, Delhi.

Transition from Neolithic to Copper Age at Kushadwip, District Bankura, West Bengal
Archaeometallurgical Studies

SUDHIR CHANDRA DUARI

AND

PRANAB K. CHATTOPADHYAY

Similarity in typology is an interesting clue in the case of *'lithic to metallic'* transition. In Bihar and West Bengal perhaps only one example had been drawn earlier by Lal (1951) where a lithic object from Ban Asuria, district Bankura, West Bengal had revealed a close similarity with its metallic counterpart. Present studies may reveal similar transition from Neolithic stone tool to its counterpart in copper and which could further be equated with Copper Hoard object. Recently the first author, has discovered a unique copper artefact - a celt from the Khalbed namely Harinmuri or Cholakandar (87°24' 30", 23°8') besides village Ramchandrapur near Kushadwip (15 km north of Vishnupur) in the same district.

The village Kushadwip and its surroundings have yielded enormous numbers of lithic tools explored by the first author from lower Palaeolithic to Neolithic, potteries from Neolithic to medieval period including black-and-red ware and other cultural assemblages of proto-historic period.

The object under review is shown in figure 1. The object is covered with thin patina of greenish black in colour. The object is found along with pits and apparently a small crack at one of its face. It is axe shaped though no handle is there. This is at present 29.15 gm. in weight. It is 41.82 mm in length, 5.44 mm thick, the cutting edge portion is 28.30 mm in breadth and top portion is 17.26 mm. in breadth. The cutting edge is a little blunt which has use marks. The object is now preserved at Kushadwip Bilasini Sangrahasala. The following are the tests performed on it to reveal its manufacturing techniques.

Chemical Analysis

This was performed in semi-quantitative measure with a Hilger-Watts Spectrograph with RU standard and selected wave length of 2700 - 4400°. Friedman *et al* (1966) method was applied to determine the probable ore for manufacturing the metal. Table 1 indicates the distribution of trace elements and table 2 indicates the chemical composition of the celt. Here copper has been shown as the base element and others in trace amount.

The specimen as revealed through the present analysis is made of pure copper (purity 99.7%). This clearly indicates no alloying was done and the other elements are the natural contamination with parent copper ore. Probable use of oxidised type of ore with malachite - azurite may be predicted having a probability of 73.71%.

TABLE 1

Distribution of trace elements in analysed metal object from Kushadwip

Ag	Fe	As	Sb	Pb	Bi	Cu	Sn	Ni	Zn	Mn	Co	Au	Al	Cr	Mo	Zr	W	Ti	Mg	V	Si	Ca
+	+	+	-	+	-	+	-	+	-	-	-	-	-	-	-	-	-	-	+	-	+	+

TABLE 2

Cu	Ag	Fe	As	Sb	Pb	Bi	Sn	Ni	Mg	Si
Base	0.01	0.05	0.01	-	0.02	-	-	0.05	0.05	0.05

Metallographic analysis

To ensure minimum possible destruction the object itself was polished at one of its edge after taking out fillings for chemical analysis and observed through a Neophot metallographic microscope. The polished surface before etching indicated only Cu-Cu_2O inclusions of moderately higher amount - which indicate perhaps no poling was conducted at all or it was not achieved properly. Later the polished surface was etched with ferric chloride, ammonium hydroxide and hydrogen peroxide solution. The etched micro-structure clearly indicates that it is in single phase where a cast structure is totally absent. Orientation of grains clearly indicates that the object was cold worked. Throughout the cross section absence of twins indicate that no annealing was conducted after cold working. In a few spots the appearance of equiaxed grains perhaps represents ageing.

The celt from Kushadwip is unique in many respects. Similarity in typology has indicated an *'intermediate'* state from Neolithic celt to Copper Hoard. This particular celt clearly indicated that it was a used one. In this respect it differs from the copper hoard as the majority was of nonutilitarian type. The crack on the surface of the same was not generated during its use but took place in course of its manufacture. This indicated that the craftsman was not knowing the annealing technique though he had learnt the art of forging (of course, in the cold working type). The object was not manufactured from native copper whereas a probable use of malachite - azurite as origin indicates that the craftsman had learnt about the extractive metallurgy of copper but was not knowing the technique of 'poling'. Poling is used to drive away oxygen from liquid copper with the help of green wood poles. The subsequent Chalcolithic objects indicate that alloying was known to the smiths whatever might be the amount of its alloying and it used to depend upon the

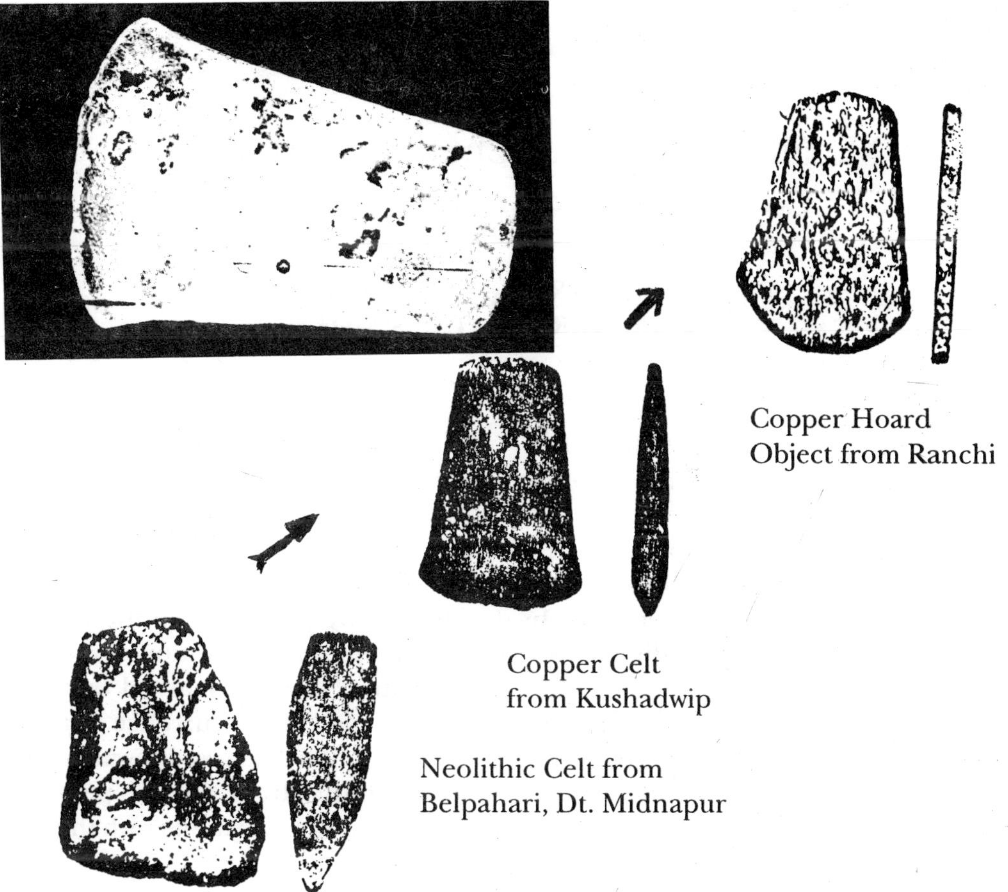

Fig. 1

availability of tin. Alloying with tin helps a better fluidity during casting and also improving thereby the mechanical strength of the material.

ACKNOWLEDGEMENT

The authors earnestly thank Shri M.L. Dutta, Shri Kanhaia Lal, Shri S.K. Banerjee and Shri N.C. Ray for their kind support to analyse the present object.

REFERENCES

Lal, B.B., 1951. Further copper hoards from Gangetic basin and a review of the problem. *Ancient India* 7: 20-39.

Friedman, A.M. *et. al.*, 1966. Copper artefacts: correlation with source type of copper ores. *Science* 152 (3728): 1504 - 1506.

Metallographic and Chemical Study of Copper Artifacts from Senuwar

PRANAB KUMAR CHATTOPADHYAY
AND BIRENDRA PRATAP SINGH

The site of Senuwar (Lat. 24° 56' N, Long. 83° 56' E) in district Rohtas, Bihar is located south of Sasaram on the right bank of the river Kudra. Excavations at Senuwar have revealed four cultural periods (Singh, 1990). Period I is Neolithic, Period II Chalcolithic, Period III NBPW Culture and Period IV, Kushan. The earliest period I, shows two sub-periods i.e. IA and IB. Period IA is metal free and it can be regarded as pure Neolithic. In the succeeding period IB all the basic culture elements of the preceding sub-period continue with the addition of metal and thus has been regarded as Neolithic-Chalcolithic. This was marked by the presence of copper and a solitary fragment of a lead rod. The occurrence of the former is noted from the earliest level of IB, while the latter is marked towards its closing phase (Table 3). Important copper objects found from this sub-period included a fish-hook, a needle (?), a pointed tool and a ring comparable with those recovered from copper hoards.

Period II (Chalcolithic in character), overlapped with the preceding sub-period IB, yielded notable objects of copper which include an ear-ring (?), needle (?), pointed tool, a miniature arrow-head, bangles and a potsherd in which a layer of copper is found still adhering throughout its inner surface. The radiocarbon date available from the lower layers of Neolithic-Chalcolithic phase reading 1770 BC±120 may be taken into consideration. The occurrence of metal to such an early date has been noted for the first time in this part of the country. The present paper seeks to examine the state of the metallurgical skill at this point of time. Both metallographic and chemical analysis of metal objects from sub-period IB and Period II have been undertaken. Five metal objects, three from sub-period IB and two from Period II have been analysed. The details of these objects are discussed hereunder.

Sub-period IB (Neolithic - Chalcolithic)

A. *Copper wire* : It was recovered from trench 2D (depth 470-496 cm, layer No. 10, Reg. No. 421). The wire under investigation has a diameter of 0.95 mm and a

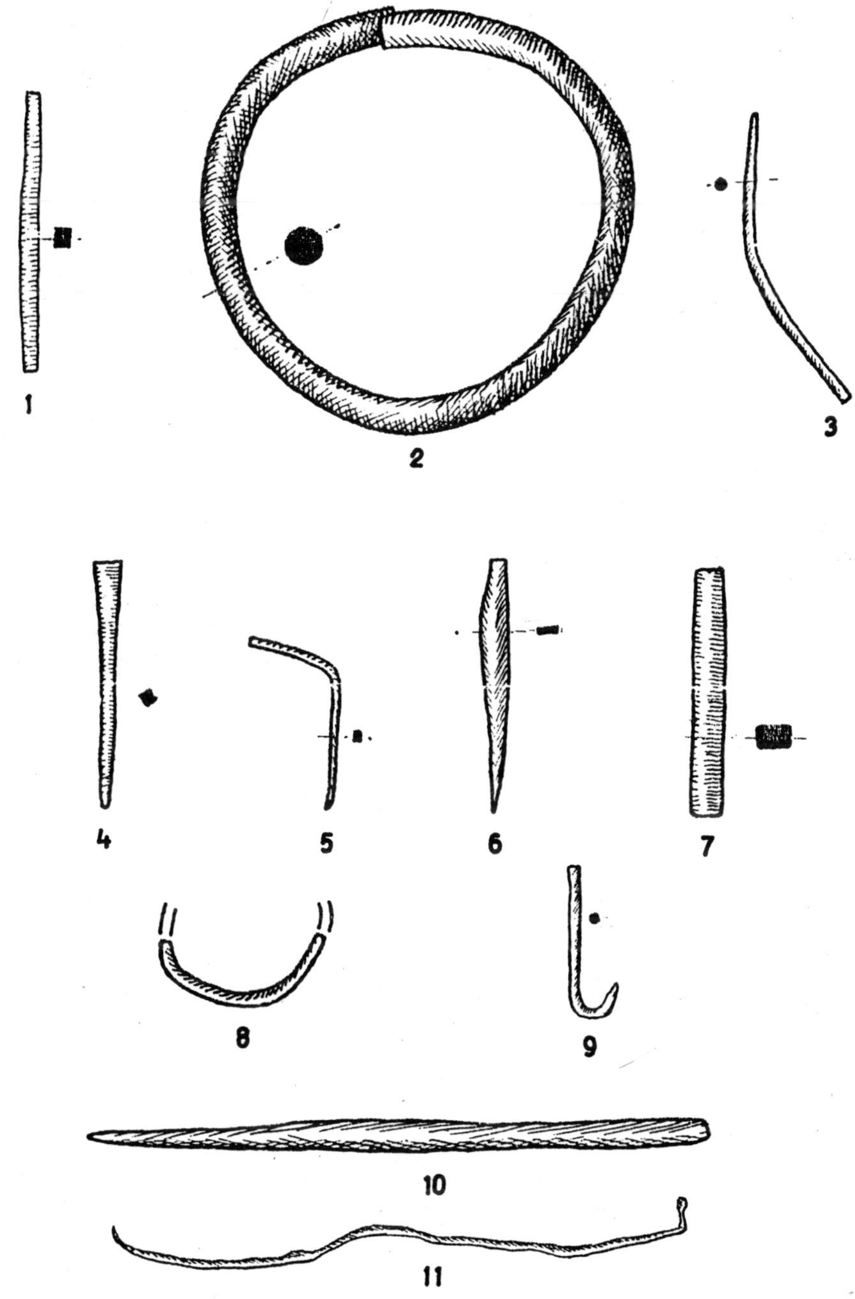

Fig. 1 SENUWAR : NEOLITHIC CHALCOLITHIC COPPER OBJECTS

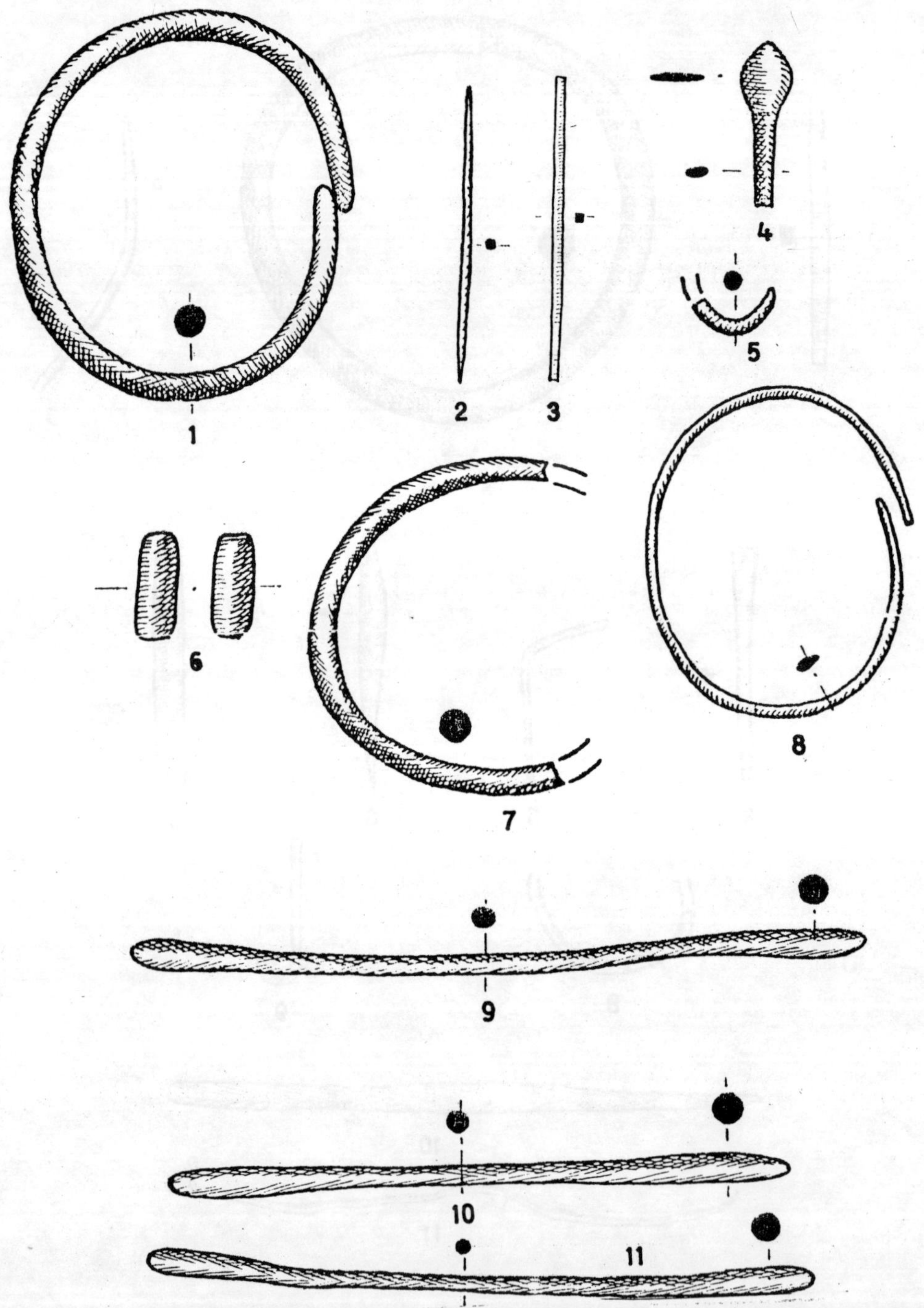

Fig. 2 SENUWAR : CHALCOLITHIC & EARLY HISTORIC

length of 11 mm. It shows a thick patina alround it which is basically made of sulphate, chloride and hydrated oxides of copper.

B. *Copper needle* (?): This object was recovered from trench 10, A'5 - A'6 (depth 693-705 cm., layer (15), Reg. No. 1875) and has a cross section of 1.4 x 2.0 mm. It was covered with a greenish patina.

Period II (Chalocolithic)

C. *Bangle*: The bangle under investigation was found in a highly corroded state and recovered from trench 2J (depth 112 cm, layer 3, Reg. No. 1022). The object had a thickness of 3.52 x 1.53 mm.

D. *Bangle*: This object was recovered from trench 9 at a depth of 480 cm from layer (8), (Reg. no. 1656).

E. The fragmentary lead rod, rectangular in section was recovered from trench 2B (depth 160-172 cm, layer No. 4, Reg. No. 162). It was found in extra corroded dark condition. Only a small bit of the metal was taken from this object for chemical analysis.

Chemical analysis of the objects have been done with Hilger Watts Emission Spectrograph, with RU standard, within a quantitative analysis of the major elements have been shown in Table 2 and the trace elements in Table 1. For investigation iron, tin and zinc objects analyses were carried with atomic absorption spectrophotometer PYEUNICUM SP 2900 in air-acetylene flame.

TABLE 1

Trace elements in Senuwar metallic objects

Object	*Ag*	*Fe*	*As*	*Sb*	*Pb*	*Bi*	*Cu*	*Sn*	*Ni*	*Zn*	*Mn*	*Co*	*Au*	*Al*	*Cr*	*Mo*	*Zr*	*W*	*Ti*	*Mg*	*V*	*Si*	*Na*	*B*
A421	+	+	+	+	+	-	+	-	+	+	-	-	-	-	-	+	-	-	-	+	-	+	-	-
B162	+	+	+	-	+	-	+	+	-	-	-	-	-	+	-	-	-	-	+	+	-	+	-	-
C1875	+	+	+	+	+	-	+	+	+	-	-	-	-	-	-	-	-	-	tr	+	-	+	-	+
D1022	+	+	+	+	+	-	+	+	+	-	-	-	-	-	-	-	-	-	tr	+	-	+	-	+
E1656	+	+	+	+	+	-	+	+	+	-	-	-	-	+	-	-	-	-	tr	+	-	tr	-	-

+ Present, - Absent, tr = 0.001

So far, four standard analysis of copper ores from Bihar are available (Agrawal et. al., 1978). Amongst those studies as mentioned in P-56, P-57, P-58 and P-188 reference is made to the reduced type ores. Senuwar copper has some similarities with Rakha ores with reference to the common presence of silver, zinc, nickel, molybdenium, magnesium, and lead and common absence of bismuth, gold, zirconium, tungston and titanium. At the same time arsenic and tin do not coincide with each other. The present knowledge, however, is inadequate regarding early mining activities in this zone in C. 1800 B.C. Besides definite information on the oxidised ores are also not available. However, there is some chance for getting ore from Rakha mines but definitely not from Singhbhum chalcopyrite (P-188).

TABLE 2
Semi-quantitative analysis (major) of metal objects from Senuwar Pd - IB and II

Elements		*A 421*	*B 162*	*C 1875*	*D 1022*	*E 1656*
Silver	Ag	0.05	0.20	0.05	0.10	0.05
Iron	Fe	0.04	0.01	0.09	0.97	0.10
Arsenic	As	0.005	0.005	0.005	0.005	0.005
Antimony	Sb	0.10	-	-	-	-
Lead	Pb	0.05	base	0.010	0.10	0.10
Nickel	Ni	0.05	-	0.05	0.20	0.05
Silicon	Si	0.50	0.10	0.10	0.10	tr
Magnesium	Mg	0.10	0.10	0.05	0.05	0.20
Copper	Cu	base	0.20	base	base	base
Zinc	Zn	0.05	-	0.007	0.008	0.003
Tin	Sn	0.07	0.002	0.20	5.30	7.60

Metallography

The four copper objects (A), (B), (C), (D) were investigated to learn about the manufacturing technique. Firstly, the objects were polished and the nature of the inclusions observed under metallographic microscope. They were then etched with ferric chloride, ammonium hydroxide and hydrogen peroxide. The presence of moderately high inclusions indicate improper poling in all these materials.

A. *Wire* : Elongated inclusions indicate a particular direction-showing its direction of working (forging). Etched microstructure indicates that the specimen was in a single phase. Total absence of dendritic cast structure indicates that it was not a cast object but was made by cold forging from a larger piece of metal alone. There is no evidence of annealing. The photomicrograph at a magnification of 200X (Pl.1A) is not a clear one.

B. *Needle*: The object was found with inclusions spherical in shape. After etching (photomicrograph, Pl.1B) was obtained at a magnification of 500X reveals a broken down cast structure. A few twins are there, though not completely recrystallised, indicate the absence of annealing process. Microstructure also indicates that it is in single phase.

C. *Bangle* (at a magnification of 500X, Pl.1C) indicates highly corroded state. A few spherical inclusions were also visible. Due to corrosion nothing has been revealed about its manufacturing technique. The only inference that can be drawn is that it has more than single phase.

D. *Bangle*: The microstructure indicates spherical inclusions. The object itself was in highly corroded state yet its photomicrograph clearly indicates a dendritic cast structure (Pl. 1D) at a magnification of 500X. From chemical analyses also it is clear that these are in two phases - an alloy of copper and tin i.e., bronze.

The results of analyses of these objects highlighted the manufacturing technique of copper objects of sub-period IB and Period II. It is now clear that metallurgy at Senuwar started in Period IB as early as around 1800 BC as the smiths learnt to smelt copper ore for procuring metallic copper. The use of native copper is not attested in manufacturing artifacts. There is no evidence of alloying in IB. The study confirmed the use of pure copper in manufacturing the artifacts. The purity comes to around 99%. Even with the presence of 0.20% Sn in one object, it clearly indicates that there is no evidence of alloying. The small amount of tin in that object may be due to the association of copper ores with tin. After reduction of metal, artifacts were made by hammering method. Lack of twins also clearly suggests that simple forging process i.e., hammering in cold condition was practiced. Microstructures also indicate this process without annealing.

TABLE 3

Periodwise distribution of copper and lead objects

Objects	*Period IB*	*Period II*
1. Wire	2	-
2. Fish-hook	1	-
3. Pointed tool/piercer	3	1
4. Needle (?)	-	1
5. Nail	1	1
6. Lead rod	1	-
7. Ring	1	-
8. Rod with rectangular section	1	-
9. Miniature arrowhead	-	1
10. Potsherd adhered with copper	-	1
11. Rod	3	1
12. Bar	1	-
13. Bangle	1	2
14. Indeterminate	4	1
Total	**19**	**10**

In the succeeding Period II alloying is well attested to which may be considered as an advancement over the preceding sub-period. During the period the content of alloying element is restricted to tin only. Thus the period may be credited with the beginning of bronze manufacturing. Craftsman had not only learnt the use of tin as an alloying element but also learnt that its presence improves the fluidity in casting practices. It has been noted that the content of tin is around 5% indicating non-availability of cassiterite in large amount. The comparatively higher content of tin in Period II against period IB clearly indicates that tin had not come as part of the copper ore itself. During this period the objects were manufactured by casting process but objects made by hammering method are not wanting. In such cases no further heating could be identified. Thus from the above study a gradual technological development from Neolithic-Chalcolithic to Chalcolithic periods is witnessed at Senuwar.

REFERENCES

Agrawal, D.P., Krishnamurthy and Kusumgar, Sheela, 1978. 'New data on the Copper Hoards and Diamabad bronzes'. *Man and Environment*, 2, pp. 41-46.

Friedman, A.M. et. al., 1966. 'Copper artifacts correlation with source types of copper ore'. *Science*, 152, 3728, 1504, 1506.

Singh, Birendra Pratap, 1990. 'Early Farming Communities of Kaimur Foot-hills'. *Puratattva*, No. 19, pp. 6-18.

Metallographic Studies of Copper Antiquities from Manjhi (District Saran, Bihar), 400 B.C. - 300 A.D.

HARI NARAIN, JAI PRAKASH
M.V. NAIR AND O.P. AGRAWAL

The development of copper metallurgy in India is only just beginning to be pieced together by Archaeologists from a number of sites and the course of this development remains the subject of heated debate among the scholars. The ancient metallurgy of Manjhi (Nagaraja Rao, 1986; Tripathi, 1987; Roy 1985-86) is particularly interesting as the iron alongwith the copper metal comes from the early period around 400 B.C. Manjhi, which is the site reported in this paper is 19 Kilometers west of the District headquarters in the Saran District of Bihar. The ancient site is sprawling along the left bank of the Ghaghra river, the size of the mound is about 526 m. from North to South, but only about 390 m from East to West. The explorations conducted at this site revealed sequence of cultures from about the beginning of the first millennium B.C. upto early medieval times. The culture content of the different periods are given below:

1. Period I (Black Slipped Ware Culture)

(From beginning to the middle of first millennium B.C.).

No metallic antiquity was recovered from this period, except pottery of different shape and size, characterized as black slipped ware, black-and-red ware, and associated red ware.

2. Period II (NBPW 600 B.C. - 50 B.C.)

This period is characterized by the introduction of the Northern Black Polished Ware (NBPW) marked with the greater use of iron. In this period, the use of copper remained subordinate to iron. A heavily corroded cast coin of copper and two inscribed sealings of terracotta indicates the prevalence of coinage and of writing during the late levels of this period.

3. Period III (50. B.C. - 300 A.D.)

The cultural equipment of this period showed marked contrast with that of the proceeding one. NBPW was completely absent. **This** period yielded the largest

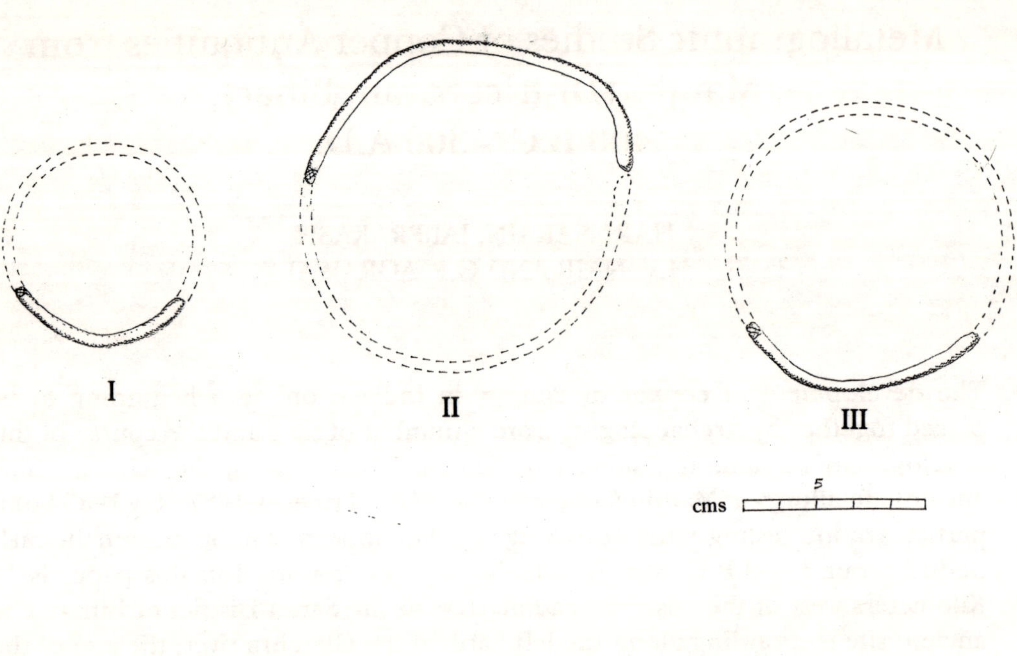

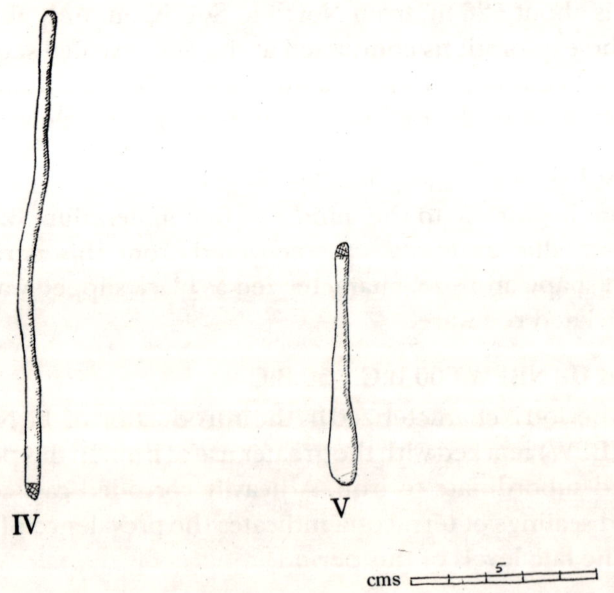

Fig. 1

number of antiquities. An inscribed terracotta sealing which, on the basis of palaeography, can be assigned to 2nd-3rd century A.D., was an important find of this period.

4. Period IV (300 A.D. - Medieval period)

A small group of people belonging to Medieval age settled over the earlier ruins after a considerable lapse of time. The present evidence shows that the site was not urbanised before 400-300 B.C. Radiocarbon dates (Tripathi, 1987) from this site for the different periods are given below:

(a)	Early phase of NBP	470 B.C. - 500 B.C.
(b)	Late phase of NBP	40 B.C.
(c)	Pd. III	230 A.D.

Twentyseven copper artifcats from 400 B.C. - 300 A.D. were received by National Research Laboratory for Conservation of Cultural Property, Lucknow, for conducting metallographic and other technical studies. Five copper artifacts consisting (i-iii) three bangles, Cu 96, 94 & 95, (iv) an antimony rod, Cu 97 and (v) a rod (possibly tool), Cu 98, Figure 1, were selected from this site for present study with the aim to determine the most likely method by which these artifacts were produced around 400 B.C. - 300 A.D.

TABLE 1

Atomic Absorption Analysis of five Artifacts from Manjhi (Bihar)

S. N.	*Description of Artifact*	*Lab No.*	*Archaeological date*	*Composition %*							
				Ag	*Pb*	*Zn*	*Ni*	*Fe*	*An*	*Sn**	*Cu*
1.	Bangle	Cu 96	500B.C.-50B.C.	0.01	1.45	0.17	0.01	0.07	h.d.	8.80	Maj.
2.	-do-	Cu 94	50B.C.-300A.D.	0.013	2.48	0.28	0.15	0.05	h.d.	6.80	-do-
3.	-do-	Cu 95	-do-	0.02	1.45	0.31	0.09	0.01	h.d.	6.00	-do-
4.	Antimony rod	Cu 97	-do-	0.013	0.85	0.12	0.05	0.04	h.d.	0.90	-do-
5.	Rod (possibly tool)	Cu 98	-do-	0.01	1.80	0.35	0.12	0.01	h.d.	1.70	-do-

*— Micro chemical analysis

Metallographic Examination

Five artifacts from the site Manjhi (Bihar) were selected for study using a Buehler Isomat low speed saw, small cross section of metal was removed from each artifacts. The details of section are shown in Figure 1. The depth of these cuts were not more than 2-5 mm. The notches in these artifacts were virtually invisible after these had been restored by a skilled restorer. The cut sections were mounted in transparent thermoplastic resin, ground smooth on abrasive wheels and polished with 0.3 alumina and 0.05 gamma alumina emulsion powder to produce an unscratched surface which could be viewed with a reflected light microscope. Each sample was then treated with an etching agent which differentially attacked the grain boundaries, thereby revealing the microstructure of the individual grain. The ferric chloride solution was used as etching agent. Further samples were drilled from each artifact with mini iron carbide drill of 5mm dia. The first drilling was discarded and second drilling was used for chemical analysis by Atomic Absorption Spectrometer, by following standard procedures. The results of the analysis are recorded in Table 1.

The microhardness of specimens was determined in the Vickers Pyramid scale after standardizing the equipment (Leitz Automatic Microhardness Tester) with 100 g. load. The photomicrographs of the prepared sections were taken on the cut film (black and white) size 8.10 x 10.6 cm. having speed 125 ASA.

Fragment of Copper Bangle Cu 96 (400 B.C. - 50 B.C.)

The bangle was heavily corroded, having deposition of green colour corrosion products and silicious material all over the surface. The bangle is in fragmentary condition, from its circular shape it appeared that approximately half of its circumference is missing. Its outer and inner dia. is 48 mm and 45 mm respectively, and weighing 2.07 gm. All the accretions and green colour corrosion products were removed mechanically with small wooden tools before drawing the sample for metallographic examination. The colour of the exterior patina was dark green. The sample was removed from one of the broken ends as shown in Figure 1 (I) measuring 3mm. in dia. Optical examination at low magnification revealed (at 50X) the corrosion attack, which has penetrated quite deep inside from the peripheral area. The severe attack of the corrosion was in the peripheral area (Pl. 2). A wide crack is present in the metal core which has divided the section into three parts. The metal was free from the cuprous inclusions but having some black colour (slightly elongated) inclusions. The badly corroded surviving metal revealed upon etching very fine grain equiaxed twinned grain structure with slip banding, (Pl. 3), (Scot, 1986; Coghlan, 1977). Core was not observed in the metal. The inter- granular corrosion was present more near the periphery, (Pl. 4), and it was also located near the cracks in the core, this indicates that the bangle piece was very deeply corroded. Average microhardness of the metal was HV 105.

All the above studies indicates that the metal used in the bangle was nicely homogenized and the measure of tin 8.8% has deoxidised the metal and made it free from cuprous oxide inclusions or probably the numerous annealing and working cycles has absorbed this into the metal itself. The metal of the bangle is having the severe attack of inter-granular corrosion on the periphery and the core, probably the cracks in the bangle has occurred due to severe cold working and the penetration of corrosion has followed the cracks. Micro-structure and microhardness indicates that the bangle which was heavily cold worked and annealed at low temperature and later on also worked for final shaping, resulting in the slight hardening.

Copper Bangle, Cu 94 (50 B.C. - 300 A.D.)

The bangle was having the soil deposition and green powdery material at few places. It is having rectangular shaped cross-section with an outer diameter of 80 mm and an inner diameter of 75 mm and weighing 7.83 gm (Figure 1.II). Almost half of the bangle portion was missing. Before drawing the sample for metallographic studies, the soil deposition and the green powdery material was mechanically cleaned. Cleaning revealed the exterior patina of the bangle in dark green. The sample removed for examination was drawn from one of these broken ends an shown in Figure 1 (II). Measuring 5 mm x 2 mm with slightly rounded corners.

Macro examination of the section revealed at 50 X, that corrosion attack was more severe near the periphery than in the core. Dove grey and black colour elongated inclusions (Coghlan, 1975; Gordon, 1986; Tylecote et. al., 1971) and corrosion cavities were present in the section. Under plane polarized light, no copper oxide could be detected in the section.

Etching revealed the fine equiaxed twinned grain structure with slip band in the surviving metal (Pl. 5) (Coghlan, 1970). No evidence of coring could be seen. The large amount of inter-granular corrosion near the periphery reveals that the bangle was much more deeply corroded than its present state suggests. Average microhardness of the surviving metal was HV 98.

Microscopic and microhardness studies lead us to the conclusion that the bangle which was severely corroded was fabricated from the good homogenized metal, having the tin content 6.8% which deoxidises the copper oxide inclusions. The dove grey and black coloured elongated inclusions which are due to the slaggy origin and lead respectively, indicate that the bangle was also hot worked slightly at some stage, but not at too much high temperature. it was not made as a casting but fabricated from one piece of metal of 2mm. thickness. The absence of the coring indicates that the metal was homogenized. In view of the small grain size (twinned crystal structure with slip) observed, it suggests that the metal was subjected to severe cold working prior to annealing and the annealing was carried out at low temperature and in the final stage of fabrication, the bangle was again cold worked, which resulted in its slight hardening.

Bangle, Cu 95 (50 B.C. - 300 A.D.)

The bangle shown in Figure 1 (III), measures 74 mm. in length, average thickness 2 mm, outer and inner diameter is 67 mm and 65 mm respectively and weight 5.6 gms. The fragment of bangle and both the ends of this piece are corroded, indicating that the break is not recent. The piece of bangle is having greenish malignant patina. All the silicious accretions and corrosion products were mechanically cleaned before drawing the sample for examination. The sample was drawn from one of the broken end as shown in Figure 1 (III) revealing cross section of a flattened 'D' measuring 4 mm x 2 mm.

Upon examination of the polished but unetched section, it was seen that the metal is not clean, containing dove grey and black non-metallic inclusions, which are probably of slaggy origin and lead respectively. These inclusions are partially elongated in the forging direction. The cuprous oxide inclusions were not detected in the section which indicates the good deoxidation of the copper metal.

The surviving metal in the section, after etching, revealed the equiaxed twinned grain microstructure with slip banding but no coring was seen in the specimen. The intergranular corrosion was present more near the periphery, (Pl. 6) and its presence decreases gradually toward core. The average microhardness of the surviving metal was HV 84.

All the above studies indicates that this bangle was fabricated in the similar manner as described for bangle Cu. 94.

Antimony rod, Cu 97 (50 B.C. - 300 A.D.)

The antimony rod shown in Figure 1 (IV) is 130 mm long having a maximum width of 5 mm. It's weight was 8.50 gm. Visual examination suggests that the rod is in fair condition, green patinated but not seriously attacked by corrosion, as we have seen in the three bangles. A cross section measuring 3 mm in dia. was drawn from the extreme portion of one end as shown in Figure 1 (IV).

Macro examination after polishing, but in unetched state, showed many dove grey and black colour inclusion in globular shape scattered over the section. The metal is free from copper oxide inclusions. The metal in the section was having some corrosion present near the periphery at some places whereas the core was free from corrosion. The metal in the section was more or less similar to coppery colour which indicates that the other metal was added in very low amount. Elemental analysis also indicates copper as major element and tin was 0.9%.

Upon etching, the metal revealed very fine equiaxed twinned grain crystal structure without slip band, (Pl. 7) (Scot, 1976; McCall et. al. 1971). No coring was seen in the section. In the section near the peripheral area, there are three places which exhibit penetrative corrosion attack (inter-granular). The average microhardness was HV 124.

All the above studies conducted on the antimony rod Cu 97 lead us to the conclusion that the artifact was very heavily forged to approximately circular section worked at low temperature and annealed at tolerably low temperature and left to cool as such. The black globular traces were due to the lead and the absence of copper oxide inclusions indicates that probably they were deoxidised by tin content or could have been absorbed in the course of numerous annealing and working cycles, but the antimony rod was not hot forged and worked after final anneal.

Rod (possibly some tool) Cu 98 (50 B.C. - 300 A.D.)

The artifact shown in Figure 1 (V) is 60 mm long having dia. of 2 mm. The weight of the artifact as received is 5.40 gm. This artifact which is possibly a tool is having green patination but is in a perfect state of preservation. For metallographic examination section was drawn from one end.

Unetched metal reveals the presence of black colour partially elongated inclusions and it also reveals the severe corrosion attack near the periphery, and in some zone stress cracks were seen running from the peripheral area toward core, light green corrosion clusters were seen in the core. Cuprous oxide inclusions were absent indicating that they might have been deoxidized by tin content which was 1.7% in the metal.

Etching revealed the moderate size equiaxed twinned grain structure with slip band, (Pl. 8) (Tylecote 1973) but no coring was traceable in the metal. The intergranular corrosion attack was very severe near the periphery which decreases gradually toward the core. The average microhardness was HV 127.

All the above studies indicate that this artifact (possibly a tool) has been given a low temperature anneal after cold working. The tool was also mechanically worked after the annealing for imparting some hardness.

Discussion

The chemical analysis of all the five artifacts comprising three bangles Cu 96, 94 & 95, an antimony rod Cu 97 and a rod Cu 98 (possibly a tool) were having the copper and tin as major elements. The other elements lead, iron, zinc, silver and nickel were present as impurities. The tin and lead varies from 0.9-8. 80% and 0.85-2.48% respectively. The other elements silver, zinc, nickel & iron varies from 0.01-0.02%, 0.12-0.35%, 0.01-0.15% and 0.01-0.7% respectively. Gold was not detected. The presence of copper and tin as major constituents in all these artifacts confirms that all these were tin bronzes having some lead content.

The presence of dove grey inclusions in these artifacts indicates that some sulfide bearing copper ores rather than oxide ores were smelted to produce the metals from which the artifacts were made.

Examination of the polished but etched metal of the sections from these five artifacts showed that the patination is thick and corrosion is severe except antimony rod Cu 97, hence the environment under the soil must have been not favourable for the preservation of the bronze.

Microscopic examination showed that all the artifacts were free from copper oxide inclusions, probably the poling done in the melting stage and the presence of tin content has deoxidized the copper metal. The absence of coring in all the five artifacts indicates that the homogenisation of the metal was done very carefully. The microscopic examination of the copper bangle Cu 96, indicates that it was fabricated from a rod having a circular section by cold working and annealing at low temperature followed by slight work hardening. The thin strips used in bangles Cu 94 and 95 having the presence of plastically deformed dove grey and black colour inclusions indicates that those may have been forged from an ingot by hot and cold working followed by annealing at low temperature, preparation of strip alone would call for skill, finally an excellent job was made of the finishing operation. The smith was clearly well skilled in the art of sheet metal working.

The antimony rod Cu 97 was forged by severe cold working and annealing at low temperature. The artifact was less corroded in comparison to others. No slip bands was seen in the antimony rod Cu 97, which indicates that after the final annealing no working was performed and the artifact left as such.

Rod Cu 98 (possibly a tool) was forged by cold hammering and annealing at moderate temperature followed by work hardening. The presence of slip lines and hardness variation from HV 84-127, indicates that the smith probably knew that some hardening over the surface area would be necessary in order that bangles and tool made of such thin bronze (average thickness 2 mm) could maintain its shape during use.

CONCLUSION

The metallographic examination of five artifacts from the ruins of Manjhi from 400 B.C. - 300 A.D. with one exception were found to be cold worked, annealed and work hardened products. The one exception was cold worked and annealed. The inclusion in the artifacts indicated that the copper was produced from sulfide ores. All the five metallic artifacts were leaded tin bronze alloy and the alloy probably was made by smelting intentional mixtures of copper and tin ores. The smith was clearly well skilled in the art of metal working around 400 B.C. - 300 A.D. and one must admire the skill of the ancient craftsman who worked his intractable metal into well designed products.

ACKNOWLEDGEMENTS

The authors are thankful to Shri Anil R. Singh for photographic work and Shri K.K. Narang, for conducting the elemental analysis and Shri Shiv Balak for making the drawings. We are also grateful to Prof. T.N. Roy, Deptt. of Ancient Indian History, Culture and Archaeology, Banaras Hindu University, Varanasi, for providing us the samples and the information regarding the archaeological site.

REFERENCES

Coghlan, H.H., 1975. Copper artifacts from Tutankhamun's tomb, *Journal of the Historical Metallurgy Society*, 9(2), 64-67.

Coghlan, H.H., 1977. A bronze strigil from Greece examined, *Journal of the Historical Metallurgy Society*, 11(2), 86-87.

Coghlan, H.H., 1980. Examination of some continental bronze age decorative objects, *Journal of the Historical Metallurgy Society*, 14(2), 80-93.

Gordon, R.B., 1986. Metallography of brass in a 16th Century artrolobe. *Journal of the Historical Metallurgy Society*, 20(2), 93-96.

McCall, J.N. and Buchheit, R.D., 1971. Metallographic Studies of Archaeological metal artifacts from South, *Arabia*, 4, 189-207.

Nagaraja Rao, M.S. (Ed.), 1985. *Indian Archaeology - A Review*, 1983-84, Archaeological Survey of India, New Delhi, 15-16.

Roy, T.N., 1985-86. Excavations at Manjhi 1983-85: A preliminary report, *Puratattava*, 16, 29-32.

Scot, D. 1986, *Metallography of ancient metallic artifacts*. Institute of Archaeology, London.

Tripathi, R.C. (Ed.), 1987. *Indian Archaeology - A Review*, 1984-85, Archaeological Survey of India, New Delhi, 12-13.

Tylecote, R.F. and Mckerrell, H., 1971. Examination of copper alloy tools from Talyahya, Iran, *Bulletin of the Historical Metallurgy Group*, 5(1), 37-38.

Tylecote R.F., 1973. Casting copper and bronze into stone moulds. *Bulletin of the Historical Metallurgy Group* 7(1), 1-5.

Iron Technology at the Chalcolithic Phase of West Bengal

ASOK DATTA

Technology distinguishes early man from other primates. It helps man in securing better living conditions. The innovation of better form of technology is always accompanied with progressive changes in the society. In sharp contrast to the earliest stone age technology which was basically concerned with human survival, the metal technology of the protohistoric period offered wide range of refinements and modifications in terms of various ways including further intensification of agricultural production. In fact the process of domestication of plants, introduced during Neolithic period, was more vigorously carried out during subsequent Chalcolithic period with the help of metal technology. Metal technology includes exploitation of copper, bronze, silver, gold, iron etc. and these minerals except silver are found to be present with varying proportion in the Chalcolithic phase of West Bengal. But among them iron was perhaps the most significant element of metal technology which brought about a drastic change in the material culture of the people of West Bengal during the formative period. The present paper aims at examining different aspects of early phase of iron technology of the Chalcolithic phase of West Bengal in terms of Archaeo-metallurgical analysis.

The origin of iron smelting technology in India is a disputed matter (Datta, 1991). The present available data suggests that instead of looking for a single origin of iron technology in India, we should look for different foci, areas where possibly iron technology was invented independently in different times and under different environmental conditions (Datta, 1991). But so far as West Bengal is concerned, the earliest evidence of iron smelting technology is found to be associated with Chalcolithic culture. Extensive explorations followed by selective excavations at different sites during the last three decades have brought about significant changes in our approach to the problems of Chalcolithic culture in West Bengal. In view of the recent discovery of large number of iron objects from stratified Chalcolithic context at a number of excavated sites, it is indeed proper and perhaps apt to change the connotation of the term Chalcolithic. The term Chalcolithic is no more valid in the present context and should be replaced by the term iron age. Present studies have shown that out of twelve excavated Chalcolithic sites, iron objects have been reported from pure Chalcolithic context in five sites while the remaining sites have yielded iron objects from period II or iron age which we consider here as transitional stage. These divisions between Chalcolithic and iron age is superficial because the matrix of this period having remained the same. The differences are found in the increase and decrease of certain elements of these periods.

Naturally the previous concept of dividing the protohistoric culture of West Bengal as period I (Chalcolithic) and period II (Iron age) does not hold good. Similarly, the proposed concept of copper-stone (early) and copper-stone-iron (late) (Ghosh, Nag, Chattopadhyay, 1987-88) also does not throw meaningful light in a wider spectrum. In fact, that distinction between iron free Chalcolithic and iron bearing Chalcolithic phases of this period is hardly tenable in the context of recent discovery of iron objects from the basal levels of a number of Chalcolithic sites.

TABLE 1

Distribution of Iron objects from different stratified Chalcolithic sites in West Bengal

Sites	*Chalcolithic Period I*	*Iron age of Chalcolithic period II or Transitional*	*Early Historic*
Pandurajardhibi	Disc, Dagger, Spearhead, Slag	Dagger, Spearheads, Points, Sickle, Sword-blade, Crucible slag etc.	Iron objects
Mangalkot	Arrowhead, Spearhead, Slag Rod etc.	Sickle, Chisel, Spearhead, Arrowhead, Nail, Peg, Rod, Slag etc.	Iron objects
Dihar	Arrowhead, Nail Slag etc.	Nail, Arrowhead, Slag etc.	—
Mahisdal	—	Arrowhead, Spearhead, Chisel, Ore, Slag etc.	—
Hatigra	—	Dagger, Rod, Lumps, Slag etc.	—
Bharatpur	—	Iron objects	—
Baneswardanga	—	Iron objects	—
Nanoor	—	Iron objects	—
Tulsipur	Iron-objects	Iron objects	—
Tamluk	—	Iron objects	—

Chakraborty and Hassan (1982) sharing the same view have suggested that the distinction is not possible at least, in case of Bahiri where iron slags have been found from the very beginning. Table 1 shows the distribution of iron objects from different stratified Chalcolithic sites in West Bengal. It shows that Pandurajardhibi, Mangalkot, Dihar, Bahiri and Tulsipur have yielded iron objects from pure Chalcolithic context. The iron objects during this phase mainly comprising arrowhead, spearheads, dagger, nail, rod, slag etc. were mainly connected with hunting and other domestic purposes. But the vast majority of iron tools belong to Chalcolithic period II or iron age or transitional stage which comprise sickle, chisel, swordblades, peg, points etc. besides the earlier types. The table further shows that except Haraipur, all the remaining excavated sites have yielded iron objects during this stage. In view of the large quantity of iron objects from different Chalcolithic sites, it is quite reasonable to assume that the iron smelting technology in West Bengal had a long history (Datta, 1991).

TABLE 2

Percentage composition and impurity pattern of Iron objects from Mangalkot, Pandurajardhibi and Hatigra

	Mangalkot			*Pandurajardhibi*		*Hatigra*
	1	2	3	1	2	
SiO_2	29.26	25.98	22.48	—	—	.102
FeO	48.59	53.67	56.39	Base	Base	Base
Fe_2O_3	—	—	—	—	—	—
Al_2O_3	3.71	2.29	3.06	—	0.05	.027
MnO	0.85	0.42	0.34	—	0.1	—
CaO	3.50	2.80	3.50	—	0.05	—
MgO	3.79	0.99	7.49	—	—	—
TiO_2	0.58	0.27	0.93	—	—	.008
P_2O_5	2.07	1.07	3.09	—	—	.51
Cu	0.20	0.10	0.10	0.153	0.001	.108
Ni	0.20	0.10	0.10	0.128	—	.036
Co	0.30	0.10	0.10	0.111	—	.037
Sn	—	—	—	—	—	.007
S	—	—	—	0.22	—	.015
C	—	—	—	—	—	.35

Archaeo-metallurgical Analysis

The role of iron technology in ushering a new era of surplus production during the formative or transitional stage in West Bengal cannot be minimized in view of the recent discovery of large number of iron tools including iron sickle from transitional stage. Metal technology, specially iron played the most crucial role in social and economic transformations.

The science of Archaeo-metallurgy in India studies the nature and character of metal technology during different protohistoric phases that turn out numerous iron objects from different stratified contexts. The pioneering work of this science has been carried out by a number of scholars including Hegde (1973, 1981), Bharadwaj (1973, 1985), Tripathi (1973, 1990), Chakraborty (1979, 1988), Chattopadhyay (1991), Datta (1991a, 1991b, 1991c) etc. In recent years, Archaeo-metallurgical studies from limited number of excavated Chalcolithic sites including Pandurajardhibi (De and Chattopadhyay, 1989), Hatigra (Ghosh, Nag and Chattopadhyay, 1987-88), Mangalkot (Datta, 1991a, 1991b) etc. are available throwing up a new horizon in the study of early iron technology of West Bengal. The study not only reveals the nature and character of early metal technology, but also throws significant light on the problem of second urbanization (Datta, 1991). Both finished iron objects and slag specimens have been examined chemically as well as metallographically from Pandurajardhibi, Mangalkot and Hatigra. The results of the analysis are given in table 2 and 3. Chronologically, examined iron specimens from Mangalkot comprising mostly rods and unidentifiable objects belong to pure Chalcolithic level while the iron specimens consisting of disc, sickle and dagger from Pandurajardhibi and Hatigra respectively belong to transitional stage, i.e. iron age or period II. Table 2 shows a comparative study of percentage distribution and composition of impurity pattern of regular iron objects. It reveals that both at Pandurajardhibi and Hatigra, the percentage distribution of impurity pattern are very low while the percentage distribution of impurity pattern of Mangalkot specimens are very high. For example the distribution pattern of silica is very high in Mangalkot specimens while silica are almost absent both at Pandurajardhibi and Hatigra. This shows that much of the iron was lost in slags in case of Mangalkot specimens. Moreover other trace elements like Aluminium, Calcium, Manganese, Magnecium and Phosphorous are found to be present in good percentages in the specimens from Mangalkot while these elements are either absent or present in negligible percentages in the specimens from Pandurajardhibi and Hatigra. In fact the presence of this low percentages both at Hatigra and Pandurajardhibi may be due to natural composition of ore. However, the pattern of percentage distribution of impurities as shown in table 2 clearly indicate that the iron smelting technology at Mangalkot was less developed than Pandurajardhibi and Hatigra. Iron lost in slags in Mangalkot may be due to inability of successful reduction. Perhaps the blacksmiths of Mangalkot during Chalcolithic phase could not produce the required temperature for a successful reduction. Moreover, a small percentage of carbon content are found to be present in two specimens one each from Pandurajardhibi and Hatigra. The respective carbon percentage is 0.22 and 0.35 which definitely indicates that the knowledge of carburization was already known to the Chalcolithic people of West Bengal during transitional or period II or iron age. Because, of the presence of this amount of carbon in iron it may be taken as low grade steel or mild steel. One thing is very clear from the above comparative study that the Chalcolithic people developed the iron smelting technology from the very beginning but since the technology was in experimental stage, initially the people could not get rid of high percentage of impurities due to not very satisfactory reduction. But subsequently the technology was modified and perfected in term of furnace management during transitional or iron age (period II) and as such better products comparatively less free of impurities are found to be present. Furthermore, the knowledge of carburization was already acquired by this time. The presence of nickel, cobalt and copper in small quantities in the specimens from Mangalkot may indicate that the source material of iron at Mangalkot during Chalcolithic stage was meteoric. Table 3 shows the percentage distribution and composition of impurity pattern of iron slags from Pandurajardhibi and Mangalkot. The table reveals a higher impurity pattern in case of Mangalkot specimens than Pandurajardhibi. Metallographic study shows that iron was shaped into bloom by using charcoal as fuel. The bloom was then hammered into different shapes with the help of further heating and forging. The study further reveals that

TABLE 3

Percentage distribution and composition of impurity pattern of Iron slag from Mangalkot and Pandurajardhibi

	Mangalkot			*Pandurajardhibi*		
	1	2	3	1	2	3
SiO_2	38.50	25.20	21.70	23.86	23.78	12.54
FeO	11.00	25.60	9.60	43.86	41.02	33.74
Fe_2O_3	34.00	34.80	47.50	24.31	20.31	43.76
MnO	—	—	—	1.26	3.52	0.22
P_2O_5	1.49	1.33	1.61	0.57	0.84	0.73
CaO	1.90	1.60	5.20	0.42	0.56	2.60
MgO	3.60	6.30	3.10	0.70	1.15	1.20
Al_2O_3	6.30	4.70	8.30	4.02	7.82	1.79
TiO_2	0.66	0.93	0.89	0.80	0.80	Trace
Cu	0.03	0.03	0.01	—	—	—

the specimens from Pandurajardhibi and Hatigra were thoroughly forged under high temperature and as such the slag particles were completely driven out. But in sharp contrast, the Mangalkot specimens were possibly forged under much lower temperature as indicated by the presence of high percentage of slag particles.

Figure 1 shows radio carbon dates of iron bearing phases of different Chalcolithic sites in West Bengal. It shows that six dates are available from Pandurajardhibi, Mangalkot, Bahiri, Hatigra, Mahisdal and Bharatpur*. It further shows that Pandurajardhibi and Mangalkot having 990 B.C. and 950 B.C. as dates for iron bearing phase seem to have had the earliest evidence of iron technology in West Bengal. Since the carbon sample of Mangalkot has been collected from the midphase of Chalcolithic culture, it is assumed that the beginning of iron technology at Mangalkot would go well beyond eleventh/twelfth century B.C. Dihar. another important site, yielding iron objects from its bottom Chalcolithic level, may also produce early date. However, the dates from Mangalkot and Pandurajardhibi are closely followed by dates from Bahiri and other sites.

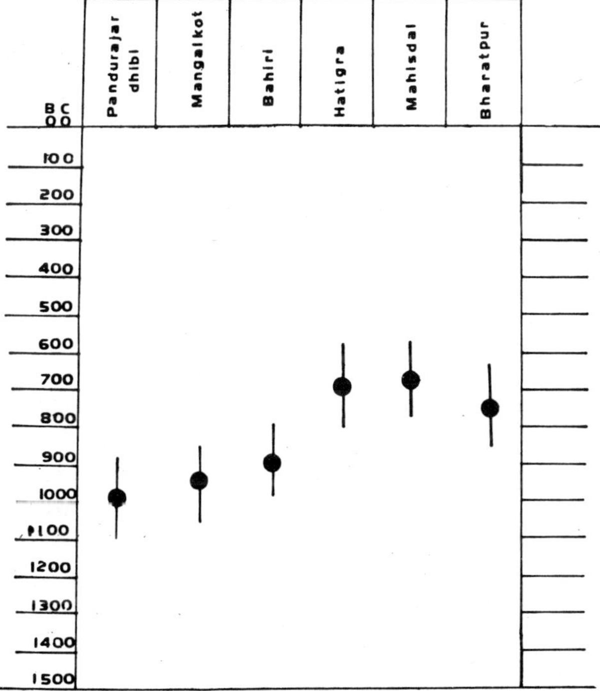

Fig. 1. Showing the Radiocarbon dates of Iron bearing Chalcolithic phase in West Bengal

* Radio carbon dates from Hatigra and Bharatpur are available from it's Chalcolithic context. But the iron objects from these sites have been found from Period II or iron age. The average dates from these sites have been calculated on the basis of Mahisdal period II date.

SUMMARY AND CONCLUSION

From the above study certain facts are very clear which can be listed as follows:

1. The beginning of iron technology in West Bengal may safely be placed to eleventh/twelfth century B.C.
2. Twofold developmental stages of iron technology can be ascertained on the basis of chemical and metallographic study.
3. The study shows that initial development of iron technology was made by the Chalcolithic people. But due to inability of successful reduction, much of the iron was lost in slag during this stage I. The specimens from Mangalkot belong to this stage which contain high percentage of impurities as compared to the specimens from Pandurajardhibi and Hatigra.
4. But soon after, the technology was modified and perfected during stage II, and as such the iron implements of this stage are completely free from slag particles due to successful reduction under sufficiently high temperature. The iron objects from Pandurajardhibi and Hatigra belong to this stage.
5. Stage II also witnessed the introduction of mild steel as evidenced from the presence of carbon contents. The deliberate addition of carbon with iron is definitely a step forward in iron technology. This high technical knowledge was acquired by the blacksmiths of West Bengal by about 6th/7th century B.C.
6. Finally, we can conclude that technology which was developed independently by the Chalcolithic people of West Bengal, in course of time (Stage II) became very sophisticated and must have played the most crucial role both in social and economic transformation of the societies during the formative stage of second urbanization.

Editor's Note

So far not a single iron of meteoric origin has been reported in India. Datta's deduction about Mangalkot iron being meteoric is interesting indeed. But the analysis results given in the paper do not substantiate the inference. The reported impurity pattern shows a small percentage of Nickel which should be at least 4.5 percent to prove it meteoritic iron. The assumption on this single count alone becomes ill founded and erroneous.

REFERENCES

Bharadwaj, H.C., 1973. Aspects of early iron technology in India. *Radiocarbon and Indian Archaeology*. D.P. Agrawal and A. Ghosh (ed.), pp. 391-401. Delhi.

Bharadwaj, H.C., 1985. Metal technology in ancient India *Archaeological perspective since independence.* K.N. Dikshit (ed.), pp. 139-145. Delhi.

Chakrabotry, D.K., 1979. Early iron age in the Indian north-west. *Essays in Indian pre-history.* D.P. Agrawal and D.K. Chakraborty (ed.), pp. 346-364. Delhi.

Chakraborty, D.K. and Jamal Hassan, 1982. The sequence of Bahiri, district Birbhum, West Bengal. *Man and Environment,* 4: 111-149. Poona.

Datta, A., 1991. Cultural migration in West Bengal (Prehistoric period). *Studies in Archaeology.* A. Datta (ed.), pp. 173-202. New Delhi.

Datta, A., 1991a. *Early stages of iron technology and development of regional pattern* (in Press).

Datta, A., 1991b. *Problem of urbanization and its antecedent state in West Bengal* (Press).

Datta, A., 1991c. Iron Technology in the Chalcolithic phase of West Bengal. (Press).

De, S. and Chattopadhyay, P.K.,1989. Iron objects from Pandurajardhibi: Archaeometallurgical studies. *Steel India* 12: 33-41.

Ghosh, N.C., Nag, A.K. and Chattopadhyay, P.K., 1987-88. The archaeological background and iron samples from Hatigra. *Purattatva* 18: 21-27.

Hegde, K.T.M., 1973. Early stages of metallurgy in India. *Radio-carbon and Indian Archaeology.* D.P. Agarwal and A. Ghosh (ed.), pp. 391-401. Delhi.

Hegde, K.T.M., 1981. Scientific basis and technology of ancient copper and iron metallurgy. *Indian Journal of the History of Science* 16 (1): 189-201.

Tripathi, Vibha, 1973. The introduction of iron in India - A chronological perspective. *Radiocarbon and Indian Archaeology,* D.P. Agarwal and A. Ghosh (ed.) pp. 272-279. Delhi.

Tripathi, Vibha, 1976. *The Painted Grey Ware: An iron age culture of northern India.* Concept Publishers, Delhi.

Tripathi, Vibha, 1990. Impact of metal technology on economic development. *Historical Archaeology of India - A dialogue between archaeologists and historians.* A. Ray and S. Mukherjee (ed.), pp. 196-207. New Delhi.

Metallographic Studies of Iron Artifacts of Domestic Use, Tools and Weapons from Sringverpura (250 B.C. - 600 A.D.) Uttar Pradesh

HARI NARAIN, JAI PRAKASH,
O.P. AGRAWAL AND M.V. NAIR

The ancient mound at Sringraur (Sringverpura) (I.A.R. & Roy, 1983) which is also known as Surya Bhita, situated 35 kms upstream of Allahabad on the Ganga. The site is being excavated since December, 1977 as a part of the National Project, "Archaeology of Ramayana site". The work was carried out under the joint auspices of the Indian Institute of Advanced Study, Simla and the Archaeological Survey of India.

The site traditionally associated with Rama who crossed the Ganga after banishment from Ayodhya has yielded as cultural sequence of seven periods, (Lal & Dikshit 1978-79) which are given below:

Period I (1050-1000 B.C.)

The period represents the fag end of the ochre coloured ware culture.

Period II (950-700 B.C.)

Black slipped, black and red, burnished grey and associated red wares. This period is further divided into II A, II B, and II C. The important features are being enumerated below:

II A. Dominance of coarse black and red burnished grey ware.

II B. Bone points, barbed bone arrowhead, a bone pendent and beads of terracotta, jasper and one gold antiquity.

Period III (700-250 B.C.)

The period is divisible into three sub periods III A, III B and III C.

III A. NBP and Black slipped ware, and few sherds of PGW.

III B. The miniature bowl and carinated handi baked bricks.

III C. Burnt brick structure, NBP was on wane and coarse grey ware became prominent.

The important antiquities of period III include copper vessels, long handle laddle, terracotta female figurine (Mauryan), balls, bone points, iron artifacts and uninscribed cast and punch marked coins of silver and copper.

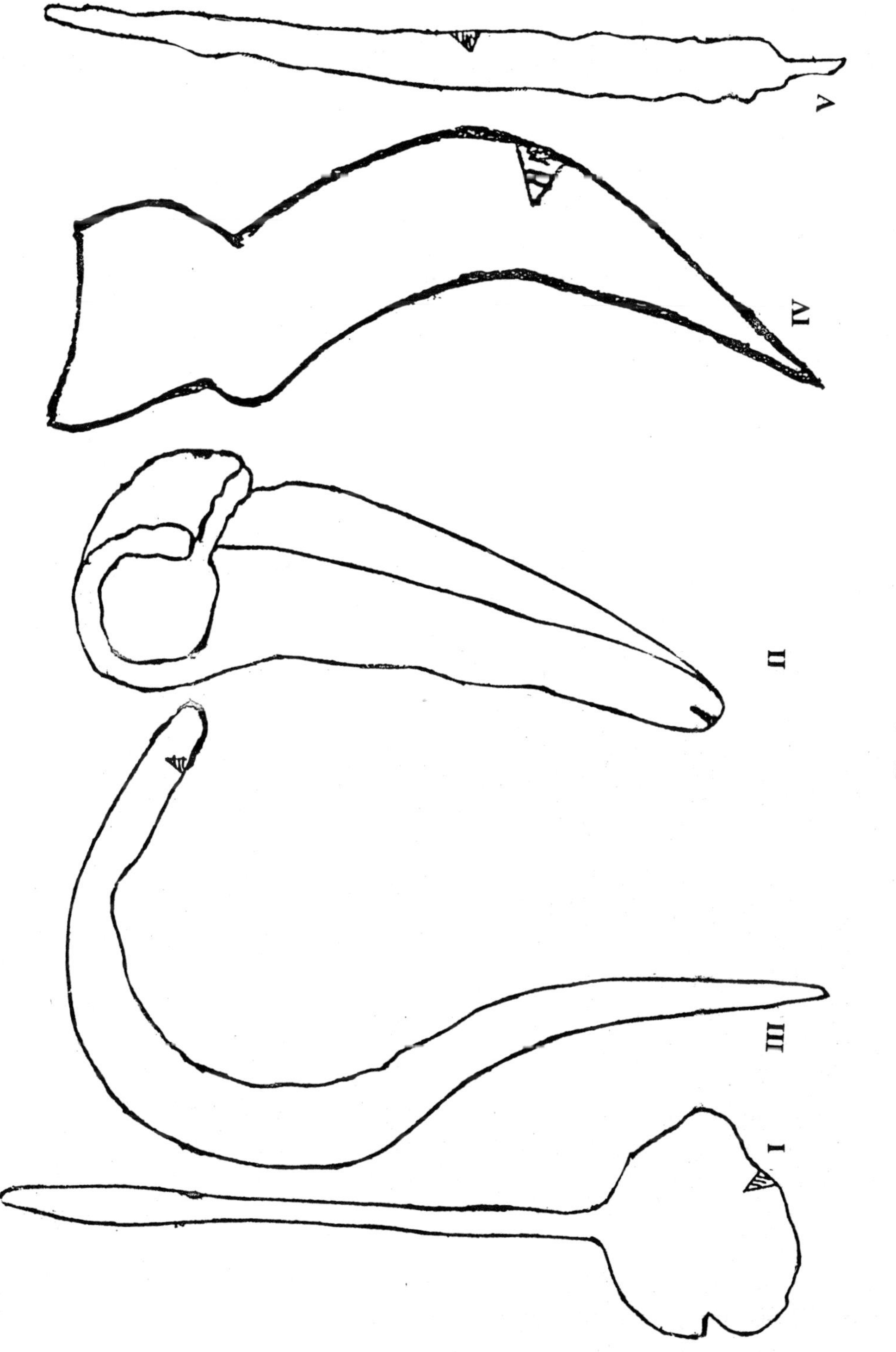

Fig. 1

Period IV (250 B.C.-200 A.D.)

The period is divisible into two sub periods; NBP and grey ware were not found. Only red ware was found.

IVA. Coins of lanky bull type, coins of the ruler of Ayodhya and Sunga terracotta.

IVB. Massive burnt brick tank of two units.

Periods V (200 A.D. - 600 A.D.)

Terracotta figurines stylistically ascribable to the Gupta times.

Period VI (600A.D. - 1300 A.D.)

Silver coins of the Gahadvala king Govinda Chandra and jewellery etc.

Period VII (1300-1800 A.D.)

A large house complex, walls had been robbed but the plan is still intact showing the outlines.

Explorations yielded bone points, barbed bone, arrowheads, bone pendant, beads of terracotta jasper and gold. Important antiquities of copper include vessels, long handled laddle, un-inscribed cast and punch-marked coins of copper and silver. Sringverpura site is very rich in iron and has yielded domestic artifacts (kitchen wares, nails, knives, scissors), tools (sickles, axes), weapons and hunting artifacts (spear and arrowhead) etc.

Fifty five iron artifacts of different types were received from this site for detailed examination. Five artifacts comprising 1. a spoon (lab no. Fe 185), 2. a sickle (lab no. Fe. 182), 3. an axe (lab no. Fe. 178), 4. a knife (lab no. Fe. 247), 5. a spearhead (lab no. Fe. 186), were selected for detailed metallographic and other examination (Figure 1) with the following main objectives.

1. To know the fabrication technique adopted by these smiths.
2. To know the development of iron metallurgy in ancient India in this area.

Metallographic Examination

The sampling for metallographic examination revealed that the artifacts were having sound metal core except for the two specimens. The exceptions are a spearhead (lab no. Fe. 186) and a sickle (lab no. Fe. 182) which were in advance stage of mineralization. All the remaining three artifacts were in a good state of preservation, although the surface appeared to be quite corroded.

Very small, V shaped and cross sections were cut from these artifacts (Figure 1) with a diamond saw. The details of these artifacts and sections are given in Table 1, the depth of these cuts were not more than 10 mm, notches in the artifacts were virtually invisible after they had been skillfully restored. The samples were embedded in a thermoplastic resin and prepared by following standard metallographic procedure (Scoot, 1986; Samuels, 1982). Etching (Samuels, 1978) of the sections was done with 2% and 5% Nital solution for 30-45 seconds. The presence of relic carbide (Curties et al., 1979; Agrawal et al., 1990) in the mineralized metal was traced at higher magnification, along with the microstructure present in the remnant metal in the sections, to help in the proper identification of fabrication technique. The photomicrographs were taken on cut film size 8.1 x 10.6 cms and speed 125 ASA. The micro hardness of these sections was determined on the vickers pyramid scale by a leitz Automatic Microhardness Tester. The equipment was standardized with 200 g load.

INTERPRETATION OF MICROSTRUCTURE

Artifacts of Domestic Use

Spoon (Kitchen Ware) (Agarwal et al., 1990) (lab no. Fe. 185)

This artifact was possibly a broken part of spoon (A kitchen ware length 470 mm, maximum width 13 mm, pan width 73 mm, average thickness 13 mm, and weight 272 gm.).

A V shaped section of size 3 mm x 2 mm was cut from the area of artifact as shown in Figure 1.I. This section of the artifact was selected and polished for examination. Polishing of the section revealed a sound metal core which was firmly embedded in mineralized metal.

The macro examination of the surviving metal showed some black colour elongated inclusions, which were probably of wüstite (FeO) and ferrous silicate. (The surviving metal was free from stress cracks etc.). The surviving metal was etched with 2% Nital for microscopic examination.

TABLE 1

Details of Iron Artifacts subjected to Metallographical Studies from Sringverpura (U.P.) (250 B.C. - 600 A.D.)

S.N.	*Description of Artifact*	*Lab No.*	*Length mm*	*Breadth mm*	*Average thickness mm*	*Weight gms.*
A. *Artifacts of Domestic use*						
1.	Spoon (Broken)	Fe. 185	470	13 Handle 73 Pan (Broken)	13	272.00
B. *Implements*						
1.	Axe	Fe. 178	255	53 nearhole 30 near cutting edge	35	520.00
2.	Sickle	Fe. 182	213	25	5	125.00
3.	Bent Knife	Fe. 247	75	15	7	74.44
C. *Weapon*						
1.	Spearhead	Fe.186	316	13	13	220.00

Optical examination aftei etching showed the homogeneously distributed ferrite and pearlite structure with carbon content (0.1-0.6%) all over the surviving metal in the section (Coghlan, 1978) Pl. 9 shows the widmanstatten pattern of ferrite, which was found in some areas. Pl. 10 is of the another area at higher magnification showing also the lammellar nature of the pearlite. The average microhardness of the surviving metal varried from 195-260 H.V. The widmanstatten structure (Brewer, 1976) would be consistent with hot forging of the kitchen ware, followed by fairly rapid air cooling. The latter is confirmed by the presence of medium lamellae in the pearlite, as coarse lamellar pearlite is only produced by very slow cooling from the austenite temperature. The rapidly air cooling and also the widmanstatten structure, make it difficult to make an accurate estimate of carbon content of a given area, from the proportion of ferrite and pearlite present and a knowledge of the iron carbon phase diagram, however the metallographic examination indicates that they varied from 0.1-0.2-0.6% C in the section. The absence of martensite in the microstructure confirms that no attempt had been made to harden this kitchen ware by rapid quenching from the austenite temperature (Brewer, 1976).

The above examinations revealed that this spoon (kitchen ware) was probably fabricated from mild steel. The steel used in the spoon was prepared by Sringverpura smith with the help of bloomery process.

Implements

Axe (lab no. Fe. 178)

This axe was in excellent state of preservation measuring 255 mm from cutting edge to head, maximum width near the whole 53 mm and 20 mm near the cutting edge, average thickness was 35 mm. Its weight was 520 gm. (Figure 1.II)

A wedge shaped section of size measuring 5 mm X 3 mm was drawn from the cutting edge of the axe as shown in Figure 1.II, Optical examination under the microscope showed that the core of the axe composed almost exclusively of ferrite. The surface layer of very variable thickness consisted of pearlite + cementite, pearlite and ferrite + pearlite and in places of widmanstatten form. The distribution of the structure in the axe is clearly visible in Pl. 11 showing both the surface layer in the cutting edge area and the core. The carbon content varied from 0.2-0.4% and 0.4-0.6% in the core.

The formation of widmanstatten structure occurs at the interface of the surface layer and the core in the region of the ferrite and pearlite structure (Pl. 12). These results indicate that the axe was not produced from all the material obtained from smelting but that tough material of ferrite structure was selected. The axe produced by this way was retreated to forging temp. in a carburising environment (Coghlan, 1977), most probably produced by wood charcoal. In this way a carburized surface layer with increased hardness was obtained. One feature which showed very clearly was that the low carbon zone contained much more slag than the higher carbon zones, which contained only a few, very thin slag stringers. The hardened cutting edge of the axe was not found to consist of the acicular or needle like structure of highly tempered martensite, the surface hardness in the axe was obtained by carburizing only. The hardness varied from 175-200 HV in low carbon and from 270-350 HV in high carbon area.

The metallographic examination of the axe shows that Sringverpura smith were having a good knowledge of the properties of the bloomery iron, which they skillfully utilised in the fabrication of this axe.

Sickle (lab no. Fe. 182)

Although heavily corroded this sickle is essentially complete in so far as the shape is concerned. The sickle is having black magnetite Fe_3O_4 in the core. Measuring 213 mm from tang to tip, maximum width of blade 25 mm, average thickness of blade 5 mm and weight 125 gm (Figure 1. III).

A wedge shaped section of size 5 mm x 3 mm was cut from the cutting edge of the sickle as shown in figure 1.III, optical examination of the polished surface of the section at lower magnification revealed that the sickle is completely mineralized and no surviving metal was traceable even on examination at higher magnification. Macro examination of the polished section showed that it was having many corrosion cracks pits and very few slag sinders. Micro examination of the section at 500 X and 1000 X after etching with 5% Nital revealed some relic carbide structure in some area of mineralized metal (Pl. 13) (Knox, 1963; Maddin, 1982).

The presence of the network of relic carbide in the sickle showed that it was forged into this shape from wrought iron and later on attempts for carburizating the artifact around 900° C and after the final forging it was allowed to cool in air.

Bent Knife (lab no. Fe. 247)

This artifact was also a broken part of a knife (length 75 mm, breadth 15 mm, weight 74.44 gm; Figure1.IV).

A V shaped section was taken from the cutting edge of the bent knife (Figure 1.IV). Examination at 125 X showed that it was forged from a large number of plates of carburized sheets and wrought iron (Agarwal; et al. Hermelin et al. 1979; William, 1974, Hogan and Rutin, 1989) (Pl. 14). The metal in the section had six un-carburized layers (wrought iron) and five carburized layers (very mild steel). Elongated black coloured slag inclusions were visible in the section in some plates.

The dark carbon bearing layers had a carbon content between 0.10-0.14% and 0.2-0.24% whereas the wrought iron layers had a carbon content around 0.05%. The average microhardness of the un-carburized and carburized layers was HV 99, 100 and 105, 115, 125, 135, respectively.

The microstructure indicated that the knife was forged from a laminated slab, prepared by forging together the sheets of wrought iron and low carbon steel. It has been examined and discussed by several scholars like Agarwal, Hermelin, William, Hogman etc. (see references, below). Finally the knife was again worked on a falling temperature and left to cool very slowly in the air. The structure and hardness do not reveal any indication of hardening and tempering treatment.

Weapon

Spearhead (lab no. Fe. 186)

Almost all the iron or steel of the spearhead had been corroded, corrosion products sticking on outer layer appeared to be hydrated ferric oxide and black megnetite Fe_3O_4 in the core. The dimension of the spearhead in rusted condition were 316 mm, maximum width 13 mm, average thickness 13 mm and weight 220 gms. (Figure 1.V).

A wedge shaped section of size 5 mm x 3 mm was drawn from the one edge of the spearhead as shown in Figure 1.IV. Macro examination of the polished sections indicated that the spearhead is completely mineralized and no surviving metal was traceable. Examination of this mineralized metal upon etching 1000X indicated that metallurgical history of this spearhead is inferred to be similar to that of sickle (lab no. Fe. 182).

DISCUSSION

The sickle (lab no. Fe. 182) and spearhead (lab no. Fe. 186) were completely mineralized and no surviving metal was available for analysis. The elemented analysis conducted on the sample drawn from spoon (lab no. Fe. 185) and axe (lab no. 178) indicated that iron was present as the major element other elements silica (SiO_2), alumina (Al_2O_3), lime (CaO) and magnesia (MgO) were present as impurities, Table 3. The silica (SiO_2) and alumina (Al_2O_3) content in these artifacts varied from 0.25-0.30% and 0.09-0.16% respectively. The percentage of lime (CaO) is in the range of 0.12-0.15% and magnesia (MgO) varies from 0.05-0.08%. Manganese (Mn) and nickel (Ni) varies from 0.010-0.015% and 0.02-0.04% respectively. Copper (Cu) was in traces and presence of chromium (Cr) could not be detected in these two artifacts. Phosphorous (P) was also present but it was in very small quantity (0.01%).

The elemental analysis and metallographic examination of the spoon and axe indicates that they were not fabricated from meteoric iron. Nickel (Ni) content in these artifacts varies from 0.02-0.04%, generally no meteoric iron contains less than 4.5% Nickel (Moso, 1975; Tylecote, 1987; Coghan, 1977).

The archaeological and metallurgical examination of early iron artifacts of the period (250 B.C.-600 A.D.) in Sringverpura (U.P.) reflects a high standard of technological skill and understanding. The metallographic examination of iron objects having metal core, spoon, axe and completely mineralized sickle and spearhead analyzed here show that Sringverpura smith were having good knowledge of the properties of the bloomery iron produced. The ferrite and pearlite part of the bloomery iron of tolerably good hardness was selected for the production of iron spoon (*Karchhee*, a kitchen ware) and the ferrite structure with increased toughness was selected for the production of the axe. In the later, the selection of a pearlite and cementite material for the production of the axe would result in failure after a very short period of use. The depth of carburization of the surface of the axe specially in the cutting edge is interesting. The high standard of the technology of production of the axe is all the more remarkable from another point of view as it relates to early stage in the production of iron and to primitive smithing technology.

The examination of the heavily corroded and completely mineralized sickle (lab no. Fe. 182) and spearhead (lab no. Fe 186) and laminated bent knife (lab no. Fe 247) reveal an understanding by the ancient metallurgist of the properties of the bloomery and carburized iron and how to select material for different purposes.

CONCLUSION

Within the limits of our archaeological and metallurgical examination with the artifacts received from Sringverpura (U.P.) having a small metal core and completely mineralized metal indicate that the artifacts were fabricated by conventional blacksmith's methods that were common in other parts of the world and India. The Sringverpura smiths were quite familiar with the concept of cementation to increase hardness with carbon content and practised it skillfully in preparing their artifacts of domestic use, tools and weapons etc. However, the absence of hardening by heat treatment in the case of axe suggest that smiths may have been aware that longer cementation would give him desirable properties and the technique of improving the mechanical properties of carburized iron by heat treatment was not known to them around 250 B.C.-600 A.D. As the control of constant conditions over long periods was probably not feasible at that time but we must admire that how skillfully only by observing the colour of flame the manupulations were conducted. They were able to control the temperature for sufficient time to produce the desireable steel. Their understanding and control of what we now know to be effects of carbon content was excellent which they utilised in the fabrication of spoon (kitchen ware), axe and laminated bent knife.

ACKNOWLEDGEMENT

We wish to thank Prof. B.B. Lal of Indian Institute of Advanced Study, Simla and Sh. K.N. Dikshit (Director Publication) of Archaeological Survey of India, New

Delhi for kindly providing the samples for these studies and for many useful discussions.

The authors are thankful to Mr. G.P. Joshi for his help in sample preparation for metallographic examination and Mr. Anil R. Singh for the photographic work.

TABLE 2

Microstructure, Carbon content and Microhardness of Iron Artifacts from Sringverpura site (Near Allahabad), U.P. (250 B.C. - 600 A.D.)

S.N. Artifact Description	*Lab no.*	*Micro-structure*	*C%*	*Metallographic estimation*	*Microhardness with 100g load*	
		Core	*Near the Core edge*	*Near the Core edge*	*Core HV*	*Near the edge HV*
A. *Artifacts of Domestic Use*						
1. Spoon (Broken)	Fe 185	FP	0.1-0.2	0.20-0.60	-195	195-260
B. *Tools*						
1. Axe	Fe 178	F.P.C.	0.12-0.4	0.4-0.8	175-200	270-300-350
2. Sickle	Fe 182	R	—	—	—	—
3. Bent Knife	Fe 247	-do-	0.10-0.14	0.2-0.24	99-100	105,115,125,135
C. *Weapon*						
1. Spearhead	Fe 186	R	—	—	—	—

F: Ferrite P: Pearlite C: Cementite R: Relic Carbite

TABLE 3

Composition (%) of Iron Artifacts subjected to Elemental Analysis from Sringverpura Site (250 B.C. - 600 A.D.)

S.N.	*Description of Object*	*Lab.no.*	*Fe*	*SiO_2*	*Al_2O_3*	*CaO*	*MgO*	*Mn*	*Ti*	*Ni*	*Cu*	*Cr*	*P*
A.	*Artifacts of Domestic Use*												
1.	Spoon (Broken)	Fe 185	Major	0.25	0.16	0.12	0.08	0.015	—	0.02	Tr	—	0.01
B.	*Implements*												
1.	Axe	Fe 178	0.30	0.09	0.15	0.15	—	0.010	—	0.04	Tr	—	0.01
2.	Sickle	Fe 182			Sample not drawn								
3.	Bent Knife	Fe 247	Major	0.82	0.98	0.59	0.39	0.01	Tr	0.06	0.01	—	—
C.	*Weapon*												
1.	Spearhead	Fe 186			Sample not drawn								

Sample not drawn as the artifact was completely mineralized.

REFERENCES

Agarwal, O.P., Hari Narain, Jai Prakash, 1988. Development of iron metallugry in Ancient India, Presented at International Colloquium on Archaeometallurgy held at Bologna, Italy, 18-21 Oct.

Agarwal, O.P., Hari Narain, Jai Prakash, 1990. *Metallographic studies of iron artifacts from* Allahapur (Meerut) U.P. 400-300 B.C. *Transaction of the Indian Institute of Metals*, 43(1), pp. 47-51.

Agarwal, O.P., Hari Narain, Jai Prakash, S.K. Bhatia, 1990. Lamination technique in Iron artifacts in Ancient India, *Journal of the Historical Metallurgy Society*, 24(1), pp. 12-26.

Brewer, C.W. , 1976. Metallographic examination of six ancient steel weapons. *Journal of the Historical Metallurgy Society*, 10(1), pp. 1-10.

Coghan, H.H., 1977. Notes on prehistoric and early iron in the old world, (ed.) T.K. Penniman and B.M. Blackwood, 2nd edition, Oxford.

Coghan, H.H., and R.F. Tylecote, 1978. Medieval iron artifacts from the Newbury area of Berkshire, Metallurgical examination, *Journal of the Historical Metallurgy Society*, 12(1), pp. 12-17.

Curtis, J.E., T.S. Wheeler, J.D. Muhly and R. Maddin, 1979. Neo Assyrian Iron Working Technology, *American Philosophical Society*, 123(6), pp. 369-90.

Hermelin, E., E. Tholander and S. Blomgren, 1979. A Prehistoric Nickel - alloyed iron axe, *Journal of the Historical Metallurgy Society*, 13(2), pp. 69-94.

Hogan L.M., and S. Rutin, 1989. Metallurgical analysis of iron artifacts from Thailand *Journal of the Historical Metallugry Society*, 23(2), pp. 99-107.

Indian Archaeology — A review, 1977-78, pp. 54-56; 1978-79, 1981, pp. 57-59; 1979-80, 1983, p. 74. 1980-81, 1983, pp. 67-68; 1981-82, 1984, pp. 66-67; 1982-83, 1985, pp. 91-92; 1983-84, 1986, pp. 84-85; 1984-85, 1987, pp. 85-86. Archaeological Survey of India, New Delhi,

Knox, R., 1963. Detection of iron carbide structure in oxide remains of ancient steel. *Archaeometry* 6, p. 43.

Lal B.B., and K.N. Dikshit, 1978-79. Sringverpura: A key site for the protohistory and early history of the central Ganga Valley. *Puratattva*, 10, pp. 1-7.

Maddin, R., 1982. *Transaction of the Indian Institute of Metals*, 35(1), p. 17.

Makkhan Lal, 1984. *Settlement History and Rise of Civilization in Ganga-Yamuna Doab* (from 1500 B.C. to 300 A.D.). B.R. Publishing Corporation, Delhi p. 91.

Moso, A.A., 1975. *Meteorites — A concise Account*, Trustees of the British Museum (Natural History), 3rd edition, London, p. 11.

T.N. Roy 1983. *The Ganges Civilization*, New Delhi, pp. 35-36.

Samuels, L.E., 1978. *Metallographic etching*, American Society for Metals, Ohio.

Samuels, L.E., 1982, *Metallographic polishing by mechanical methods*, American Society for Metals, Ohio.

Scoot, D., 1986, *Metallography of ancient metallic artifacts*, Institute of Archaeology, London.

Tylecote, R.F., 1987. *The early History of Metallugry in Europe*, 1st edition, London.

William, A.R., 1974. The metallographic examination of a Burgkmair etching plate in the British Museum. *Journal of the Historical Metallurgy Society*, 8(2), pp. 92-94.

Iron Technology : Its Role in the Historical Development in the Morar Valley, M.P.

R.P. PANDEY

The use of Iron has led to urbanization in the Ganga valley as well as in several other river valleys in the northern and central India. Now some fresh evidences are coming as a result of excavations being conducted at historical settlements.

In the recent past, some sites have also been excavated in the Gwalior region in the Morar valley, revealing unique evidence including some of the urban centres coming up in the area.

A continuous cultural sequence right from Lower Palaeolithic period has been discovered in Gwalior region. The Morar valley has also evidenced the sites from Middle Palaeolithic to the historical periods. There are a large number of historical sites located on this river (*IAR*, 1971-72). The important ones among them are Mehra, Jaderua, Sooron and Baretha. Among these Jaderua and Sooron have been excavated by the Department of Ancient Indian History, Culture and Archaeology, Jiwaji University, Gwalior (M.P.) and have revealed the antiquities associated with urban centres on its right and left banks. (Fig. 1)

The Morar river rises from the Santhau hills about 20 km. south of Gwalior in Gwalior district. After traversing a variety of landforms in the uplands and plains, the river joins Vaisuli, a tributary of the Sindh near Jhawalpura (village in Bhind district). The total distance covered by the river is about 100 km. The Morar river possesses an unique situation. After flowing for a distance of about 5 km, the river enters in the plains and is joined by several small and big nalas, on both its banks. Several smaller hillocks are also seen rising in the plains. The hills of the source region and hillocks of the plains generally contain quartzite shale rocks and laterite deposits. The laterites are generally used for road construction in the villages in the vicinity. The lateritic pellets are also encountered in the trenches of the excavation at the sites. A better exposure of the laterite is recently observed in the Moti Jheel area, opposite to the Petrol Pump. These smaller hillocks sometime bear mesolithic material over them. The better examples can be seen at Putalighar (very close to Gwalior) and at Sooron. The geological formations of the area are the Archaeans, Vindhyan series, laterites and alluvium. The formation of laterites in the area indicates source of iron. Other rock types, rich in iron contents could not be located

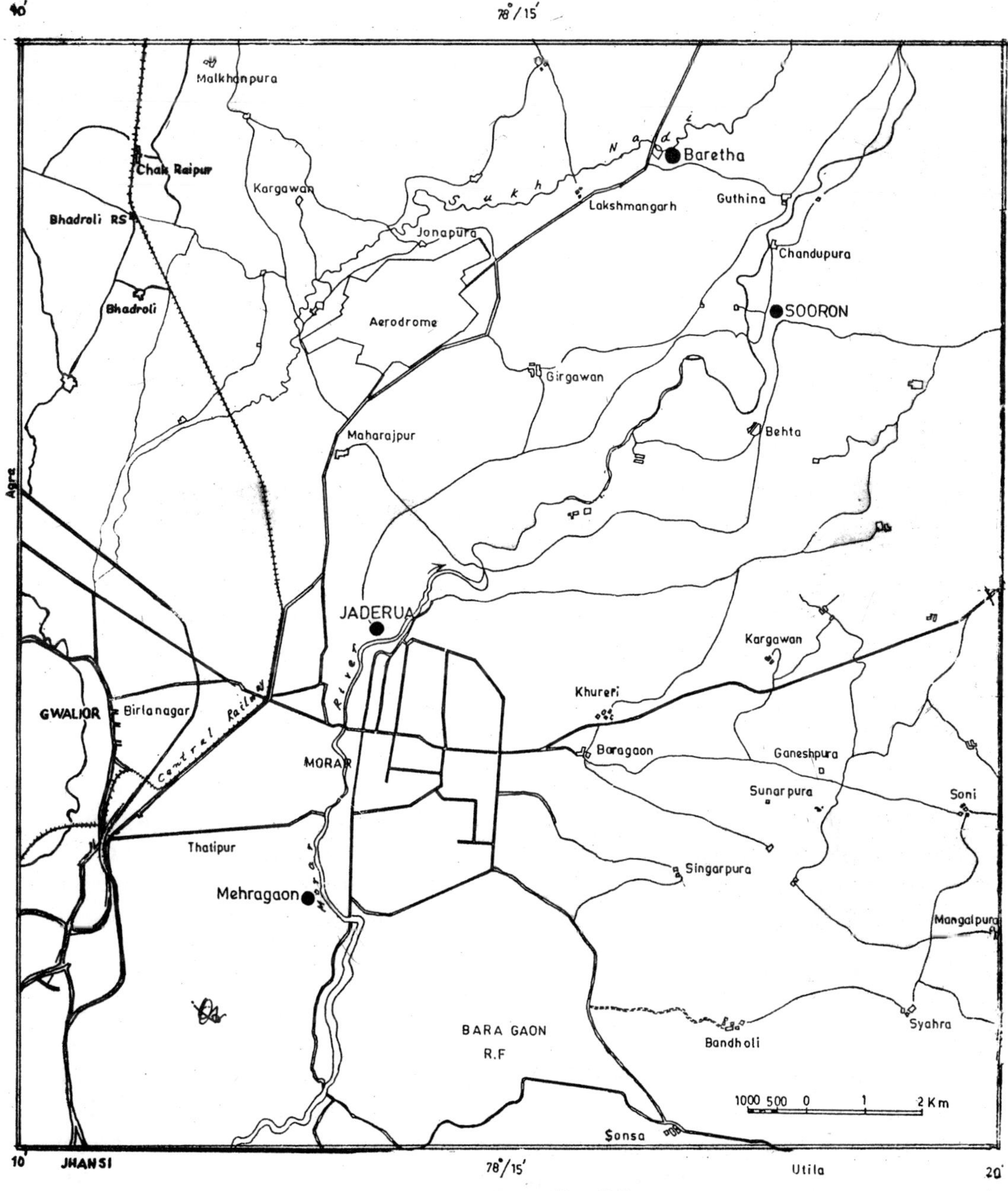

Fig. 1 : Locality Map

by the writer during the course of explorations. The Vindhyan quartzite has contributed in many ways. The alluvial deposits on the Morar valley are measured upto 20 m while no gravel deposit could be located. The accumulation of the alluvium mainly belongs to younger alluvium group, thereby indicating later developments. All the historical sites along this valley are situated on river laid alluvium thereby indicating that before settlement of the early historical cities, the Morar river was active and could support the early historical cities. Even today, the river water is used for irrigation purposes. The area is covered with reddish and yellowish soils and nourishes thorny scrub type of vegetation.

The area experiences a typical monsoonal rainfall of about 700 mm per annum and can be called semi-arid. June is the hottest month of the year and December the coldest.

THE SITES

Both the excavated sites are situated on the Morar river and are being discussed below.

Jaderua

The site of Jaderua is situated on the left bank of the Morar river and is about 2 km north of Morar on the Morar-Maharajpura road. The site has been excavated in the seasons 1971-72 and has yielded evidence of early historical culture ranging from 5th century B.C. to 10th century A.D. The site is presently covered by the modern village.

Sooron

The site of Sooron is situated on the right bank of the Morar river, 25 km north-east of Gwalior and can be approached by a cart tract from Maharajpura airport. The mound was excavated in 1979, 1984 and 1987. The mound site measures 500 m north-south and 300 m east-west directions. The trenches were dug on the western and northern slopes. The site yielded the cultural sequence from 800 B.C. to circa 100 A.D. The top of the mound is covered by modern village at present.

Cultural Sequence

A cultural sequence ranging from 800 B.C. to 900 A.D. is unearthed from both the sites. Sooron, however was occupied earlier, yielding Red ware from the lower most strata and the settlements at Jaderua continued upto circa 9th century A.D. Thus human occupation continued from 800 B.C. to 100 A.D. at Sooron and 600 B.C. to 900 A.D. at Jaderua. Sooron shows continuous, though a gradual development while Jaderua remained unoccupied from A.D. 300 to 800 A.D. The period and antiquity wise distribution is mentioned in Table 1.

Iron Tools

The iron objects having been used for day to day activities are found right from the earliest periods of these twin sites. The limited excavations on these sites have yielded iron tools, implements and weapons like chisel, spearheads, a large sickle from Jaderua and iron pieces, nails, axes, spears and arrowheads from the site of Sooron.

TABLE 1

Period and antiquitywise distribution of the Jaderua and Sooron

Period	*Pottery*	*Intiquity*	*Date*
		SOORON EXCAVATIONS	
IV	Red-Ware & Brick structure	Ball, Beads, Iron-Arowhead, Glass Bangle, Stone weight, Copper-Cast coins	100 B.C. to A.D. 100
III	Red-Ware, N.B.P.	Sunga T.C. Bull, Ram, Beads, Bangles, Ball, Stylus, Stone weight, Semiprecious stone Beads, Punch Mark coins, Gamesmen	300 B.C. to 100 B.C.
II B	Red-Ware, Coarse Grey Ware	T.C. Ball, Pendent, Kaudi, Animal figures, Iron Nails and T.C. Wheel	450 B.C. to 300 B.C.
II A	Red Ware	Ball, Beads, Ear-ring, Stone weight, Iron-Axe	600 B.C. to 450 B.C.
I B	Red-Ware, Black-Ware with Mica, Red-Ware few pieces of Grey-Ware	Iron Nail, Broken Beads	700 B.C. to 600 B.C.
I A	Red-Ware, Black Slipped Ware, Black & Red Ware Grey-Ware, Burnished Grey-Ware & Micacious Black Ware	T.C. Toy, Flesh Rubber and Iron	800 B.C. to 700 B.C.
		JADERUA EXCAVATIONS	
III	Brick structures	not mentioned	900 A.D. to 1000 A.D.
II B	N.B.P. Red Ware	Coins, T.C. female figurines,	100 B.C. to 100 A.D.
II A	N.B.P. Red Ware	Chisel, Spear head arrowhead, large sickle, beads of carnelian agate and T.C.	200 B.C. to 450 B.C.
I	Black & Red Ware Grey-Ware	Muller, Saddle quern Crystal beads	600 B.C. to 500 B.C.

Metallographic Studies of the Implements and Weapons

The metallographic studies of iron implements and weapons from Sooron ware carried out by the NRLC, Lucknow (Narain *et al.*, 1990), (Table 2). Out of ten artifacts, four artifacts were selected for metallographic studies. They are axes, a knife and spear. The elemental composition works out to be, as iron content being maximum while S_2O (0.58%), Al_2O_3 (0.45%) CaO (0.32%) MgO (0.18%) very small percent. Titanium and Copper percentage were only in traces.

Presence of silica, manganese and nickel in these tools are suggestive of iron extraction from the ore by direct process. These studies also show that the people of the Morar valley knew the concept of cementation due to which carbon content could be increased to 0.05%-0.6%. The iron was first extracted from the ores by smelting in a furnace. To get a desired shape of the implements and weapons they were, further heated in the charcoal fire due to which carbon from the combustion gases used to diffuse into the metal but further improvement of the mechanical properties by heat treatment (quenching and tempering) and lamination techniques were not known to the people of the Morar valley, (Table 3 and 4) while the people of the Ganga valley had the knowledge of carburization, heat treatment and lamination for developing the mechanical properties in the iron implements as early as 60 B.C. (Narain *et al.*, 1991 and personal communications).

As mentioned earlier, iron is found right from 800 B.C. in the area and has certainly played an important role in the urbanization in the Morar valley. In the earlier phases only iron nails and bits of iron appeared but subsequently various tools used for different purposes make their appearance. Among them are agricultural implements, tools of domestic use and a few objects which could be classified as weapons.

On both the sites iron slags are found in plenty on the surface as well as from the excavated trenches. The iron slags at the site are an indication of large scale iron smelting and iron smithy in the area during different cultural periods. The sites of the Morar valley lie not far from the hills. These hills must be very rich in iron ore and other minerals. The inhabitants during early historical times, must have discovered such resources in the hills and exploited them for their use at the site. Such iron rich rocks and minerals could not be located by the writer so far in the hills while at a site namely Dhuan near Ghatigaon in Gwalior district, lying about 30 km from these sites, iron was being extracted till the last century. A thick laterite deposit was also noticed by the writer during cutting for the construction near Moti Jheel, opposite Petrol Pump very close to Gwalior.

The settlements on the Morar valley as stated above are discovered from Middle Palaeolithic, Upper Palaeolithic and Mesolithic. The microliths found on a hillock near Sooron are an indication that the area attracted early inhabitants

TABLE 2

Artifacts subjected to metallographical studies from Sooron

No	Artifact	Lab No.	Archaeological date	Length mm	Breadth mm	Thickness mm	Weight mm	Section size mm
I	*Implement*							
1.	Axe	Fe 188	600 B.C. to 450 B.C.	80	60	14	109.00	4x5
2.	-do-	Fe 187	100 B.C. to 100 A.D.	150	28	26	355.00	4x6
3.	Knife	Fe 189	100 B.C. to 100 A.D.	113	88	12	50.15	3x5
II.	*Weapon*							
1.	Spear	Fe 190	300 B.C. to 200 B.C.	191	14	12	125.00	3x5

(After Narain *et al.*, 1991)

TABLE 3

Microstructure, Carbon content and Hardness of Iron artifacts from Sooron

No.	Artifact	Lab No.	Microstructure Carbon %	Metallographic estimation		Hardness 100 gm H.V.	
				Uncarburized area	Carburized area	Uncarburized area	Carburized area
I	*Implement*						
1.	Axe	Fe 188	F/P(M)	0.05	0.3-0.4	100	180
2.	-do-	Fe 187	F/P(ME)	—	0.5-0.6	—	200-210
3.	Knife	Fe 189	F/P(ME)	0.05	0.2-0.3	195	150-200
II	*Weapon*						
1.	n	Fe 190	F/P	0.05	0.2-0.3	180	230

F=Ferrite M=Mild Steel P=Pearlite VM=Very mild steel ME=Medium Steel (After Narian *et al.*, 1991)

TABLE 4

Composition of Iron artifact from Sooron

S.N.	Artifact	Lab No.	Composition %								
			SiO_2	Al_2O_3	CaO	MgO	Mn	Ti	Ni	Cu	Cr
1.	Axe	Fe 187	0.58	0.45	0.32	0.18	0.01	Tr	0.04	Tr.	—

Tr = Trace only

= Not reported by analyst (After Narian *et al.*, 1991)

probably due to its natural resources. The metallurgy of these sites further suggests the importance of this area.

The early historical cities sprang up with the use of iron. The river has laid down thick alluvium and has lateral extent of alluvium throughout its stretch. These agricultural tools must have been used in large scale in the fertile alluvial plains due to which agricultural surplus was available, giving constant all round progress leading to urbanization. This is indicated by a variety of evidence. The area has also yielded evidence of cultivation of *gram* and *urad* and an unique evidence of granery. The availability of the grains from the excavations are suggestive of large scale agriculture (of variety of cereals) was practiced in the recent alluvium of the Morar valley and surplus was properly stored in the burnt brick graneries with beautiful insulating system to ensure proper supply of food grains throughout the year (Fig. 3). For preparation of the food, various tools and implements were utilized. In the beginning mullers, saddle querns must have been used and at later stages stone mill must have come into prominence. Such evidences have come up from the Period II of Jaderua and Periods III and IV of Sooron.

The iron technology and its exhaustive use also changed the living standard of the inhabitants of the Morar valley. The people at these sites gradually started living in the brick built houses. Several burnt brick floors were exposed during excavations. It seems that some of the burnt brick structures were still supported by wooden posts as many post holes were recorded near the floors. It looks that roof was still not being made of the bricks or stone slabs. The tiles were in use for roofing as the evidence of tiles was found from the excavations. To the north of granery, a floor having 12 post holes was recorded. This is considered to be the weavers' workshop (Fig. 4). The cities of the Morar valley were laid out in a planned manner. This is indicated by sewage disposal system due to the presence of ring wells. Another important element of the cities of the early historical period was also found during excavations. A defence wall measuring about 16 m in north-east south-west direction is noteworthy. The moat near the defence wall could not be determined due to limited area of excavation.

Due to constant progress and metallurgical contribution, variety of crafts also sprang up, showing diversified progress. A number of desired iron tools were made and metal work developed. Some of the semi-precious stones, viz. agate and carnelian in making ornaments and beads were not available in this area and must have been imported. This exchange system must have forced the people of this area to send their surplus in the other areas too. Some of the shapes among the beads like tortoise shape are interesting and indicate special attachment towards water living animals. Ivory bangles also indicated the import from other areas. The making of the animal and human figurines and also gamesmen for their amusement indicate the urbanized economy. The finds of coins at Sooron and Jaderua

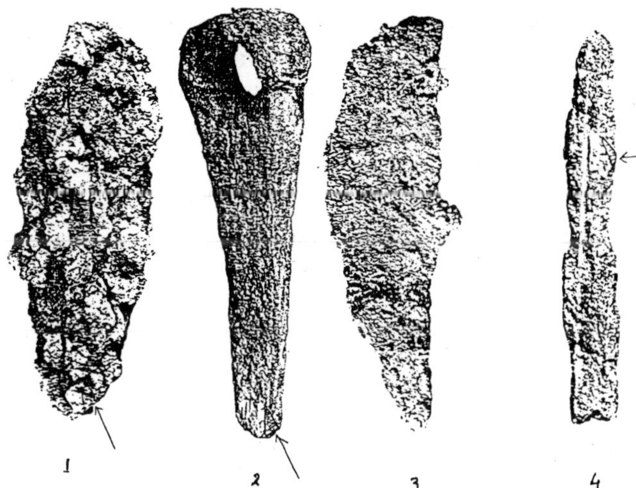

Fig. 2: Appearance of the four artifacts after removal of the metallographic samples. (After Narain et al 1991)

METER

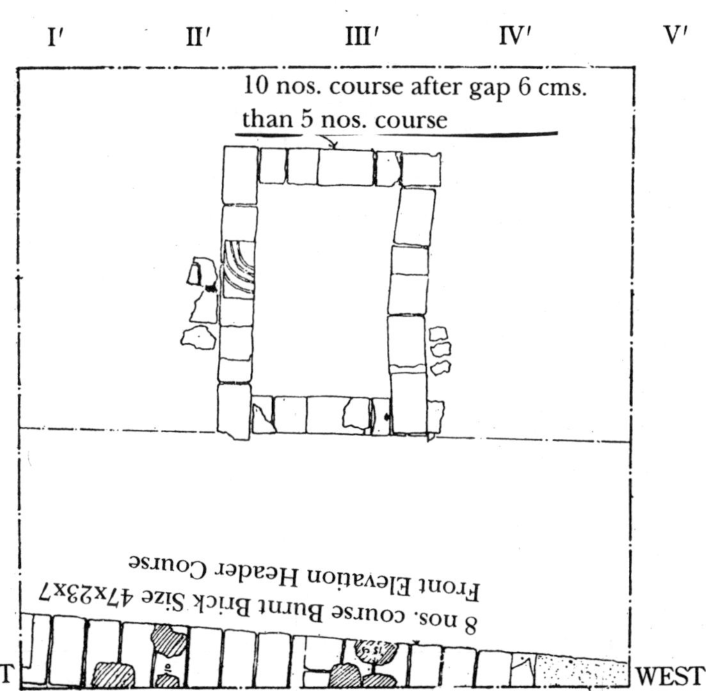

Top course, depth from 0 = 1.46; Below course depth from 0 = 2.55; Total Height = 1.09

Fig. 3: Plan Structure No. 6 (Quadrant 2nd Brunt Brick Grainary Sealed by Layer1)

are suggestive of central rule, proper revenue system, and developed economy during early historical periods in the Morar valley.

Thus, the iron technology played an important role for development of the urban centres in the Morar valley which had linkages with the contemporary cultures of the northern and central India.

Sealed Layer (1)

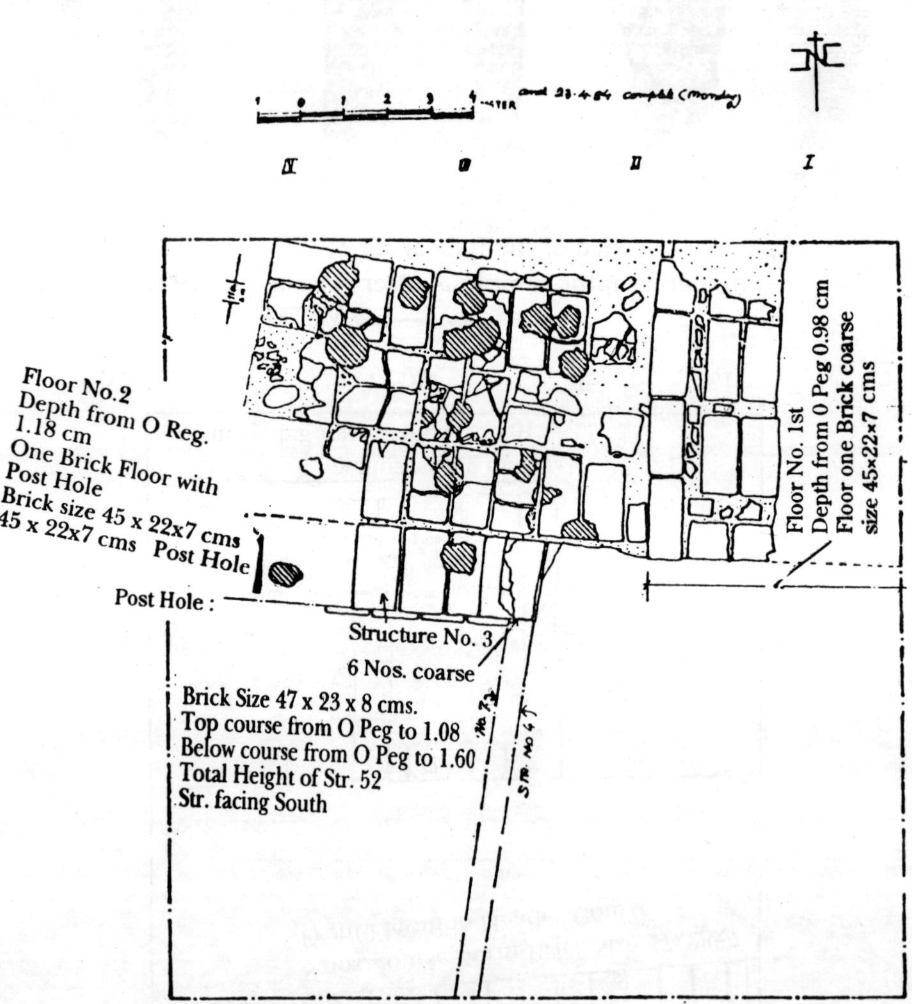

Fig. 4: Plan Floor Structure No. 1st and 2nd. Trench Quadrent I O-V, Sooron Excavation.

ACKNOWLEDGEMENTS

The writer is grateful to Prof. R.N. Misra for fruitful discussions, suggestions and also providing unpublished brief report of Sooron. He is also thankful to Mr. D.D. Mishra for preparing the line drawings and Mr. U.S. Sharma for various helps.

REFERENCES

Indian Archaeology — A Review 1971-72, pp. 29-30.

Narain, H., A.K. Srivastava, J. Prakash, I.K. Bhatnagar and O.P. Agarwal, 1990. Metallographic studies of Iron Implements and weapon from Sooron 600-100 A.D. *Pragdhara,* No. 1, pp. 166-171.

Metallurgical Study of Ancient Swords

B. PRAKASH

The Indian heritage of independent development of iron and steel technology has been proved beyond doubt and it dates back to earlier than 1000 B.C. Mention of iron and steel has been made in Vedas at many places. Ghosh and Chattopadhyay (1982) have found a hardened steel sickle at Barudih dating back to 810 B.C. Sushruta Samhita (1982) has mentioned in detail the design and processing of many surgical tools made of Tikshna Loha (steel) and its heat-treatment to obtain razor sharp edge. Hadfield (1912) has mentioned the possibility of the use of the Indian steel tools and craftsmen in the erection of the Egyptian Pyramids. During the Medieval period Indian 'wootz' steel and the swords, daggers etc. made from it had gained world wide fame and the famous watering pattern or Damascusizing structure visible with naked eye on the steel blades had become a hallmark and a matter of technological curiosity. This paper describes the characteristic features of Damascus or Indian swords and compares it with the European 'Pattern Welded' sword and other objects. It also discusses the results of some of the experiments done in reproducing these ancient structural designs.

Tikshna Loha is essentially Fe-C alloy containing carbon in the range of 0.7 to 1.9%. In the literature at least four methods of manufacturing this type of alloy have been mentioned, i.e.

(i) Solid state carburization of the cutting edge of the blade.
(ii) Liquid state carburizing in crucibles to produce wootz steel ingot.
(iii) Liquid state decarburization of white cast iron to produce wootz steel cake.
(iv) Melleablization of white cast iron in presence of iron oxide (Fe_2O_3 or Fe_3O_4).

All these processes have been described in detail and reviewed by Buchnan (1807), Voysey (1832), Prakash (1990), Rao (1989), Yater (1982), Bronson (1986), Mahmud (1988) and many others. Pl. 15 shows the macro-structure of cross-section of 'wootz' steel ingot as reported by Tylecote (1984). This structure consists of dendritic platelets of cementite in a matrix of transformed ledeburite as shown in Pl. 16 giving the microstructure of hyper eutectoid steel (1.33% C). These ingots were given elaborate heat treatment and a number of such ingots weighing about

1 to 2 kg each were skillfully forge welded to produce various war implements. The heat treated Dasmascus or Indian sword was recognised by the 'watering' or Damascusizing mark visible on the steel blade as shown in Pl. 17.

Recently more detailed microscopic examination of two 'shamshir' and one dagger from the Bharat Kala Bhawan, BHU was conducted by the author to gain first hand knowledge of such etch patterns. Pl. 18 shows a picture of these objects. Micrographic study of the surface of these objects has revealed that, while 'Shamshir' No. 11/1604 and dagger No. 3/9172 belonging to early 18th century showed the typical 'watering' pattern, sword No. 11/1603 showed lamelar arrangement of cementite globules in a matrix of tempered martensite. Pls. 19-21, show the etch patern seen on the dagger and 'shamshir' No. 11/1604 respectively. Such pattern added beauty as well as excellent edge hardness combined with flexibility and high strength to the weapon which no other sword could match.

Pattern Welding Technology

At least for the past 900 years European blacksmiths have been trying to develop the Damascusine structure but without any success. In this attempt they have evolved a new technique of 'Pattern Welding' which consists of forge welding of steel rods and strips of different carbon content and then giving them several folds and twists, followed by repeated forge welding. This technique of hot forge welding and finally shaping into war weapons developed a beautiful pattern in black and white after final polishing, heat treatment and etching with organic and inorganic acids. Fig. 1a and 1b show a knife and a gun barrel made by this technique whose surface is beautifully decorated with the pattern welded design. These patterns are definitely different from the Damascusized steel structure and there is marked difference in the edge hardness and other properties. Fig. 2 shows the vickers hardness plot of an old pattern welded knife from Bohemia as reported by Pleiner (1984).

A large number of such pattern welded swords and other weapons have been found all over Europe and a number of detailed reports on their metallurgical examination as well as method of manufacture have been published in journals on archaeology and blacksmithy. Yater has conducted a large number of experiments on developing such patterns. Fig. 3 shows some of the arrangements of the strips and rods of varying C-steel to obtain pattern welded swords and knife. Recently a large number of papers on this subject have appeared in a conference proceedings on 'The Craft of the Blacksmith' edited by Scott and Cleere (1984). The latest experimental study of this technique has been by Denig (1990) who has produced axe and other items having beautiful etch patterns.

Thermo-mechanical Treatment of Damascus Swords

Yater and Piaskowski (1974) have published a detailed review on the structure

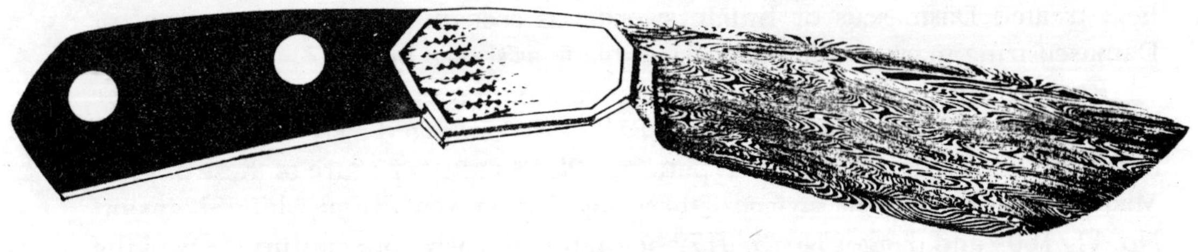

Fig. 1a William Fiorini, La Crescent, MN. Skinner. W1-A203E Damascus steel, chevron twist, wight strand (48 Layer each strand) ebony, and brass. Engraving by Bruce Le Page. Photo: David V. Scadlock

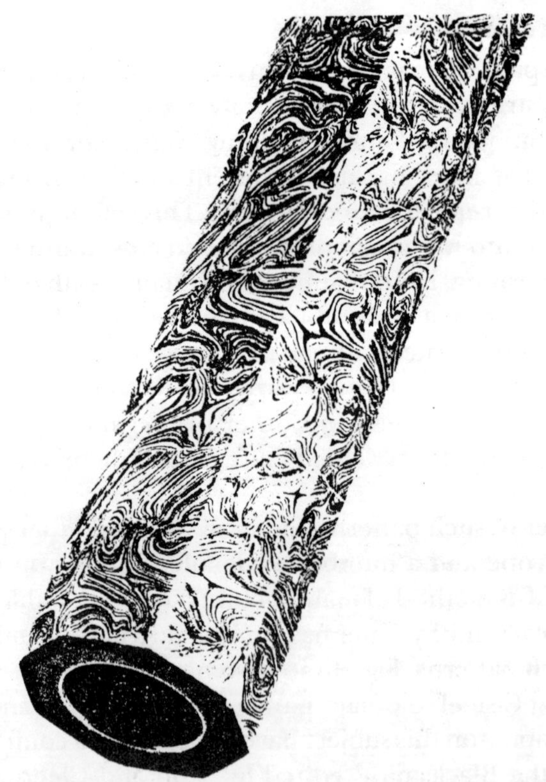

Fig. 1b William Fiorini, La Crescent, MN. 53 cal. smooth bore black powder pistol barrel (and detail). Mild steel-wrought iron Damascus steel, double twist, 32

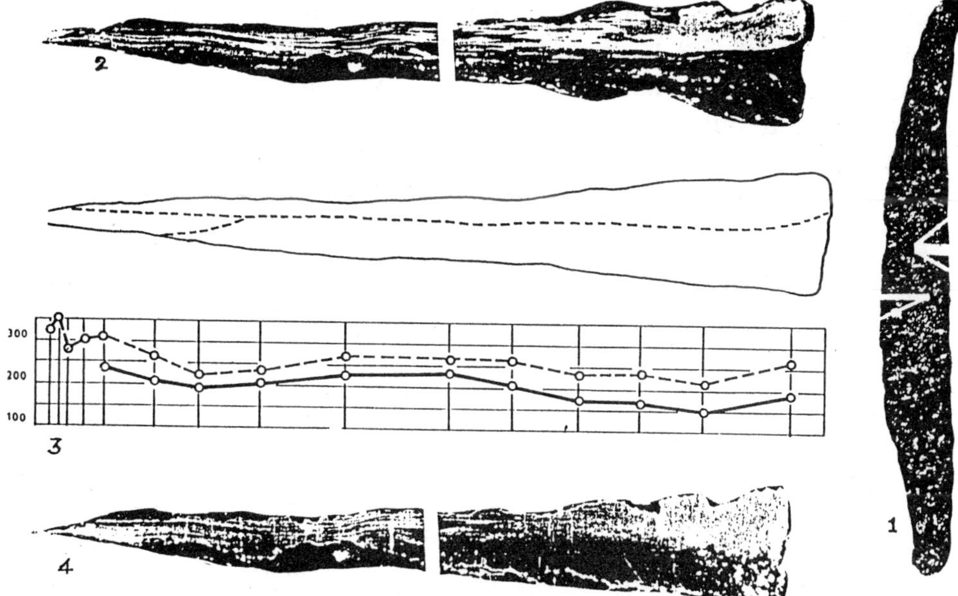

Fig. 2 Radovesice, Bohemia, flat-tanged knife 626, house 164A, HD/LA; (1) location of specimens: (2, 4) transverse sections etched with Oberholfer's regent and 5% nital (7x); (3) microhardness mHV 30g. in pearlitic (dotted) and ferritic areas.

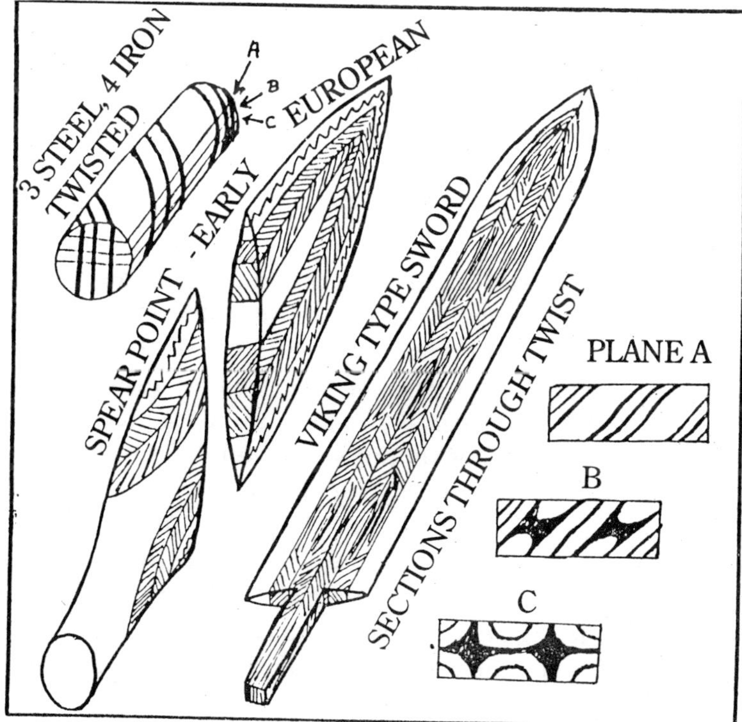

Fig. 3 Pattern Welded Blades

of Damascus steel swords and they have referred to the early publications of Anossoff (1974) and Balaiev (1974) who made a detailed study of the microstructure of watered Damascus blades. The microstructure of such a blade shown in Pl. 17 clearly indicates that the massive cementite plates shown in Pl. 16 have been broken into globular cementite before the ingot is hot forged to make war weapons and then given the final heat-treatment. In recent years Verhoeven *et al.*, (1987) have conducted a series of experiments to reproduce this famous pattern and the nearest structural pattern produced by them is shown in Fig. 4, where Fig. 4(a) shows the original structure of watered steel blade and Fig. 4(b) gives the pattern produced by them by 'Austenitizing' spherodizing and rolling of 1.65% carbon steel. This figure Fig. 4(b) shows the distribution of globular cementite at the elongated grain boundary in a matrix of fine pearlite, but this structural pattern is very much different than the original Damascusized structure shown in Fig. 4(a).

The sequence of changes occurring in any Fe-C alloy can be understood from the study of the Fe-C equilibrium diagram shown in Fig. 5(a). This figure shows the change in microstructure of L-iron with the increased 'C' content of the metal viz. from low carbon steel containing network of L-Ferrite (C in solid solution of Fe) to 100% Pearlite at 0.8% C and the separation of excess cementite at the grain boundary as the 'C' content increases to 1.4% and beyond. On further increase in 'C' content cementite phase starts separating on Widmannstatten planes along with that at the grain boundry giving rise to the structure shown in Pl. 16.

On reheating of the steel to temperatures beyond 723° C the room temperature phases viz. Ferrite, Pearlite and Cementite change to Austenite whose quantity and distribution can be controlled by controlling the thermo-mechanical parameters. The major structural change is the dissolution of 'C' from cementite into Austenite and the change of Body centered cubic structure (b.c.c.) of L-Ferrite into Face centred cubic (f.c.c.) structure of Austenite. The rate of heating as well as cooling of Fe-C alloys have marked effect on its structure producing many new phases and hence changing the property of steel. The temperature ranges for some of the heat-treatment viz. annealing, normalizing etc. are also shown in Fig. 5(a).

The brittle columnar and dendritic structure of the cast ingot has to be changed before it could be forged. Verhoeven *et al.* had given the austenitizing treatment to 1.65% C steel by reheating the cast bar to the austenite range and soaking it there for several hours followed by slow cooling or annealing. This changed the dendritic platelets of cementite into a network of cementite on the grain boundary and on hot rolling and spherodization at ~ 700° C the structure shown in Fig. 4(b) was produced. The other possible heat-treatment is to first give a prolonged annealing treatment at cherry red temperature 750-850° C when a large percentage of the pearlite and cementite will get converted into Austenite and then slow cooling so that the platelets of cementite are broken to increase the percentage of pearlite grains both by transformation of Austenite as well as decomposition of massive cementite and ledeburite. The ancient Indian blacksmiths have been known to cover the ingots with clay (to prevent oxidation) and given prolonged annealing of 48 hrs. or more before hot forging. The thermal decomposition of cementite is probably responsible for the presence of graphitic carbon as reported by Percy (1964) in the following chemical composition of 'wootz' steel.

Combined C	1.33%
Uncombined carbon	0.31%
Silicon	0.045%
Sulphur	0.180%
Arsenic	0.04%

If the high carbon (> 0.8% C) steel is heat-treated between 650 to 750° C the massive cementite as well as cementite present in pearlite get converted into globular particles present in a matrix of ferrite. This heat-treatment was also known to the ancient blacksmiths as indicated by the microscopic study of the 'watered' steel blades by Yater and Verhoeven *et al.* (1987).

In order to produce swords and other weapons a number of rods prepared by heat-treatment and hot forging were reheated to 1000 to 1100° C and forge welded before giving the final shape. After the forging and shaping of the blade it was filed and cleaned free from oxidized scale and polishing to give semifinish. Then it was reheated to the cherry red temperature 750-800° C for final conversion to the Austenite, containing spheroids of cementite, and it was quenched into water, brine or oil etc. to transform Austenite into Martensite which has hardness of > 60 Rc. In ancient times the hardening was done in several ways as described by Rao (1989) and Prakash (1991) Some of these quenching techniques are mentioned below:

1. Rao (1987) has mentioned that in Damascus the blade was heated to a colour just like morning sun rising in a desert and then it was cooled to a colour of king's purple, and then it was thrust into the body of a muscular slave's belly. It was believed that by this method the strength of the slave was transferred to the blade giving it the desired strength.
2. Zschokke (1924) has described another method where the red hot blade was carried at a furious gallop by a horseman mounted on a horse with the sharp edge forward. This was a process of quenching in chilled air of winter.
3. Sushruta and Varahamihira (1956) have mentioned the use of alkaline sap and ash of banana tree for quenching these blades for a short period followed by self-tempering. Varahamihira has given a full account of this in 'Kharaglakshnam'.
4. Another method of hardening and tempering prefered by the Indian craftsmen was to thurst the red hot blade into the green trunk of banana,

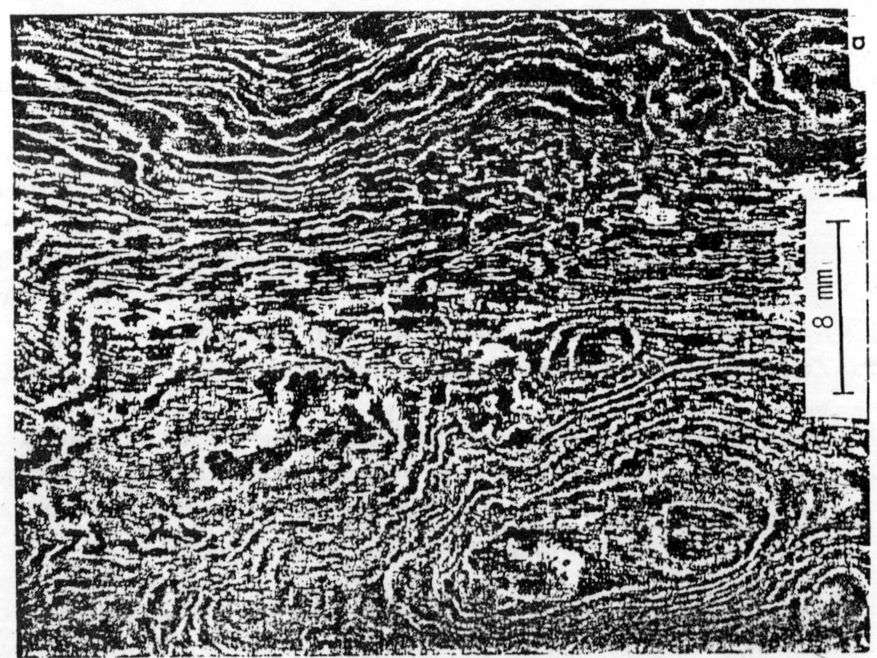

Fig. 4 Polished and etched faces of (a) the Damascus sword and (b) the Moran blade

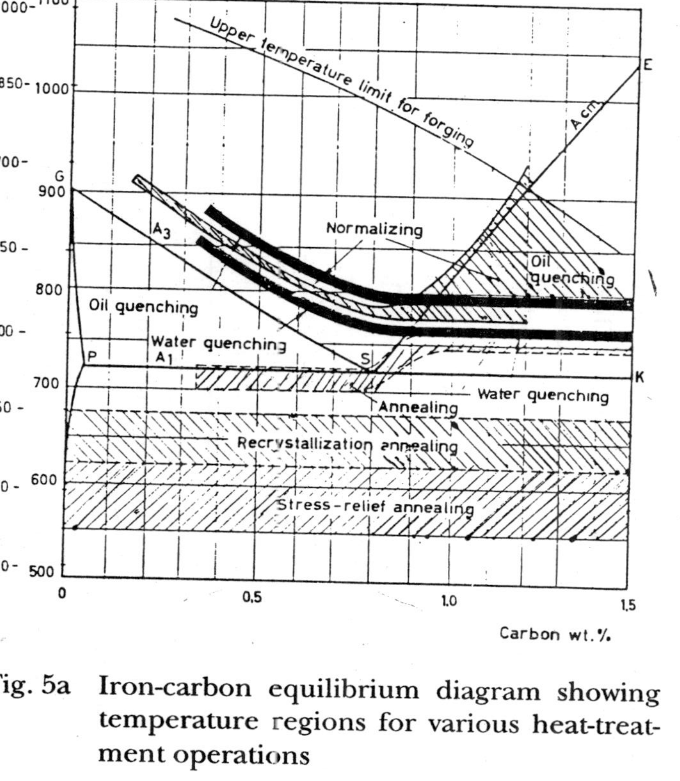

Fig. 5a Iron-carbon equilibrium diagram showing temperature regions for various heat-treatment operations

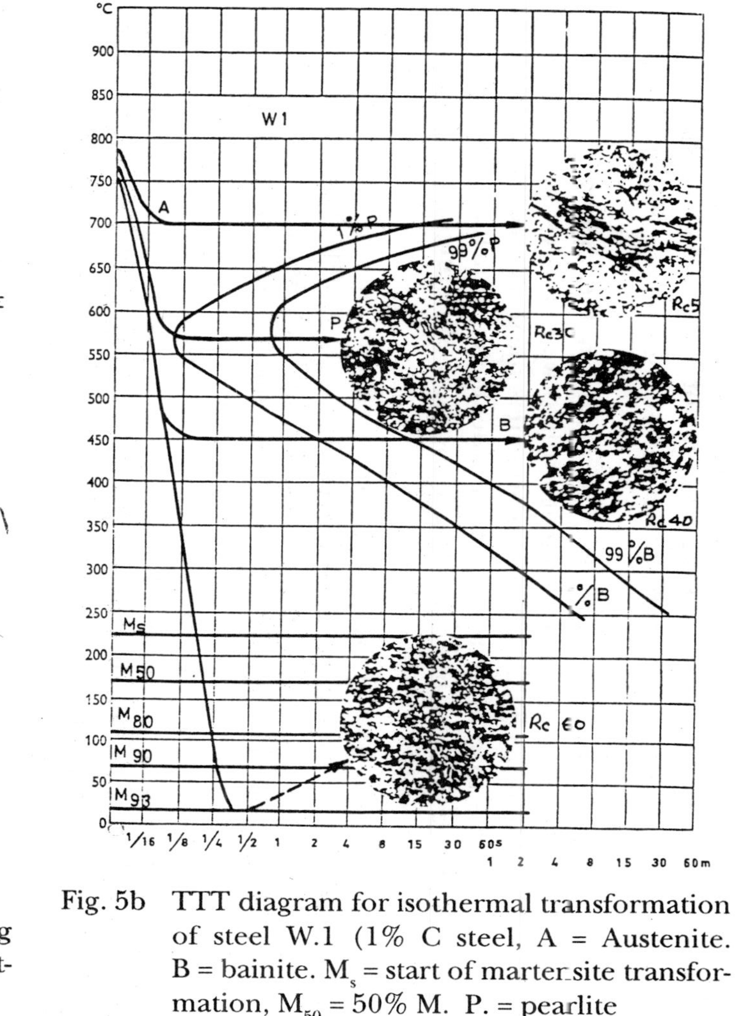

Fig. 5b TTT diagram for isothermal transformation of steel W.1 (1% C steel, A = Austenite. B = bainite. M_s = start of martensite transformation, M_{50} = 50% M. P. = pearlite

where its alkaline sap was sufficient to quench the blade to martensite and then the dried sap acted as heat insulator helping in the tempering of the thin cutting edge by conduction of heat from the thicker back part of the blade.

5. The most common method of hardening and tempering of high carbon steel implements was to quench it into water for few seconds and then withdrawing it so that the heat present in the thicker edge (retained heat) is able to convert martensite into tempered structure. This process of hardening steel has given the generic name 'Watering of Steel', and it is still prevalent in Indian villages.

The main purpose of this watering treatment and the effect of cooling rate on the structure of steel can be understood from the study of the Time-Temperature-Transformation Curve (TTT Curve) shown in Fig. 5(b). As it can be observed from this figure, in order to obtain the hardened Martensite phase the cooling rate of the hot steel blade must correspond to the curve 'A', and if the cooling rate is slow the resulting structure will be Bainite or Pearlite having lower hardness. It is obvious that neither these structures nor such diagrams were known to the ancient craftsmen but by experience and keen observation they had gained experience to produce swords and other weapons of steel having excellent cutting property, flexibility and surface design.

Experiments on Pattern Welding and Watering Treatment

In the Department of Metallurgical Engineering, Institute of Technology, B.H.U. Rakshit *et al.* (1990-91) have conducted a series of experiments on 'Pattern Welding and Watering' of two steels of the compositions given in Table 1, and studied their microstructure and hardness attained after quenching and tempering.

TABLE 1

Material	*% C*	*% Si*	*% Mn*
L.C. Steel	0.3	0.09	0.47
H.C. Steel	0.6	1.26	0.62

The chemical composition of these steels does not correspond to that of wootz steel but first steel having low carbon (L.C.) has its hardenability very much similar to wrought iron. The other steel having 0.6% C and 1.26 Si (H.C.) is known for it high hardenability similar to that of 'wootz' steel. Pls. 22-23 show the microstructure of the two steels. The lower 'C' steel (Pl. 22) consisted of pearlite colonies in a matrix of Ferrite whereas the high 'C' steel (Pl. 23) consisted of ferrite grain boundary in a Pearlite matrix.

5 to 7 mm thick plates of these steels were arranged in double and triple layer as shown in Fig. 6 and they were rivetted together. These assemblies were coated with clay and dried in Sun to prevent their excessive oxidation during heating. These plate assemblies were heated to ~1150° C and forge welded under a 2 tonne power hammer by applying the force in the direction shown in Fig. 6.

The hot forging operation was continued till a uniform plate thickness of ~2 to 3 mm was obtained. In order to make the thinner cutting edge further cold forging was done by hand hammer. This forging operation was done at ~700° C (cherry red temperature) till the cutting edge having a thickness 1 mm was obtained.

The forged sample was given spherodization treatment at 800° for 1 hour and then it was cooled in air. This was followed by surface grinding to remove the oxide layer formed on the surface as well as clean and polish the surface.

The quenching and tempering of these samples was done by reheating them in a electric muffle furnace to cherry red temperature ~850° C and inserting the red hot blade into a green banana tree trunk and leaving it there for 24 hours. The alkaline sap of the tree first quenched the sharp edge to convert the Austenite fraction into Martensite and then when the sap had evaporated away the sharpened edge was tempered due to the flow of heat from the thicker back edge.

Pl. 24 shows the forged shapes of the samples. Out of these the hardness of the three quenched and tempered samples (No. 3, 4 and 5) were measured and their microstructure was studied.

Hardness Profile and Microstructure of the Blades

Sample No. 3: This double layer sample consisted of one L.C. layer forge welded to another layer of H.C. The Rockwell hardness measurement on the L.C. side varied between Rb. 73 to 85 (~Rc 8 to 10) as indicated by hardness mapping in Fig 7(a). The hardness mapping on the H.C. side as shown in Fig. 7(b) showed hardness variation between Rc 30 to 37 on the thicker edge and Rc 39 to 44 on the thinner edge. Pl. 25 shows the microstructure of the two sides as well as a Macroscopic picture of the etched cross section of the forge welded piece. Pl. 25(c) clearly shows the bond between the two layers.

While the microstructure of L.C. side Pl. 25(a) shows large grains of pearlite in ferrite matrix, the H.C. side Pl. 25(b) shows the presence of globular cementite in a matrix of tempered Martensite. It also shows the decrease in the size of the carbide towards the edge i.e. area with higher degree of cold work.

Sample No. 4: In this double layer forging was done in such a manner that the cutting edge consists of H.C. steel whereas the thicker noncutting edge is made from L.C. steel. Fig. 8(a) shows the hardness profile of this sample after the hardening and tempering treatment. In this case while the thin H.C. edge has a

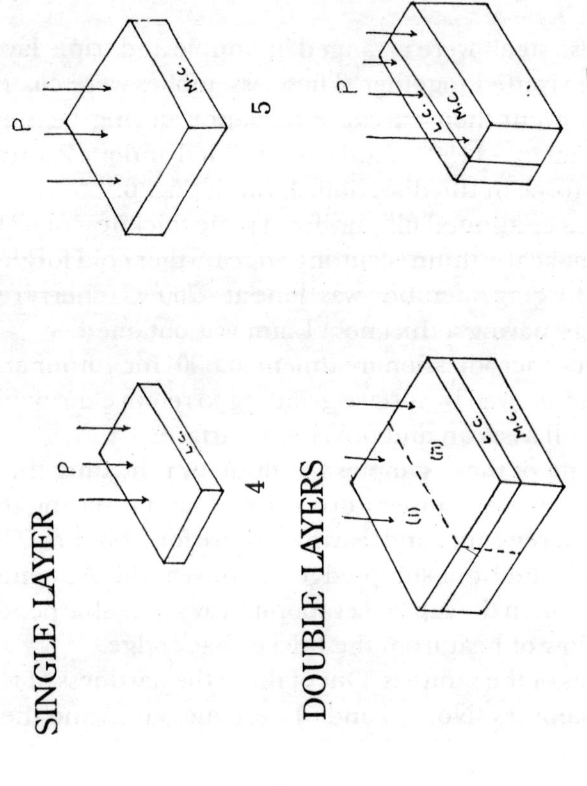

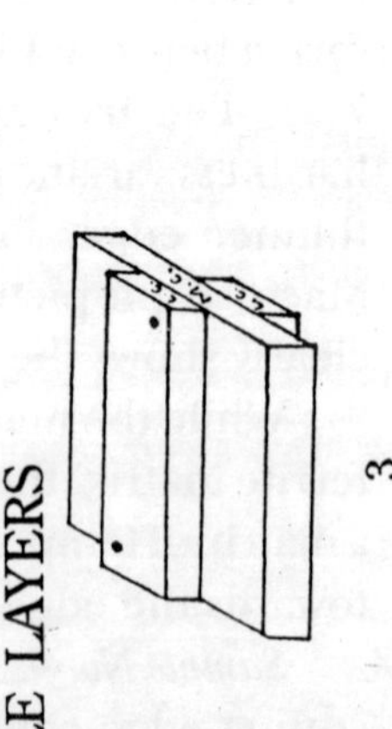
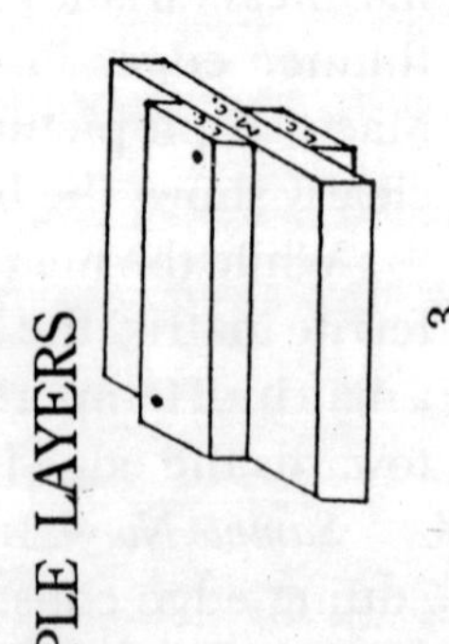
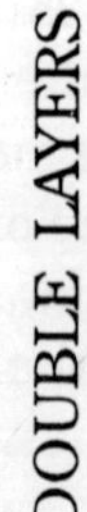
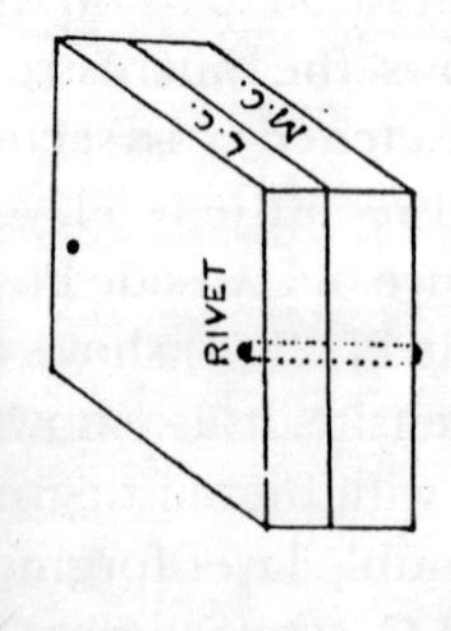

Fig. 6 Sample Preparation

SAMPLE-3
Side 1

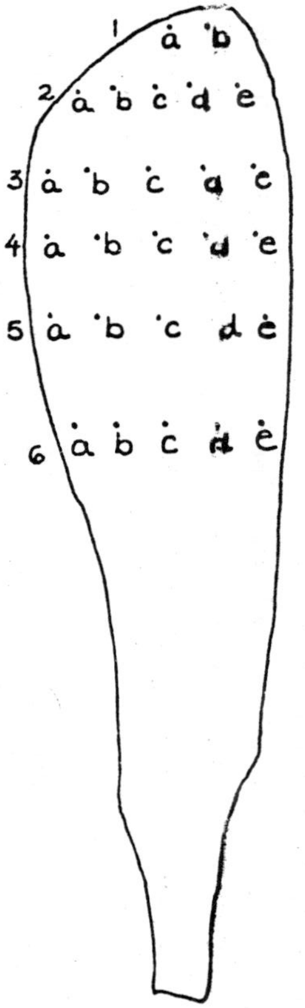

Rc Scale

	a	b	c	d	e
1	80	78			
2	85	81	83	76	73
3	84	80	80	75	74
4	85	78	77	75	73
5	85	81	77	76	73
6	84	79	77	75	73

Fig. 7a Low C steel side of a quenched double layer sample. No hardening seen.

SAMPLE -3
Side 2

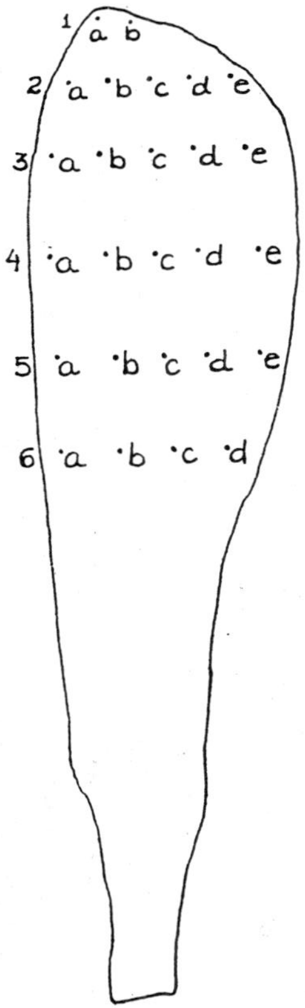

Rb Scale

	a	b	c	d	e
1	37	37			
2	33	34	37	40	43
3	31	34	37	39	44
4	32	36	39	39	43
5	32	35	37	41	43
6	31	34	38	39	43

Fig. 7b Medium C-silicon steel side of a quenched double layer sample pronounced hardening at thin edge (1, 2, 3, 4, 5, 6).

hardness of Rc 42 to 49 the hardness of the L.C. edge is Rc 8 to 12. The microstructure of the cross-section is shown in Pl. 26 and the H.C. part clearly shows the carbide particle distribution pattern. It is obvious that this carbide pattern has great similarity to that of original 'wootz' steel blade but it is not the same. Also the surface of the blade did not show any such watering pattern.

Sample No. 5: The Rockwell hardness mapping of this sample on the averse and reverse sides are shown in Fig. 9. In this case the variation in the hardness on both the sides of the blade decreases from the arrow shaped edge to the centre, because in this case L.C. steel plate has been sandwitched between two H.C. plates. The hardness at the sharp edge varies between Rc 39 to 45 whereas the hardness in the central part is between Rb 73 to 85. Pl. 27 shows the microstructure of the hardened and tempered edge. It clearly shows the presence of carbide network and globular cementite in a matrix of tempered Martensite. The etched microstructure of the cross-section of this triple ply steel blade is shown in Pl. 28. It clearly shows poorly welded edges as well as the watering pattern of carbide in the H.C. plate. This pattern is very much similar to that shown in Fig. 9 but it also differs from the 'watering pattern on the wootz' steel blade in many respects.

All the samples and specially sample No.5 have gained a shining silvery lusture after hardening and tempering treatment and they have not shown any rusting even after their storage for the past six months. This indicates that the alkaline sap of the Banana tree reacts with the steel surface forming a rustless coating on the steel blade. More detailed study is required on the nature of this corrosion resistant coating.

CONCLUSION

The experiments conducted by Denig (1990) have shown that 'Pattern Welding' technique does not produce the same structure as the 'Watering Pattern'. Further, the experiments done by Verhoeven *et al.* and Rakshit *et al.* (1990-91) on rolling and forging of steel blades have shown that:

1. Steels of different carbon content could be forge welded at temperatures 1100-1150° C to give a variety of patterns revealed after differential etching of the blades surface. It is not very sure whether such blades were given any hardening and tempering treatment.
2. In the case of longitudinal forge welding of H.C. plate to L.C. steel so as to make the cutting edge of the sword from H.C. and the back edge of L.C. steel as in the case of sample No. 4 (Fig. 8) the cutting edge attained a hardness of > Rc 43 whereas the back remained soft giving flexibility to the blade.
3. By this technique Damascusizing or watering pattern could not be developed on the surface of the steel blade.

4. Hot rolling of H.C. steel followed by spherodizing treatment as described by Verhoeven *et al.* does not produce the carbide distribution pattern similar to that of 'wootz steels' Watering Pattern.
5. Hot forging under power hammer as described by Rakshit *et al.* has also failed to produce this structure on the surface of the blade although the structure of the cross-section of H.C. part did show a flowery pattern of carbide globules.
6. The hand forging of the blades using a light hammer at 700 to 800° C has produced much finer carbide globules. The skilled use of hand forging technique at cherry red temperature (~ 800° C) might have been the process resulting into the flowery distribution of carbide globules and the typical watering mark might have been produced due to uneven flow of the carbide. This needs further experimentation.
7. The hardening and tempering experiment has shown that it is possible to give these treatments to the steel by thursting the hot blade in to the green Banana tree trunk. This technique imparts a corrosion resistant coating on the surface of the blade.

REFERENCES

Anossoff, P.P., 1974. in J. Piaskowski edited *O Stali Damascenskiej,* Wydawnictuo Polskeej Akademi Nauk, Warszaw, Poland.

Anvils Ring, 1990. *Showcase,* 18, 1, 33.

Balaiev, N.I., 1974. in J. Piaskowski edited *O Stali Damascenskiej,* Wydawnictuo Polskeej Akademi Nauk, Warszaw, Poland.

Bronson, B., 1986. *Archaeomaterials,* 1, 1,14.

Buchnan, F.A., 1807. *Journey from Madras through the countries of Mysore, Canara and Malabar,* London.

Denig, H., 1990. Herstellung von Damassenerstahl, International Conference on *'Iron Palaeometallurgy and Culture'*, Sevenans, France.

Ghosh, A.K. and P.K. Chattopadhyay, 1982. *MASCA Journal* 2,2,63.

Hadfield, R., 1912. Sinhalese Iron and Steel of Ancient Origin. *JISI,* 85, 34.

Mahmud, S.J., 1980. *Metal Technology in Medieval India,* Daya Publishing House, Delhi.

Percy, J., 1964. *Metallurgy of Iron and Steel,* London.

Piaskowski, J., ed. 1974. *O Stali Damascenskeej,* Wydawnictuo Polskiej Akademi Nauk, Warszaw, Poland.

Pleiner, R., 1984. *The Craft of the Blacksmithy,* Ed. B.G. Scott and H. Cleere, Belfast.

Prakash, B., 1990. Some Aspects of Process Control of Ancient Iron Making. International Symposium on Iron Palaeometallurgy and Culture, Sevenans, France, 1-3 Nov.

Prakash, B., 1991. *Metallurgy of Iron and Steel* and Blacksmithy in Ancient India - paper communicated to *Ind. J. History of Science,* INSA.

SAMPLE - 4

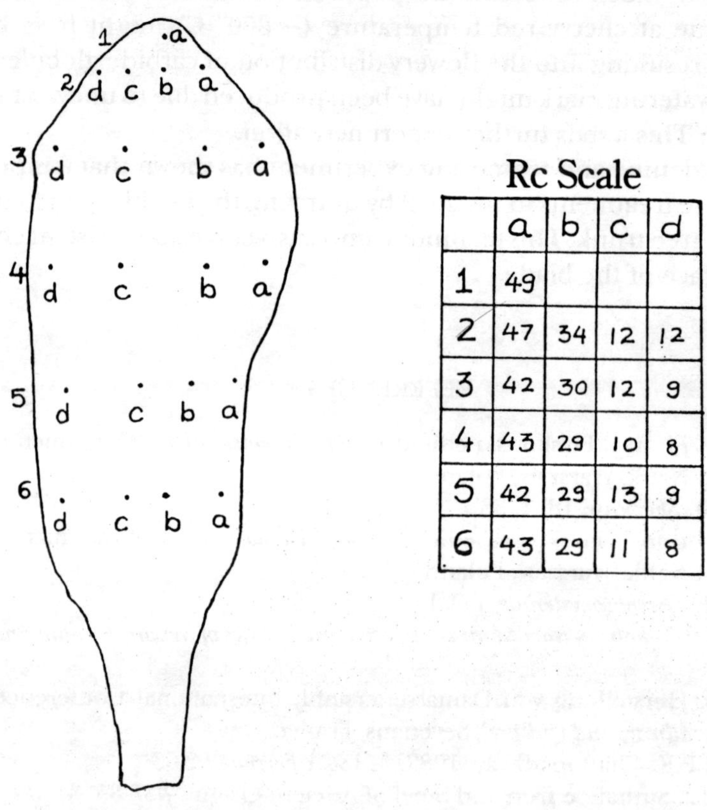

	a	b	c	d
1	49			
2	47	34	12	12
3	42	30	12	9
4	43	29	10	8
5	42	29	13	9
6	43	29	11	8

Fig. 8 A quenched double layer sample surface shows two parts - hardened thin edge a (1, 2, 3, 4, 5, 6) of medium C-silicon steel welded to a thick edge of low C steel.

SAMPLE - 5
Side 1

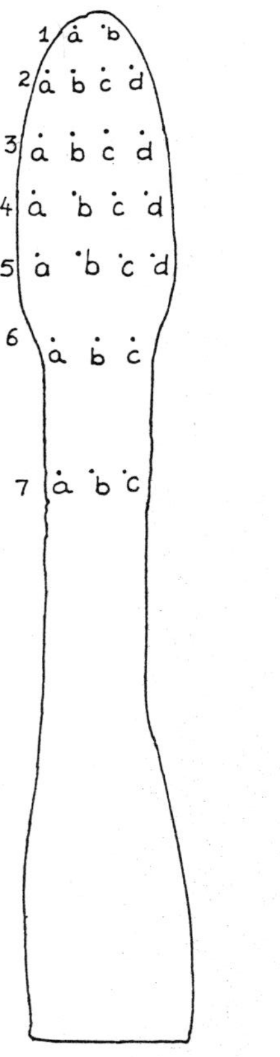

Rc Scale

	a	b	c	d
1	45	45		
2	42	43	43	43
3	39	41	40	41
4	39	38	38	39
5	38	37	38	37
6	36	34	33	35
7	29	30	26	

Fig. 9a Medium carbon silicon steel side of triple layer sample. Hardening all along the edges of the sharp arrowhead.

Sample 5
Side 2

Rc Scale

	a	b	c	d
1	85	84		
2	80	82	81	80
3	78	76	78	78
4	77	77	77	74
5	75	76	75	
6	73	73	73	
7	71	71	70	

Fig. 9b " Low C steel" side of triple layered sample. Negligible hardening along the edge of the sharp arrowhead.

Rao, K.N.P., 1989. *Metals News,* 11, 3.

Rakshit, A., P.Y. Bhopalan and B. Prakash, 1990-91. Pattern Welding and Watering of Steel Blades - B. Tech. (Met. Engg.) Project, Banaras Hindu University (1990-91).

Scott, B.G. and H. Cleere (eds.) 1989. *The Craft of Blacksmithy,* Symposium vol. Belfast.

Sushruta Samhita, 1982. Chaukhamba Orientalia, Varanasi.

Tylecote, R.F., 1984. *The Metallurgist and Materials Technologist,* July p. 343.

Verhoeven, J.D., *Metallography* 20, 2 (1987) 145; *ibid.*, Verhoenen J.D. and L. Jones (1987) 153; *ibid.*, Verhoeven, J.D. *et al., Materials Characterization* 24 (1990) 205.

Voysey, H.W., 1832. *J.A.S. Bengal,* I. 249.

Vrihatsamhita of Varahamihir, 1956. *History of Chemistry of Ancient and Medieval India,* Indian Chem. Soc., Calcutta, *op. cit.*, Kharaglakshanam.

Yater, W.M., 1982. *The Anvils Ring,* 10;11, 4 (1983-84) 2.

Zschokke, B., 1924. *Revue de Metallurgie,* Paris, Nov. 639.

Glory of Indian Iron and Steel Technology in Archaeology

SYED JAFAR MAHMUD

Although metal-working was not the first craft known to mankind, our present material civilization is derived from our knowledge and use of metals. Metals and metal-working have conditioned the development of human culture throughout the ages. Generally, utilization of metals affects the society in two ways. From one standpoint the technology of metal assisted in transforming the economy and livelihood style. In this respect mention may be made of its influence on agriculture and methods of warfare. Control over the use of harder type of metals empowered ancient societies to overpower their neighbouring communities. We also know that conquest of all kinds help the spread of ideas. From the realm of ideas we could easily move to other aspects in early civilization. It may be remembered that change of materials does not confine itself to a change in technology. Introduction of new materials provides the artists and craftsmen with new opportunities to enlarge the scope of their art. The same thing happened in case of metals. Metals provided and opened up new avenues of artistic ideas in a given society.

From the time of Tubal Cain onwards, the early users of metals would probably have described them as bright substances which were hard but could be hammered into various useful shapes. All the metals they knew were heavy—iron, for instance, being nearly three times as heavy as granite. If the metal-worker was also a hunter or warrior, he could speak with satisfaction of the sharpness of metal weapons and of their endurance through many combats. Craftsmen in wood or in stone would praise the new metal cutting-tools which made their jobs so much easier. Thus, even in the very early days, there grew an appreciation of the value of metals.

Iron in Antiquity

Iron finds mention in ancient Indian texts. It has been universally accepted that the Rig Veda is the earliest collection of hymns in Sanskrit. In the Rig Veda, the word 'Ayas' has been mentioned several times. It would appear that this word meant one of the metals, namely, iron, copper or gold in different contexts. It has, however, been pointed out that Ayas would only mean iron and not copper or gold, and also Loha would refer to copper and Shyam to iron. The latter word has been used in the Atharva Veda also. In order to differentiate between copper and iron, the latter metal has been referred to as Krishnayas. In the later Vedic age, the word Ayas was used with different adjectives prefixed to mean either copper or iron. Thus, Lohayas or Lohitayas meant copper, whereas Shyamayas or Krisnayas signified iron. Gold was looked upon as a mark of nobility and iron as a mark of the common people. In the light of the reference to iron in the Vedas, it would be evident that iron was known in the early Vedic times in India and came to be identified clearly as time progressed.

Archaeological Findings

Based on a study covering various sites in India using radiocarbon technique, it has now been possible to establish approximate chronologies for the different cultures which prevailed in this country. In India, the last deglaciation started around 15,000 B.C. in the Kashmir area and the Holocene period could be placed around 11,000-10,000 B.C.

The available C^{14} dates show that the Mesolithic culture in India started around *c.* 6000 B.C. The Neolithic cultures here are marked by a settled life, agriculture, hand-made pottery and ground stone tools. These cultures have been distinguished under three phases. Phase I, where no metal is found, has been assigned to *c.* 2450-1900 B.C. In Phase II, copper makes its first appearance along with the wattle and daub houses with a time bracket of *c.* 1900-1350 B.C. Phase III, *c.* 1400-1000 B.C., is marked by more extensive use of metals. Iron appears in the beginning of the first millennium B.C. The Harappa culture of Indus civilisation appears suddenly without a visible slow evolution. Though Harappa has not been fully C^{14} dated so far, and its script not yet deciphered, the beginning of Harappa culture has been assigned to *c.* 2300-2400 B.C. The Chalcolithic cultures, which are customarily treated separately from Harappa are characterised by the use of copper and stone tools. The C^{14} dates extend over a wide range from *c.* 2000 to 800 B.C. With the advent of iron, the Gangetic plains were colonised, leading to urbanisation. In the North, the Painted Grey Ware (PGW) and the Northern Black Polished Ware (NBPW) were the early Iron Age cultures. The C^{14} dates for PGW fall within a bracket of *c.* 800-300 B.C. except for one with 1025±110, while the NBPW sites cover *c.* 600-50 B.C.

The recent findings in Indian archaeology with radiocarbon dating, reveal that, in general, from the Mesolithic to the Iron Age, Indian culture started later than the West Asian counterparts. The advent of copper technology in India is about 1000 years younger, and that of iron, a few hundred years later than West Asia. Whether Indian technologies were indigenous or borrowed, their dates are certainly later than West Asia. It has been said that the earliest Iron Age sites in the Ganges valley, which date to the early centuries of the first millennium B.C., suggest that the iron industry was in an experimental stage and slowly developed in the south around 1100-900 B.C. In the light of modern archaeological evidence, it can be definitely stated that iron was in use in India around the first millennium B.C.

Location of Iron Mines

There is hardly any iron bearing region in India where we do not find old working and slag heaps. On the basis of these remains a list of regions has been drawn up where iron was worked and sources open to smelters during this period:

1. *Sind* : Ranikot area (magnetite, red and brown haematite).
2. *Baluchistan* : Bolan area (clay iron-stone)
3. *Northwest Frontier Zone* : Bajaur (black magnetic iron sand), Bannu (earth haematite), Hazara (magnetite and haematite) and Waziri hills (limonite).
4. *Kashmir* : Different parts of the state (magnetite, haematite, limonite and other sedimentary ores).
5. *Salt Range and Kot Kerana Hills* : Earthy haematite.
6. *Punjab and U.P. Himalayas* : Kangra (magnetic and micaceous, Shale near Shimla (magnetic and micaceous), Mandi (magnetic and haematite micaceous schists), Sirmur (magnetite, probably mixed with specular ore), Kulu, Almorah, Garhwal (red and brown haematite).
7. *Patiala* : Brown haematite and magnetite.
8. *Rajasthan* : Alwar, Jaipur, Udaipur, Ajmer, Bharatpur, Bundi, Jodhpur, Kota (basically haematite, magnetite, limonite).
9. *Gujarat* : Surat, Panchmahal, Rewa Kantha, Ahmedabad, Kutch, Kathiawar (lateritic ore in Kutch, Kathiawar; ore unspecified in the rest).
10. *Central India* : Almost the whole of it; lateritic ore in Bhilsa, Ujjain, Shajapur, Shivpuri, Mandasore, the ore of the Vindhyan System principally in Mandasore and that of the Gwalior series in Gwalior, the ore of the Bijawar, series in Indore. Jhabua and Dhar; also, haematite in Lalitpur, Hoshangabad, Narsingpur, Nimar; important deposits in Bastar, Chanda, Durg; other areas: Jabbalpur, Bilaspur, Banda, Mandla, Rewa, Bundelkhand.
11. *Maharashtra*: Konkan laterite, Ratnagiri, Kolhapur, Mahabaleshwar, Goa.
12. *Mysore*: Laterite of south Konkan; haematite quartizites and titaniferous ores, magnetic sand; areas: Tumkur, Mysore, Chitaldurg, Kadur, Kolar, Shimoga.
13. *Kerala* : Magnetite and laterite, also black magnetic sand; area: Malabar and Travancore.
14. *Madras*: Magnetite, haematite, laterite; spectacular ores; areas: Tinnevally, Madura, Balem, Trichinopoly, Coimbatore, Nilgiris, Pudukottai Chingleput, north and south.

15. *Andhra*: magnetite, limonite etc., areas: Cuddapah-Kurnool, Guntur, Bellary, Nellore, West Godavari, Krishna, Vizagapatnam, Hyderabad.
16. *West Bengal*: Different ore-types including the magnetic haematite of Borai-Keonjhar, Talcher, Sambalpur, Mayurbhanj sand in the western and northern parts of the state.
17. *Assam*: Clay iron-stone and an impure limonite in upper Assam; titaniferous magnetic iron sand in the Khasi Jaintia hills.
18. *Bihar*. Various ore-types in the entire Chhotanagpur block.
19. *U.P.*: (except Nainital - Almorah - Garhwal) : magnetic and magnetite sand in Mirzapur.

Mining

Indian miners, as known primarily from ethnographic evidences and medieval texts could recognise by the nuance of the soil and the type of flora, a suitable site for procuring iron ore: Very often the ore was found on the surface.

Open Cast Mining

Mining was generally restricted to surface out-crops. With the help of a pick axe trenches were made with stairs in sides. Ores were dug out and taken out of the trench filled in cane baskets.

Shaft Mining

Mining was done by narrow adits or stopes driven from the strike of the ore body. Openings of the mines were 1 metre (approx.) in diameter, just big enough for a person to crawl inside with the support of ropes. Openings were made with the help of pick axe. Elliptical shafts were used for giving roof supports which reflects the sophistication of mining technology of that period. Timbering was also resorted to. Small pillars of ore were left to support the ground and in some cases timbers 15-18 cm in diameter were used to support the stoped out areas.

The labourers while at work sat on their heels with the lamp upon their heads, holding a chisel in left hand, struck with a hammer held in the right hand and collected the chipped fragments in the basket held upon knees. The bamboo baskets of ore were passed along a chain of persons usually 4 to 12 in number, who sat upon their haunches in the galleries or stood on the niches cut in the shafts; when the whole reached the foremost man, they moved upwards, and again arranged themselves, and this was continued until the mouth of the mine was reached.

Tools of Miners

Pick Axe: The handle was made of a piece of wood with a knob at one end, into which a piece of hard iron was thrust and sharpened at the point.

Iron Hammer: Special type of hammers made of iron 10 inches long were used.

Chisels: Iron chisels pointed at one end 3 inches (approx.) in length were used for following the strike of the ore body.

Spade: Spade made of iron was used to collect the ore from the ground.

Pitchers: There are evidences to show that earthen pitchers were used as it has been found in the old workings.

Smelting

Beneficiation of iron was done to enrich the metal content and also to improve the physical characteristics of the ore charged in the furnace. It was achieved by crushing, sizing and agglomeration of resultant fines by putting a large hard stone on the ground on which they laid the ores, and then either with a stone or hammer, pulverized the ore at places with the help of water mill which was washed in water, which had a profound effect by decreasing the silica and alumina content. Indian metallurgists knew that the preparation of sized iron ore burden could achieve,

(a) Reduction in Charcoal rate by reducing the metallurgical load in the blast furnace by better utilization of available heat, and

(b) Increased productivity and better quality of Iron.

Smelting of the beneficiated ore was done in a blast furnace, which was one of the first steps in the transition between the original method of reducing the ore in an open fire and its reduction in a modern blast furnace. Doubtless that the Indian furnace was the descendent of such a structure. Blast furnace consisted essentially of a vertical cylindrical furnace, built of clay, circular in shape ringed by bellow (wind box) or air chamber from which air blast entered the cupola with fuel and the materials making the charge. There were mainly two types of blast furnace used in India :

1. Fosse Type
2. Aerial Type

Furnaces

Their smelting furnaces, though crude in appearance, are nevertheless very exact in their interior proportions, and it has often surprised me to see men who are unquestionably ignorant of their principle, construct them with precision, in so simple a manner; their unit of measure is the breadth of a middle sized man's finger; 24 of which constitute their large and 20 their small cubit; thus there is a constant ratio of 6 to 5 prevailing throughout these furnaces, nor is it of the least consequence, that their dimensions are larger or smaller, so long as all the parts are in the same proportion; the length of these measures is on an average 19-20 English inches for the large cubit, and 16 English inches for the small one.

As they have no standard measure, their fingers, their span, and their arm are substituted by which a piece of stick is measured which they use in practice; neither is the division of the cubit necessary though the large one supposed to be divided into six parts and the small one into five, of four fingers each-as the measurement is invariably ascertained by their fingers; the length of these parts is on an average 3.20 English inches.

Steel

The method for making steel was to convert wrought iron in crucibles, resembling modern cementation process.

Wrought Iron

Whether smelted in open hearths or in small and large blast furnaces, the end product was wrought iron. Indigenous method of iron production during the period was that of direct reduction. Indians took the advantage of the fact that despite the high melting point of iron (1540° C) iron oxide could be reduced to metal at a low temperature of 800° C. At a temperature of 1100-1500° C, this reduced iron flows together forming a semifused porous mass called as bloom. The bloom can be easily forged and worked. Flux was not used, save in the case where the ore was self fluxing. This had the disadvantage that lot of iron oxide passed into slag but helped in removal of phosphorus fusible.

This low temperature smelting avoided the reduction of any other metal present in the ore and also diminished the danger of carburization of iron. The product was slag bearing wrought iron. It was fibrous in nature and could be easily welded and forged. Metal was very pure and no other metal including manganese was present. Sulphur was very low and consequently metal had high resistence to corrosion. For chemical composition of wrought iron refer to tables.

However, it must be admitted that the yield from best ore rarely exceeded 50 per cent of iron in the ore. Ordinarily the yield were of the order of only 20 per cent and the remaining was lost in the slag. Rich iron slags littered all over the country prove the point. (See Table 1).

TABLE 1

Analysis of Iron Slags from various Smelting Centres

Description of sample	*SiO_2*	*Al_2O_3*	*FeO*	*CaO*	*Loss on Ignition*	*MnO*	*MgO*	*S*	*P_2O_4*
Rewakanta Bombay	53.64 —	5.39	28.96 —	10.49	1.52				
Madhya Pradesh	10.33	1.85	8.13	2.49		0.23	1.07	0.03	0.35
Rajdoha			73.94						

TABLE 2

Analysis of Iron Ore by Harris

Ferric Oxide	(Fe_2O_3)	69.55
Ferrous Oxide	(ReO)	19.50
Silica	(SiO_2)	5.83
Manganese Oxide	(HnO)	0.22
Alumina	(Al_2O_3)	0.51
Lime	(CaO)	0.36
Magnesia	(MgO)	Trace
Sulphur	(S)	0.02
Phosphoric anhydride	(P_2O_5)	0.03
Moisture at 100° C		0.60
Combined water and loss		3.28
Total		100.00

TABLE 3

Percentage Composition of some Ancient and 19th Century Iron Objects

Description of specimen	*Fe*	*Carbon*		*Mn*	*N*	*S*	*Si*	*P*
		Graphitic	*Combined*					
1. Delhi Iron Pillar 5th cent. A.D.	99.768	—	0.23	—	0.0065	tr	0.026	0.180
2. Konark Iron 1250 A.D.	—	—	0.11	—	—	0.024	0.100	0.015
3. Rajdoha crude iron 19th cent. A.D.	98.180	0.66	—	0.013	—	0.005	1.110	0.028
4. Rajdoha Hammered and reheated iron, 19th cent. A.D.	99.947	—	0.03	—	—	—	0.010	0.030
5. Modern Tribal Smithy iron from Koraput, Orissa	—	—	0.23	—	—	0.006	0.100	0.015

It is the power of wrought iron to resist outside atmospheric influences better than steel that has caused the production in indigenous wrought iron to survive side by side with modern steel works.

Process of Making Wootz Steel

Indian steel generally referred to as wootz was made by a method resembling the modern cementation process or crucible process. Small pieces of wrought iron weighing 1/2 to 2 Ilbs were packed into crucibles with 1/10th of their weight of chopped dry wood and leaves of specific plants e.g. Avaram (*Cassia auriculata*). Crucibles were sealed with clay and up to two dozen of such crucibles were staked into the furnace which was maintained at high temperature by blowing air over charcoal fire. The process was completed within 2-1/2 to 4 hours. The cake so formed was forged at a very low temperature. Composition of wootz steel given by Percy is :

Combined carbon	1.33%
Uncombined carbon	0.31%
Silicon	0.45%
Sulphur	0.181%
Arsenic	0.037%

Source : Percy J., *Metallurgy* : *Iron and Steel*, London, pp. 254-278.

This composition agrees with high carbon steel, slowly cooled from above its melting point, which would be nearly 1450° C.

Malleable Iron

Malleable iron is a cast alloy consisting principally of iron and carbon which is rendered tough and ductile by a controlled heat-conversion process. Because of its unique metallurgical structure, it has the valuabe properties of strength, great toughness, remarkable resistance to heavy and repeated impact, excellent ductility, high resistance to corrosion, and easy machinability. Its versatile castability makes possible sound castings, accuracy to pattern, in complicated as well as simple forms over an extensive range of weights and sizes. The combination of these properties in this one metal has established it as the ideal material for wide range of applications.

The first use of iron is lost in the mists of prehistoric time. It is clear that primitive man learned that heating a type of reddish earth would give him a red hot mass of material which he could pound into shapes of his own design. The reddish earth was iron ore—oxides of iron. The wood fire which heated it formed charcoal. This served as the reducing agent to deoxidize the ore, leaving essentially pure iron which could be 'worked' or 'wrought'.

In later recorded time, the ore was reduced in Catalan hearths, then by their successors, the early stack furnaces. In these, temperatures were sufficient to produce only an incandescent spongy mass of iron. The metal, however, was almost pure chemically, for two principal reasons: First, the temperatures were still below the ranges where chemical combination could occur between the iron and any other elements present - notably the carbon of the charcoal; and, second, at the temperatures, attainable the charcoal ash, resulting from combustion of the fuel, sintered or partially fused with the clay and other minerals found in nature with the ore, to form a slag. Some of the slag worked through and out of the hot mass, but because the iron was pasty in texture, much slag ramained entrapped. The glowing masses with which the early iron masters worked were known as 'blooms' and their furnaces as 'bloomeries' which in their later built-up forms became the 'high furnaces' and were the fore-runners of the modern cupola and blast furnace. The blooms were forged by hammering, to shape and form the articles desired, and to force out as much as possible of the entrapped slag.

Improvements in furnaces and their performance led in time to complications; the iron produced became more difficult to work and when cooled, it lacked strength and could easily be broken. Reheating the iron, it was found, at least partially restored to more ductile conditions. This rule-of-thumb technique relied on the basic chemical principle that with sufficient temperature carbon will migrate in iron. If the carbon content is higher in the materials or atmosphere, than in the iron itself, carbon will enter and migrate through the iron. The first process — absorption of carbon — is called carburization; the second — removal of carbon — is decarburization. During carburization, carbon combines with iron to form iron carbide, harder and more brittle than pure iron. Decarburization frees carbon from the carbide, restoring the iron to a chemically purer, softer state.

Up to this time the iron obtained by the direct reduction of the original ore could be wrought, for the temperatures had not been high enough to cause any appreciable carburization. Now the higher temperatures provided by improved furnaces had introduced the hardening influence of carbon into iron-making. As the ore was reduced with charcoal at these more elevated temperatures, the carbon-rich fuel and its gases of combustion supplied the excess carbon that carburized the formerly pure iron of the blooms, making it hard and unworkable. It therefore became necessary to go to the indirect or two-stage process of first producing a carburized hard bloom by reduction of ore, and then by a heat-treatment decarburizing it to prepare it for ultimate forging. When the hard blooms were reheated, preparatory to being formed into desired shapes, much less fuel was needed than that required in the original reduction of the ore, and so the furnace atmosphere was now decarburizing in nature. Hence the carbides were partially broken up, carbon migrated out of the iron, and the iron was chemically repurified, its hardness lowered, and its workability restored.

As the height of the slag furnaces increased, the time of contact between ore and charcoal lengthened, temperatures grew hotter and carbon absorption increased. Ironmasters, who were now making blooms so large that they were unwieldly and had to be pulled apart into pieces of convenient size for forging, were beginning to hope that finally they would be able to melt iron and cast it as other metals were cast. The blooms, they expected, would soften progressively and ultimately liquefy. The time came in the first decade of the fourteenth century when, in some furnace, well before seemingly high enough temperatures were reached, iron actually meltered. Breaking into this furnace for the removal of the expected bloom, the iron-workers found that the fluid metal had run down into furnace hearth, collected into a molten pool and rid itself of the lighter slag, which floated on the surface. The now well-known capacity of carbon for lowering the melting point of iron had come into play and had produced an iron rich in hard iron carbide, but relatively slag-free and clean.

The introduction of melting had several profoundly important effects. Since working out slag during forging now became almost unnecessary, the indirect process proved to be more efficient and economical than the direct, which involved both a much longer reduction period at lower temperatures to avoid carburization and then laborious forging to drive out the slag. Melting, therefore, was the greatest single influence in the adoption of the indirect process, which became general practice during the seventeenth century. Moreover, once molten iron was available, coupled with experience in the casting of other metals, iron founding was at hand. Castings made of the newly available material had to be confined to a limited field, since they were of a highly carburized iron, and therefore very hard and brittle. The advent of molten iron, too, permitted a time lag in the progression from raw material to finished product, through the device of running the molten metal out of the furnace into shallow, thinly connected troughs to cool into pieces that were easily broken apart. The troughs, when filled, had a most realistic resemblance to a sow with suckling pigs, hence the term 'pig-iron'. For the infant foundry industry of this period, with meagre demand for its brittle, hard castings, relatively little melting stock was necessary. Of the two means for obtaining it — the direct reduction of ore or the melting of pig iron — the second was clearly cheaper and more practical and was adopted where possible, thus contributing further to the establishment of the indirect process. No significant amount of gray iron, malleable iron, or cast steel is made today by a process deriving its basic iron direct through the reduction of ore.

The process of malleablization consisted of melting iron, pouring castings, packing them in pulverized iron ore, and heating them to bright redness for many days. Lengthy heat-treatment in this temperature range slowly removed the carbon present as iron carbide in the metal as cast. This decarburization was partial to complete, according to the thickness of the casting and the temperature and time of heating. The metal thus produced was almost free from carbon and for the first time provided ferrous castings possessing some toughness and ductility.

At elevated temperatures, the chemical affinity of oxygen for carbon is stronger than its affinity for iron. As carbide gave up carbon, the iron ore packed around the castings released oxygen to combine with the freed carbon to form gaseous carbon monoxide and dioxide, which escaped from the furnace. The packing performed three functions: First, it supplied the oxygen for oxidizing the carbon, second, it supported the castings that otherwise would have softened and lost shape under the applied heat, third, it acted as a 'buffer', shielding the castings from direct contact with the hot furnace atmosphere—a contact which in the long period of heating would have caused undue oxidation of iron as well as carbon, creating a heavy scale on the castings and perhaps even destroying thin sections.

This process, with only minor deviations, has been followed in India to the present day in those foundries still producing 'white-heat' malleable iron. To understand this descriptive term, one must remember that this iron is decarburized in the sense that most of the carbon freed from its carbide has been eliminated, though not all the carbide may have been broken up into iron and carbon. The constituents of the annealed metal, therefore, are mainly iron and iron carbide, although some of the freed carbon may also remain. Because of the presence of the iron carbide, the grains of the metal, when under tension, stretch to a limited extent or not at all, and then shear through and across their sections along minute and fairly flat planes which reflect light. These minute reflecting planes give a whitish or grayish colour to the ruptured surface, and hence its description as 'white-heat'.

Since decarburization was the basis of the medieval process, only small castings could at first be made malleable by it. Larger ones were liable to be hard and brittle in the centre. Though present day manufacture of white-heat malleability has advanced beyond these limitations, decarburization continues to be basic to it. Some graphitization—the precipitation of carbon from iron carbide and its retention in the casting as nodules of free 'temper' carbon does take place in white-heat malleable castings during the heat-treatment, but small castings, say upon a half-inch section, are almost entirely decarburized. Heavier sections consist of a white rim and a core in which some of the carbon remains in the combined form and which contains some nodular graphite. The properties of the rim differ greatly from those of the core, and so the size of casting which can be made successfully in white-heat malleable is still limited.

This survey of iron technology from its inception to high level steel production is an attempt to place the story of indigenous iron manufacturing in India.

REFERNECES

Allchin, B. and R., 1968, *The Birth of Indian Civilization.*

Alwin, Verrier, *Agarias,* London.

Banerjee, M.N., 1929. Iron and Steel in Rigvedic Age, *Indian Historical Qrly.,* Vol. 5.

Bhardwaj, H.C., *Aspects of Ancient Indian Technology.*

Buchanan, F., 1807. *A journey from Madras* (London).

Caley, E.R., 1964. *Analysis of Ancient Metals,* Pergamon Press.

Chakraborti, Dilip, 1976. The beginning of Iron in India, *Antiquity*, L.
Crackoft, W., 1833, Iron Smelting in the Khasi Hills. *J.A.S.B.*, vol. II.
Kosambi, D.D., 1963, The Beginning of Iron Age in India. *J. Econ. and Social Hist. of Orient*, Vol. 6, pp. 309-318.
Mahmud, Syed Jafar., 1988. *Metal Technology in Medieval India.*
Rahman, A. and B.V. Subbarayappa, Note on the native method of bar Iron Production in South India (Salem Region). *I.J.H.S.*, Vol. No. 2, 1966.

Studies and Archaeometallurgical Significance of Some Gold, Silver and Copper Coins from Rajghat (Varanasi) Excavations — A Brief Review

SOMNATH MISRA AND T.N. TIWARI

India is a land with one of the most ancient civilisations in the world, and we Indians can be justly proud of our past achievements, especially in the realm of the intellect. In fact, the famous nineteenth-century scholar and Indologist Prof. Max Müller made some very significant remarks on this matter, in his book *India: What can It Teach Us?* When he wrote,

> "If I were to look over the whole world to find out the country most richly endowed with all the wealth, power and beauty that nature can bestow-in some parts a very paradise on earth — I should point to India. If I were asked under what sky the human mind has most fully developed some of its choicest gifts, has most deeply pondered on the greatest problems of life, and has found solutions of some of them which well deserve the attention even of those who have studied Plato and Kant — I should point to India".

Archaeological investigations have revealed to us a rich heritage of India and given a new perspective to our ancient culture and civilisation. There has been an enormous spurt in archaeological excavations and studies since the Independence. These excavations have brought to light a very rich material heritage in the form of antiquities from different regions of India.

Since no written records are available, it is only through the scientific studies of material remains excavated from various archaeological sites that we can learn about the scientific and technological knowledge of ancient Indians. That is how the study of archaeology has been greatly enriched in recent years by the active collaboration between archaeologists and scientists. Current archaeological researches tend to draw heavily from recent advances in physical, biological and earth sciences. A happy outcome of this fruitful collaboration is the development of *archaeometallurgy* — a fascinating discipline devoted to a better understanding of ancient communities (especially their technological knowledge) involving extraction, processing and applications of metals and alloys.

Archaeometallurgical researches in India during the last two decades have revealed the stupendous intellectual effort made by early Indians in exploiting natural mineral resources and making a variety of materials using copper, silver, gold and iron and their alloys. These researches have, moreover, thrown some light on the probable sources of ores, methods of ore dressing, techniques of smelting and extraction of metals from their ores, the temperatures obtained, the role of fluxes, nature of slags, methods of purification of metals, processes of fabrication, casting and forging techniques, and many other aspects of metallurgical art, science and technology.

This paper presents a brief review of the chemo-metallographic studies and archaeometallurgical significance of some gold, silver and copper coins obtained from Rajghat (Varanasi) excavations (ca. 600-200 B.C.). A brief description of Rajghat excavations is given in the following section.

Rajghat (Varanasi) Excavations

Varanasi is one of the most ancient living cities in the world, which has a continuous history since 1000 B.C. Since time immemorial, Varanasi has been among the most important centres of religion, philosophy, literature, music, art and culture. Rajghat area of Varanasi (Latitude 25°18' N and Longitude 83° 1' E) represents the oldest part of this eternal city.

Varanasi enjoys a commanding position, situated on the crescent-shaped left-bank concave side of the river Ganga. The crescentic nature and character of the river bank has made the position of Varanasi so stable that it has shown little shifting in its site during the last three millennia. The highest level in Varanasi is found in the north-eastern part of the area, known as Rajghat plateau, which is 82 m above sea level and 15 m above the surrounding country side.

Archaeological excavations of the extensive mounds of Rajghat at the confluence of the rivers Barna and Ganga, undertaken in the early sixties by the Department of Ancient Indian History, Culture and Archaeology of Banaras Hindu University, have brought to light a continuous cultural sequence at Varanasi from about 1000 B.C. to the modern times (Bhardwaj, 1979).

Besides other important antiquities, the excavations at Rajghat brought to light a number of gold, silver and copper coins belonging to the period 400 B.C. to 900 A.D. The study of such ancient coinage is of great interest to the numismatists, historians of science and technology, archaeologists and archaeometallurgists. Because the coins could have been made with greatest care, utilising all the metallurgical knowledge and skill, these would be the best representation of the contemporary metallurgy of the period to which such coins belong (Bhardwaj and Misra, 1968).

The metallographic examination of these coins could give information about their micro-structure and the amounts of the principal metal and trace impurities. This information would throw much light on the art and science of smelting of the principal metal, its casting, fabrication, processing, heat treatment, stamping, embossing and other metallurgical details. These, supplemented by hardness measurements, can throw further light on the metallurgical processes involved. Even the macro-examination of the superficial features of such coins is helpful in elucidating the degree of perfection of the final finishing operation.

Gold Coins

Background

Gold was among the first metals to be explored and used by man. During the Harappan period (ca. 2300 B.C.), for example, gold was commonly used for jewellery.

Emergence of regular coinage in the West is dated back to around 700 B.C., and this coinage is in gold. No gold coins contemporary to this period have been found in India; though there are some indications, in the Vedic and post-Vedic literature, of some sort of metal pieces minted in gold, electrum and silver.

Gold coinage as a clear monetary system in India first emerged during the Kushana period. The Gupta gold coins succeeded the Kushana coinage and bear their influence to some extent. Nevertheless, the Gupta coins are supposed to have their own evolution, and they are considered one of the finest expression of Indian art on the one hand, and technological excellence on the other.

Excavations at Rajghat have yielded four gold coins of the Gupta period. The archaeological and general details of these coins (Bhardwaj and Misra, 1971) are given in Table 1.

TABLE 1

Archaeological and general details of four Gold coins from Rajghat excavations (Bhardwaj and Misra, 1971)

Coin No.	*Reg. no.*	*Weight (g)*	*Archaeological details*
1	130	8.9677	Skanda Gupta, 455-467 A.D., Archer type
2	131	7.7504	Chandra Gupta, 375-414 A.D., Archer type
3	132	7.7834	-do-
4	133	7.7204	Kumar Gupta, 414-455 A.D., Archer type

Chemo-metallographic Studies

The four gold coins were examined for their gold content. One of them (Coin No. 1, Reg. No. 130) was studied metallographically. The main results of these studies (Bhardwaj and Misra, 1971) are shown in Table 2.

TABLE 2

Density and chemical composition of four Gold coins from Rajghat excavations (Bhardwaj and Misra, 1971)

Coin no.	*Reg. no.*	*Density (g/cm^3)*	*Wt. (%) Gold*	*Wt (%) silver**
1	130	16.001	75	25
2	131	15.804	74	26
3	132	16.250	77.5	22.5
4	133	15.807	74	26

* Iron is present in traces.

On account of the great antiquarian value and archaeological importance of these coins, it was not possible to take their samples for detailed chemical analysis. They were, however, tested qualitatively for the alloying elements present. Silver was found to be the only significant alloying element. The percentage of gold and silver was determined by the measurement of specific gravity.

These gold coins show excellent details of embossing on both sides, indicating that striking technique was well developed. Both sides of the coin examined metallographically (Reg. No. 130) show fine elongated cavities on the surface. The micro-structure of the edge of this coin shows equiaxed fine grains of size approximately 0.002 cm. Micro-hardness measurement gave a value of 160 kg/mm^2, which is unusually large for a gold coin and further confirms that the coin has been hardened by alloying with silver. Its composition by weight (75% gold and 25% silver) corresponds roughly to the composition which undergoes an ordering transformation in the solid state resulting in edge hardening (Smoluchowski, 1951). This composition also maintains some of the favourable mechanical properties of pure gold, viz., low annealing temperature, high ductility and extremely easy working characteristics.

Traces of iron detected in the coin may have been due to the usual contamination in the parent gold. This, incidentally, has a desirable property of inhibiting grain growth.

Archaeometallurgical Implications

From the available evidence of macro-and micro-structures, it may be concluded that the material has been either hot worked or cold worked, followed by annealing above the recrystallization temperature, as is clear from fine equiaxed grains.

Dies of steel or even bronze could have been sufficient to emboss the coin blanks. On the basis of recovery of a clay mould of Archer type of Chandra Gupta II from Rajghat excavations, there could be the possibility of the Gupta period gold coins having been cast in the clay mould; but considering the small size of grains, absence of dendrites and fine embossing, the coins were probably die struck rather than cast. Moreover, as the mould in question is not from a stratified dig, one cannot attach much importance to it; it could even represent a forger's mould prepared from a genuine gold coin!

Silver Coin

Background

Early history of coinage in India is obscure. Certain small ingots of silver with three circular dots are probably the example of the earliest existing Indian coins. These are followed by heavy bent bars of silver with certain devices punched on one side. These coins came into circulation probably prior to 600 B.C. A little later (around 500 B.C.), we come across Punch-marked coins, reported in thousands from all parts of India. These are rectancular and circular flat pieces, based on alloys of silver and copper. These are cut from a sheet of metal and clipped to proper weight.

We review here the studies conducted on a silver coin (of about 900 A.D.) obtained from the Rajghat excavations. The archaeological and general details of this coin (Bhardwaj and Misra, 1968) are presented in Table 3.

TABLE 3
Archaeological and general details of the Silver coin from Rajghat excavations (Bhardwaj and Misra, 1968)

Item	*Detail/Value*
Shape	Round
Surface attributes	Geometric design
Approx. date	900 A.D.
Weight	1.8 g
Diameter	1.8 cm
Other macrofeatures	Cracked, rounded edges

Chemo-metallographic Studies

A small portion of the coin (about 0.1 to 0.2 g) was clipped for chemical analysis. It was dissolved in concentrated nitric acid and the residue was filtered off. Silver was precipitated as AgCL and determined gravimetrically. Lead was determined as $PbSO_4$. Copper was precipitated as cuprous thiocyanate. Nickel was determined as dimethyl glyoxime complex. Qualitative tests were made for the identification of As, Sb, Bi, Fe, Ni and S.

The coin was ground and polished by usual methods and polished surfaces were examined under a microscope to reveal non-metallic inclusions. Then the surfaces were etched with aqua regia and examined for various microstructural constituents under a metallographic microscope.

The coin was indented in a Vicker's hardness tester with a diamond pyramid indentor under 5 kg load. The results of chemical analysis, metallographic examination, hardness measurement and characteristic microstructures of this silver coin are presented in Table 4.

TABLE 4

Results of chemo-metallographic analysis of the Silver coin from Rajghat excavations (Bhardwaj and Misra, 1968)

Item/Parameter	*Detail/Value*
Chemical composition	Ag - 61.54%, Cu - 32.14% PB - 0.50%, Insol. - 3.4%
Impurities	Fe 0.54%, Ni - 0.50%, Sb-trace, As-trace, S - 0.20%, Sn - 0.57% CaO - 0.52%, Bi-trace
Hardness	89 V.H.N. Kg/mm^2
Microconstituents	Cu (dark) globules and eutectic, compressed, elongated, interconnected crack
Approx. melting point	810^0 C
Fabrication technique	Cast, moderately cold worked without subsequent annealing

Archaeometallurgical Implications

The ninth century silver coin from Rajghat excavations contains 0.50% lead and this is also detected metallographically as small globules. This indicates that the silver used for the coin might have been extracted from lead-bearing natural ores of silver. These are mostly sulphide ores of which argentiferrous galena is an example. Native silver, occurring as almost pure virgin silver, does not contain any lead and may be ruled out (Gowland, 1917).

The extraction of silver from the sulphide ore must have been through processes of roasting the ore to oxide, followed by reduction with carbonaceous materials such as charcoal, and finally removal of lead (and other easily oxidisable impurities) by cupellation. The recovery of silver by cupellation requires a temperature of about 1000° C.

The coin under consideration reveals the presence of bismuth (see Table 4), which is believed to be responsible for embrittlement of this coin as evident from the low hardness value (89 kg/mm^2) and the crack in the microstructure. The notorious embrittlement of the noble metals copper, silver and gold by bismuth is due to the latter being insoluable in the solid noble metals and its easy spreading as thin liquid film along the grain boundaries. Addition of lead counteracts this easy spreading of bismuth as a thin film. Lead, moreover, is insoluable in solid silver so that there is no possibility of precipitation hardening of silver by lead (Cotterll, 1959).

This coin contains 32.14% copper by weight. There is no natural ore of silver which can yield, on smelting, such a high copper content directly. Hence the copper must have been introduced as an intentional alloying element. Studying the phase diagram between silver and copper, one finds that the melting point of silver decreases from 960.5° C to 779° C on increasing the copper content from zero to 28.1% by weight. Most silver-based alloys of copper will show a mixture of two microconstituents at room temperature, e.g., almost pure silver and an eutectic (intimate lamellar mixtures) of silver and copper rich solid solutions (Ag and Cu), respectively.

The microstructure of this silver coin consists of primary copper - rich solid solution dendrites and fine eutectic lamellae. The entire structure has been compressed possibly by cold work on the embrittled materials (due to grain boundary film of bismuth) and subsequently enhanced by stress corrosion.

In conclusion, the composition of the silver coin from Rajghat excavations is such as to give it optimum hardness and wear resistance, and lowering of melting point. The low percentage of lead shows that cupellation was carried out with great perfection. Comparative absence of slag inclusions shows that the metal was cast from completely molten metal. The coin might have been clipped from a cast sheet, which might have been slightly hammered to give uniform thickness. Cast sheet might have been obtained by pouring the super-hot melt on flat substrate. The debasement of silver by copper is deliberate, as hinted earlier, and its purpose might be both economising on silver and providing the coin strength and wear resistance.

Copper Coins

Background

Indus Valley people had mastered the metallurgy of copper around 2350 B.C. They made commercially pure copper (99% Cu) and copper-rich alloys with tin, lead, arsenic and even nickel. Both forging and casting techniques of fabrication were developed. This ancient civilisation had even developed a system of coinage but no copper coins had yet emerged.

The earliest copper coinage in India is represented by the Punch-Marked Coins dated Ca. 500 B.C. These were followed by the so-called local or tribal coins, which have been reported from many archaeological sites such as Taxila, Mathura, Ayodhya, Kannauj, etc.

We review here the investigations conducted on a copper coin obtained from the Rajghat excavations. The archaeological and general details of this coin (Bhardwaj and Misra, 1969) are presented in Table 5.

TABLE 5

Archaeological and general details of the Copper coin from Rajghat excavations (Bhardwaj and Misra, 1969)

Item	*Detail/Value*	
Coin type	Uninscribed, rectangular	
Surface attributes	*Obverse*:	Tree in railing, hollow cross, crescent on hill and taurine;
	Reverse:	Elephant with rider, standard and two unidentifiable symbols.
Approx. date	400 - 200 B.C.	
Weight	2.10 g	
Dimensions	15.9 x 14 x 3 mm	

Chemo-metallographic Studies

For chemical analysis, a small portion of the coin (about 0.1 g) was clipped, chemically cleaned, washed in distilled water and then dissolved in nitric acid after drying at 110° C. The residue was filtered off. Lead was estimated as $PbSO_4$, copper as cuprous thiocyanate, iron as Fe_2O_3, sulphur as $BaSo_4$ and nickel as dimethyl glyoxime complex. Arsenic was detected by the well-known Marsh test.

For metallographic examination, the coin was mounted in bakelite at a maximum temperature of 150° C and a maximum load of half a tonne. The mounted coin was ground and polished for examination under a metallurgical microscope. Then the surfaces were suitably etched and examined under a Leitz Panphot metallograph. The coin was indented under a diamond pyramid indentor under a 5 kg load. The results of these chemo-metallographic studies are summarized in Table 6.

TABLE 6

Results of chemo-metallographic analysis of the Copper coin from Rajghat excavations (Bhardwaj and Misra, 1969)

Item/Parameter	*Detail/Value*
Chemical composition	Cu - 86.90%, Sn - 7.83%, Fe - 1.50%, Ni - 1.07%, CaO - 1.00%, Pb - 0.83%
Impurities	S, As, P and other oxides - 0.83% (by difference)
Hardness	66 V.H.N. kg/mm^2
Macrofeatures	Evidence of cast marks
Microconstituents	Coarse grains with coring, grain boundary network of low melting constituents, which are brittle and dislodge during fabrication
Fabrication technique	Cast in sand or clay moulds

Archaeometallurgical Implications

The presence of sulphur in the copper coin under consideration indicates that the copper must have been extracted from a sulphide ore. The simultaneous presence of iron (see Table 6) leads to the conclusion that this ore must have been chalcopyrite (Bhardwaj, 1965).

The significant amount (7.83%) of tin in this coin must have been an intentional alloying addition. The other oxides and insolubles (such as CaO, MgO, SiO_2, etc.) must be due to entrapped slag particles inherited from the fluxing and slagging operations (Newton, 1963). The chemical composition of copper slag from the Rajghat period is shown in Table 7. The small traces of lead and arsenic must have been inherited from the corresponding sulphide contamination of the parent chalcopyrites.

TABLE 7

Chemical composition of Copper slag from Rajghat period Ca. 400-200 B.C. (Bhardwaj, 1979)

Compound	*Weight (%)*	*Compound*	*Weight (%)*
SiO_2	39.17	CaO	5.39
All_2O_3	12.02	MgO	0.62
Fe_2O_3	38.72	SO_3	1.20
CuO	1.05	Others	1.83

The coin is a cast product as evident from both macro and micro-structures. Its composition corresponds roughly to a basis of phospher bronze, but the full utility of tin addition (e.g., increased hardness and strength with sufficient ductility for cold working) has not been realised.

The presence of coarse grains with segregation, the dendritic structure and the grain boundary network of low-melting constituents indicate that the melt was cast in sand or clay moulds where cooling was low. The brittle and non-adhering network also explains the unusually low hardness (66 kg/mm^2) of this coin. The surface symbols and the figures in this coin have been incorporated by providing corresponding designs in the mould walls.

CONCLUDING REMARKS

The present paper is a brief review of the metallurgical achievements of early Indians as evident from laboratory studies of some gold, silver and copper coins from the Rajghat (Varanasi) excavations. This review barely scratches the surface of ancient Indian metallurgy of noble metals.

Important technological changes took place in India as early as 2300 B.C. when the Harappans successfully exploited the natural mineral resources. After the collapse the Harappan civilization, the Indian technology resumed its momentum around 1000 B.C., with the appearance of superior quality ceramic, silver and copper coinage, and profuse use of iron for weapons, household pots and pans. These new materials would require proper selection of raw materials, fuels, furnaces, crucibles, workshop tools, control of temperature and conditions of combustion (oxidizing/reducing) through the regulation of furnaces.

In the preceding sections, we have mentioned some emerging facts on ancient India metallurgy of gold, silver and copper. Many problems, however, remain to be solved by further researches in Indian archaeometallurgy. There is a great need, for example, to fix major metallurgical development in India (such as use of native gold and copper, annealing and melting of native metals, alloying of copper and tin, separation of gold and silver, preparing of brass, etc.), both in time and space. For this, it would be necessary to take up studies of metal artifacts from various archaeological regions and phases in the Indian subcontinent.

REFERENCES

Bhardwaj, H.C., 1965. Some technical aspects of Indus Valley copper. *Bharati*, 9: 57-80.

Bhardwaj, H.C., 1979. *Aspects of Ancient Indian Technology*. Varanasi, Motilal Banarasidass.

Bhardwaj, H.C. and Misra, S., 1968. Early Indian silver coinage—a metallurgical review. *J. Num. Soc. India*, XXX: 1-11.

Bhardwaj, H.C. and Misra, S., 1969. A metallurgical review of some early Indian copper coins. *J. Num. Soc. India,* XXXI: 194-204.

Bhardwaj, H.C. and Misra, S., 1971. A metallurgical note on Gupta period gold coins from Rajghat (Varanasi) excavations. *J. Num. Soc. India,* XXXIII: 123-127.
Caley, E.R., 1948. On the application of chemistry to archaeology. *Ohio J. Sc.*, 48: 1-14.
Caley, E.R., 1949. Archaeological chemistry. *Chem. Engg. News*, 27: 2140-2142.
Cotterll, A.H., 1959. *Theoretical Structural Metallurgy*. London, St. Martins Press.
Gowland, W., 1917. Silver in Roman and early times. *Arch.*, 69: 121 160.
Hall, E.T., 1959. Chemical investigations of museum objects. *Archaeometry*, 2: 43-52.
Newton, J., 1963. *Extractive Metallurgy*. New York, John Wiley & Sons, pp. 226-227.
Smoluchowski, R., 1951. *Phase Transformation in Solids*. New York, John Wiley & Sons.

The Development of Zinc Technology in Ancient Gujarat

SWARNA KAMAL BHOWMIK

Introduction

The origin of zinc smelting is not definitely known. However, certain evidences lead us to belive that the method of extraction of metallic zinc from zinc-ores was known to the metal craftsmen of ancient Gujarat long before Muslim conquest of India.

The discovery of three important metal-hoards from Gujarat has provided the writer of this article an opportunity to examine the development of zinc-technology in ancient Gujarat.

Large number of metallic works of art have been produced in medieval Gujarat before 14th century A.D. The centres of metallic art production had flourished in the north-eastern Gujarat, north Gujarat and central Gujarat. In early medieval period, Gujarat had witnessed enormous development in the manufacture of metallic objects of art. The discovery of three important hoards of metallic art objects at Mahudi of north Gujarat, at Lilva-deva of north-eastern Gujarat and at Akota of central Gujarat has established beyond doubt the existence of highly organised tradition of non-ferrous metallurgy and well-organised metal-working communities in ancient Gujarat. Large number of metallic objects of art and worship belonging to Jainism have been recovered from these places. The examination of their chemical composition and their metallographic studies clearly demonstrated that zinc as an individual metal was known and it was manufactured from zinc-ores by metal craftsmens of ancient Gujarat before 10th century A.D.

Technological Investigation

Our technological investigation of the Mahudi Metallic images which are dated from 7th to 12th century A.D., revealed that the efforts of metal workers of Mahudi towards making zinc-bronze alloy was successful (Bhowmik, 1973-74). Metal-workers of Mahudi were acquainted with three varieties of metal, namely, (1) copper, (2) zinc-bronze with high tin content, and (3) zinc-bronze with low tin content. They were successful in developing an alloy of copper-zinc-tin. During the time, they probably carried out several experiments with metallic zinc for manufacturing an alloy which is better than bronze. For this, different quantities of metallic zinc were added deliberately to bronze alloy for getting desired effect. This was possible with the knowledge of extraction of metallic zinc from the zinc-ore. On heating and reducing, the calamine (a type of zinc-ore) sublimates in the form of white smoke which has a tendency to disappear in the air unless it is condensed on the iron gratings above the roasting pit. The white smoke of zinc is condensed into metallic zinc in a form resembling egg-shells.

From the critical technical study of metal images discovered from Lilvadeva, which are dated from 8th-12th century A.D., it may be observed that the metal workers of Lilvadeva adopted a very efficient method for the extraction of metallic zinc from zinc-ores. The knowledge of advanced technique of metal extraction and skill in metallurgy was probably evolved by experiments over a long period. Such persistent efforts for experiments combined with long experience probably enabled the metal working communities of Lilvadeva to evolve an ornamental alloy known as lead-brass which is a superior alloy in comparison to the bronze alloy and the zinc-bronze alloy. The very fact that the metal-workers of Lilvadeva were successful in manufacturing leadbrass alloy itself suggests a tremendous progress in the scientific extraction of zinc as a metal from the zinc-ore known as calamine. And this break-through took place as early as 9th-10th century A.D., in the ancient Gujarat (Bhowmik, 1980)

From the technical investigation of important metallic works of art from Akota of central Gujarat, which are dated from 6th-14th century A.D., it is possible to infer that ancient Akota (near Baroda city) witnessed continuous activities in the manufacture of metals and metallic alloys from 5th-6th to 14th century A.D. During the long period of nearly 800 years, metal-workers of Akota carried out several experiments. They were acquainted with four varieties of metallic alloys, namely (i) bronze, (ii) zinc-bronze, (iii) lead-brass and (iv) real brass (Bhowmik, 1976).

Extraction of metallic zinc from zinc-ores played a very important role in all these scientific developments. The extraction of metallic zinc from calamine (a type of zinc-ore) was carried out successfully in ancient Gujarat. Zinc was used as a metal from 8th century A.D. onwards and it was extracted successfully from zinc-ore during 10th-11th century A.D.

It may be noted that zinc occurs hardly more than 20 per cent and lead occurs approximately 13 per cent; while tin occurs 2 per cent in the "lead-brass" alloy used in manufacturing metallic objects of art from Lilvadeva as well as from Akota. There are definite evidences to prove that during the 11th century A.D., zinc as a metal was available in Gujarat. On the other hand, zinc as a metal was not known to the west during this time and it was known to the western world only during the late 18th century A.D. The total brass technology was, however, known in ancient Gujarat especially to the metal-workers of Akota of central Gujarat as early as 13th century A.D. It may be noted that the success of real brass technology largely depended on the commercial extraction and production of metallic zinc from zinc-ore. Metallic zinc was commercially available in Gujarat from 12th-13th century A.D. onwards. The Aravali region of Gujarat especially the "Amba-Mata" copper belt of north Gujarat is associated with rich mines of ores which formed major source of zinc for the metal working communities of ancient Gujarat. The possibility of importing zinc-ores from Zawar mines of Rajasthan, however, cannot be ruled out. An examination of the large number of clay-retorts at Zawar-zinc-mines shows that these retorts belong to a period prior to 14th century A.D. Moreover, some of these retorts were found to contain carbonacious reducing agents and they are, significantly marked by the presence of pure zinc metal condensed in the neck of these retorts (Mahmood, 1978).

Literary References

The *"Rasaratnakara"* of Nagarjuna mentions that copper could be converted into brass by heating it with calamine (a zinc-ore) and carbonacious matter (Ray, 1958). The method of preparation of brass as mentioned in the "Rasaratnakara " of Nagarjuna (8th century A.D.) has been repeated in "Rasarnavatantra" (12th century A.D.) which also mentions the methods of the extraction of zinc from calamine (zinc-ore), (Ray, 1958). The "Rasendracudamani" of Somadeva (12th century A.D.) mentions about a process for making a brass-like-alloy of copper, tin, zinc and lead (Ray, 1958). These literary evidences and other ancient texts (Sikdar, 1983) clearly demonstrates that zinc as a metal was known to the metal-working communities of India as early as 11th-12th century A.D. and it was extracted from calamine (a zinc-ore) for commercial use.

The fact that brass is an alloy of metallic copper and metallic zinc is crystal clear from the statement in Bhavamisra's "Bhavaprakasa". The "Bhavaprakasa" is a famous treatise of the 15th century A.D. It states that brass is a semi-metal (upadhatu) of metallic zinc and copper; because, it is derived from these two metals. Bhavamisra in the first part of his treatise, Bhavaprakasa, devoted one complete chapter on the subject of Dhatu-Upadhatu (i.e., metal and metal-alloys). He states that there are seven Upadhatus and he has included bronze and brass amongst them. Bhavamisra mentions several terms synonymous with brass. He mentions that brass has four Sanskrit names, viz., Pittala, Arkuta, Ara, and Riti. He explains that when the heated alloy in the crucible turns red in colour the brass is known as "Rajatiri" and when it assumes a yellow colour, the metal is known as Brahma-Riti. Explaining the properties of brass Bhavamisra states that this metal is an alloy of copper and zinc.

CONCLUSIONS

The above quoted literary references clearly demonstrate that in India, real brass was known as an alloy of metallic zinc and copper as early as the 14th century A.D. and during the 15th-16th century A.D. the brass was a popular alloy which was commercially produced. This brass as an alloy of metallic copper and metallic zinc was, however, vaguely known to the western world towards the late 18th century A.D. (Bridge, 1920) and until 19th century A.D., calamine-brass was prepared (Gowland, 1912). In Gujarat, real brass as an alloy of copper and metallic zinc was known as early as 13th-14th century A.D., and this fact has come to the surface of our knowledge from the technical studies of the metallic objects of art discovered from Mahudi, Lilvadeva and Akota of Gujarat.

In view of the above studies, it may be inferred that ancient Gujarat was probably ahead of other countries including Europe in the field of non-ferrous metallurgy and practical chemistry which no honest thinker can ignore.

REFERENCES

Bhavamisra, 1955. "Bhavaprakasa Purva Khanda", translated by Shri Girija Sankar Maya Sankar Sastri, published by Sastu Sahitya Vardhakakaryalaya, Ahmedabad and Bombay, pp. 409 and 112.

Bhowmik, S.K., 1973-74. "Metal images from Mahudi". *Bulletin of the Baroda Museum and Picture Gallery,* vol. XXV.

Bhowmik, S.K., 1976. "Art and technology of some metallic images from Akota". *Bulletin of the Baroda Museum and Picture Gallery,* vol. XXVII.

Bhowmik, Swarnakamal, 1980. "Ancient Metallic art and metal technology of Gujarat". The Museum and Picture Gallery, Baroda.

Bridge Port Brass Co., 1920, *Seven centuries of brass making,* p. 15.

Gowland, W., 1912. Copper and its alloys in early times. *Journal of the Institute of Metals,* VII, p. 44.

Mahmood, J., 1978. "Zinc metallurgy in medieval India", Paper read at Seminar on Science and Technology in India (1400-1800) INSA, New Delhi.

Michael, H.V. and Asaro, F., 1979. "Chemical study of the plate of brass". *Archaeometry,* vol. 21, Part I, February, pp. 3-14.

Ray, P. (editor), 1958. *History of Chemistry of Ancient and Medieval India,* Culcutta, pp. 129; 138-139; 152.

Sikdar, J.C., 1983. "Some light on metallurgy as revealed in Suvarna-Raupya-Siddhi-Sastra of the Jainacarya : Jinadattasuri (v.s. 1210)". *Reflection of Indian Art and Culture,* Museum and Picture Gallery, Baroda, pp. 215-220.

Ancient Industry of Lead, Silver & Zinc in Rajasthan

P.T. CRADDOCK, I.C. FREESTONE, LYNN WILLIES,
H.V. PALIWAL, L.K. GURJAR AND K.T.M. HEGDE

Introduction

India's cultural heritage, as reflected in her art, architecture, exotic stone sculpture and iconography of God and Goddess has received much appreciation. However, the technical infrastructure necessary to carve, cast or build these marvellous monuments in the past has received little attention. In fact, there is little appreciation of Indian technological heritage, mainly due to inadequate studies on this subject.

Vast heaps of spent retorts and slag, extensive ancient mine workings and ruins of exotically carved Brahmanical and Jain temples in and around Zawar, located 44 km south of Udaipur, stand as testimony to the flourishing zinc mining and smelting industry in India in the past.

These remains have attracted the attention of many visiting metallurgists and geoscientists ever since Zawar mines reopened in 1940s, have carried retorts with them to conduct scientific investigations to unravel the mysteries of zinc smelting being practiced at Zawar. Carsus (1960), Somarlatte (in Warner, 1972), S.W.K. Morgan (1976), Straczeck and Srikantan (1967) were among few who have reported the evidence of early zinc distillation process, and speculated how zinc ore was distilled in clay retorts to extract metallic zinc. Their observations were, however, published only in appropriate mining, metallurgy and earth science journals which do not have a widespread circulation with the result that people in other disciplines, particularly those working in relevant fields, namely the archaeologists and science historians seemed to have remained ignorant of the significance or even existence of the remains of a flourishing lead-zinc mining and smelting industry in India at Zawar.

Findings of Mr. Morgan and others have drawn the attention of archaeometallurgists of British Museum and they had proposed a multi-disciplinary joint archaeological investigation campaign to uncover the mysteries of ancient zinc extraction technology practiced at Zawar. The investigations were carried by a team comprising Dr. Paul Craddock and I.C. Freestone of the British Museum, London, UK, Dr. Lynn Willies of Peak District Mining Museum, Derbyshire, UK, L.K. Gurjar of Hindustan Zinc Limited, Udaipur, India, and Dr. K.T.M. Hegde of M.S. University, Baroda, India. (Dr. Hegde could not participate in subsequent excavations due to pre-commitments elsewhere). Mr. H.V. Paliwal, Director (Mining Operations), Hindustan Zinc Limited, was the main guiding force behind the entire work. The investigations commenced in December 1983, after a reconnoitrery survey in 1982. The unearthing of two sets of ancient zinc smelting furnaces, still intact with their full charge of 36 retorts turned out to be the most significant aid to decode the technology of zinc smelting during the investigations. Subsequently, investigations were carried out at Dariba and Agucha during 1986 and 1987 season.

Remains of similar magnitude in the form of extensive mine workings, heaps of mined out debris, slag, furnace refractories and large timber revetments are also found at Dariba, District, Rajsamand and Agucha, District Bhilwara, Rajasthan. Realising the importance of these remains, Hindustan Zinc Limited (HZL) has obtained the radio-carbon dates of these remains of Dariba in 1976, which show that the industry was in operation 3000 years before.

The remains of early mining and metallurgy studied by the team are at least as sophisticated as anything further west, and there are no parallel or analogies for the zinc smelting processes that investigative teams have uncovered. This was the sharp edge of technical innovations, taking place on a major scale, well away from Europe, the Mediterranean and the Middle East.

Geology

The Aravalli hills run for over 1000 kms. approximately northeast from Gujarat to Delhi principally through the state of Rajasthan. Although the hills are rarely mountainous they are ranged in a series of ridges running parallel rather in the manner of a grid iron and in historic times at least they have formed a considerable barrier to communication.

These ridges were formed by the repeated isoclinal folding of the highly metamorphosed Precambrian rocks. Seven principal shear zones have been located in the Aravallis and all are rich in minerals, the northeast zone tends to be rich in copper (and includes the famous copper mine of Khetri), whereas the southwest zone is rich in lead-zinc ores (the mines of Ambaji, just inside the state of Gujarat for example); although exploitable quantities of all the metals mentioned above are found throughout (Sychthenovong). The repeated folding means that the ore deposits usually dip steeply.

The copper industry in the area has long been recognised as being of considerable antiquity, probably dating back to Harappan times and evidence of copper smelting was found at the settlement of Ahar near Udaipur dating to the early second millennium BC.

Mineralization and Topography

The mineralization at Zawar is principally of argentiferous mixed galena/ sphalerite with varying amounts of pyrites and to a lesser degree of marmatite, a solid solution of sphalerite with the pyrite. The galena and sphalerite are microscopically distinct and capable of being separated manually, and almost certainly explains why zinc smelting developed only at Zawar. The silver content is rather low and difficult to extract.

At Zawar the host rock is sheared and metamorphosed dolomite in a country rock of greywacke and phyllite. The sheared dolomites occur now as steeply dipping isoclinal folds forming ridges flanked by more rounded hills of phyllite and greywacke (Figure 1).

The hard calc-silicate rocks form long ridges which run for several kilometres north from Dariba, rising dramatically from the otherwise featureless plain (Figure 2). In the south this forms a conspicuous ridge of gossan, and this is the only substantial deposit of gossan to survive in the three mines. Although, there are early trench and quarry workings along the ridge, the majority of the workings are below the level of the plain either from the major open cast pit or from the deep mines.

The primary orebody at Dariba is a complex, polymetallic (zinc-lead-copper [silver]) stratiform deposit, occurring within highly metamorphosed, steeply dipping Precambrian rocks. The main orebody being about 800 m long with a maximum width of ca 40 m. The main ore minerals are sphalerite, galena and chalcopyrite but other sulphides (including pyrite) and complex sulphosalts also occur together with copper zone which was certainly worked in the 19th century if not earlier. The main enrichment of silver is confined to the zone of coarse grained galeno-chalcopyrite and complex sulphosalts.

At Agucha (Figure 3) the mine is in a level plain with very little surface indication of mineralization other than the now very eroded open pit mine in the form of shallow depression, indeed it was only very recently rediscovered during the investigation for garnet by the Rajasthan State Department's Geologist. The main ore body, which comprises massive and finely disseminated sphalerite and patches, veins and stringers, galena, argentiferous, is lens shaped, reaching a maximum thickness of ca 100 m (average ca 60 m) and having a length of ca 1500 m and is the single largest and richest deposit in India. The calc-silicate bands are in the hanging wall of the deposit have acted as the locus for the concentration of coarse, recrystallised argentiferous galena

The area of ancient workings appear to fall into two categories:

(1) Working of the putative gossan overlying the main ore body, by means of a large open cast trench, and
(2) Underground mining of the pods and stringers of coarse galena associated

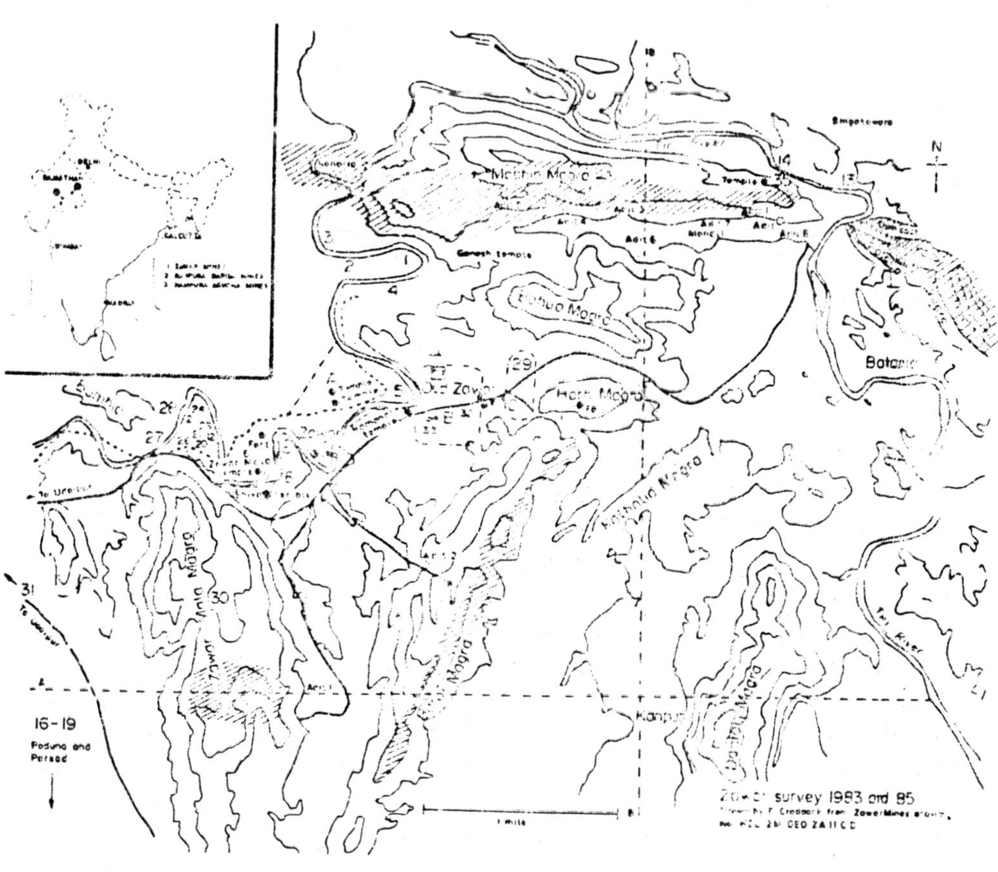

A Main area of small retort heaps
B Main area of intermediate or large retorts
C Main area of large retorts
TI Furnace for large retorts
1 Tap slag
2 Tap slag, and retorts
3 Large retorts
4 Several slag heaps
5 Tap slag, large retorts and perforated plates
6 Tap slag
7 Tap slag
8 Tap slag and retorts
9 Tap slag
10 Tap slag and large retorts
11 Tap slag
12 Large retort bank
13 Tap slag
14 Tap slag
15 Scattered tap slag in river gravels
16 Tap slag
17 Tap slag
18 Large tap slag
19 Large tap slag
20 Large retorts
21 Large retorts
22 Large retorts
23 Large retort wall
24 Large retorts
25 Large retorts
26 Lagre retorts
27 Large retorts
28 Large retorts
29 Tap slag and large retorts
30 Small retorts and furnace
31 Tap slag
32 Tap slag
33 Intermediate retorts
34 Intermediate furnace
35 Large retorts
36 Primitive retorts in dam fil

Fig. 1: Plan of Zawar and inset showing location of sites in India

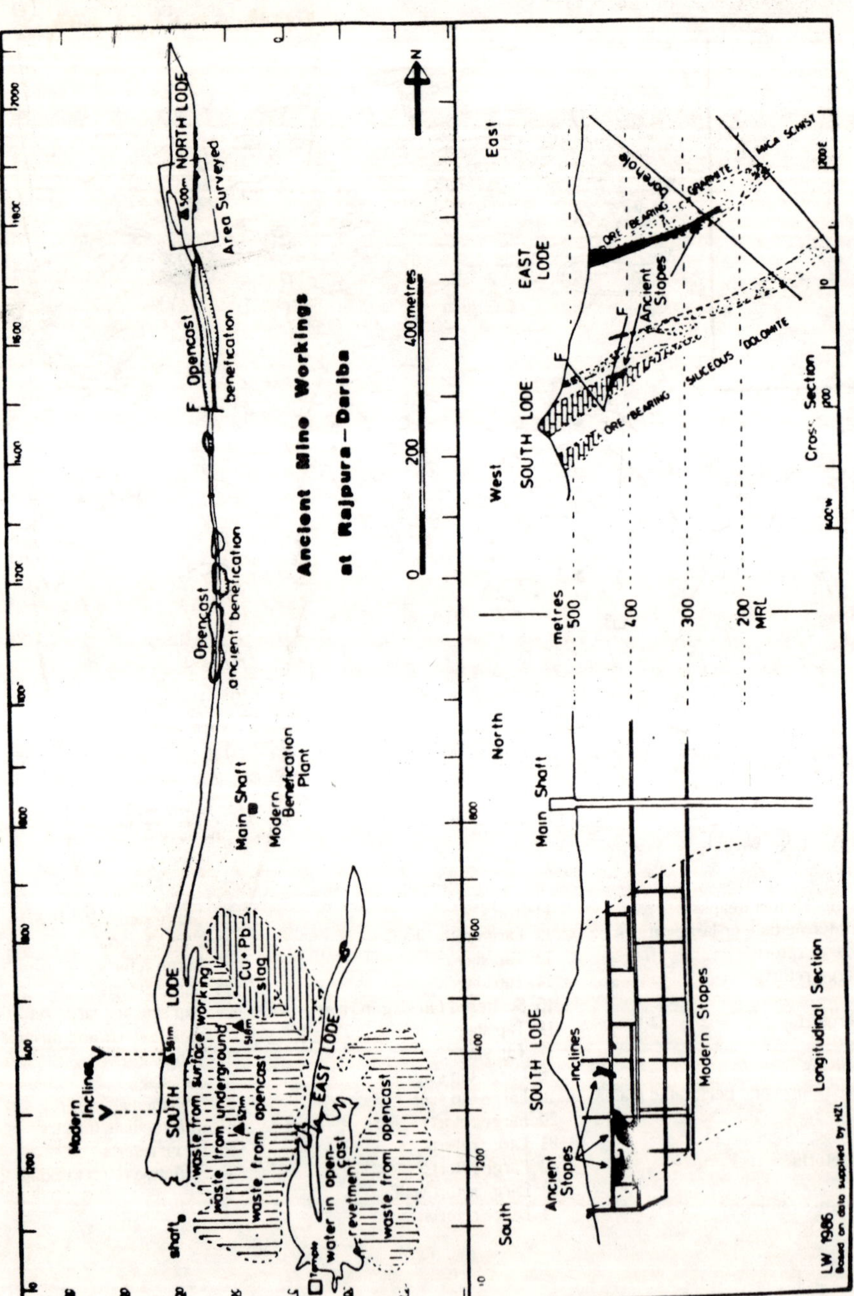

Fig. 2 Rajpura-Dariba. Plan and sections of ancient mine workings

with the lenses of calc-silicate rock on the hanging wall (i.e. east) side of the main orebody and within it.

It is likely that silver was the chief metal sought, although there is little evidence of discarded sphalerite in the heaps of beneficiation debris which surround the great trench.

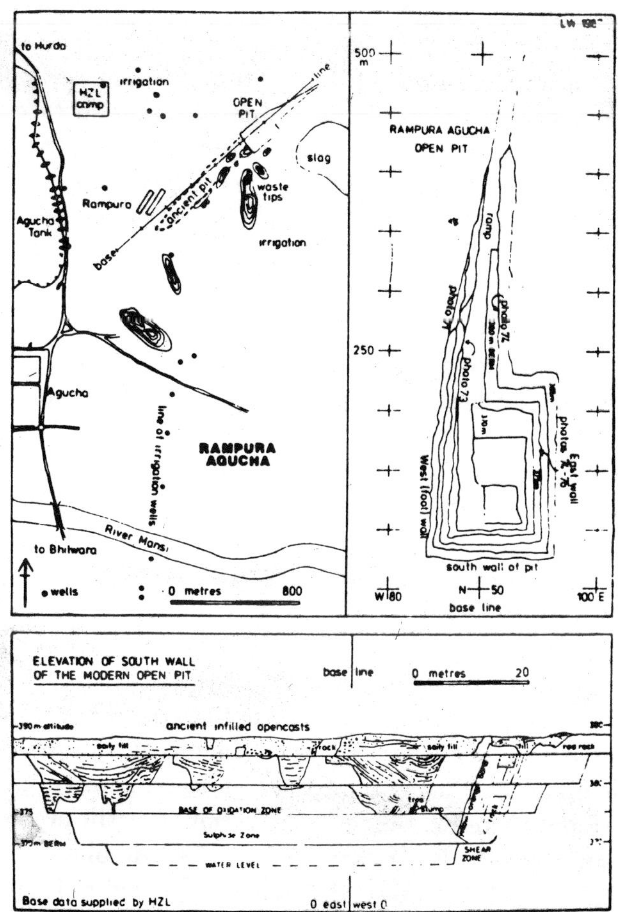

Fig. 3: Rampura-Agucha. Plan of ancient mine workings

Mining

In general the ore was followed down from where it outcropped at the surface, and at all three mines there are very numerous surface and near surface workings. Method of mining was more or less same at all the three mines and Zawar was studied in detail, hence mining at Zawar is described here.

Exploration and surveying of ancient mines at Zawar reveals that major mining systems were developed in the later half of the first millennium BC at both Mochia and Zawarmala hills. The absence of either later working or collapse at the latter mine has resulted in the virtually intact survival of one of the most extensive and sophisticated mining systems known anywhere (Willies *et al.*, 1984 and 1987).

Traces of early mining are visible along strike of the ridges for at least 12 km. These include major excavations forming the top of stopes, shafts for ventilation, access and haulage routes leading vertically down to intersect the inclined stopes.

Once the ore was located on the surface based on the presence of gossan or mineralized veins, the ancient miners followed the downward extension along dip and pitch of the ore shoots (Pl. 29). Thus numerous near surface workings were developed in the form of small inclined galleries for exploiting the very richest ore. These workings in relatively weathered rocks near the surface were often worked with pick and chisel (gad) and the distinctive marks made by the tools can be clearly recognised. However, once the weathered and fractured rocks had been left behind as the work progressed deeper workings were developed using fire-setting method to break-up the rock. Thus almost all the underground workings have the distinctive smooth continuous profiles characteristic of fire-set work, with few evidences of tool marks, and everywhere the walls are still blackened with soot and the floor are buried under deep layer of ash, charcoal and partially burnt brush-wood. The galleries and stopes are often of considerable size due to the great thickness of the orebody and the strength of the host rock which enabled very large chambers to be excavated without fear of collapse (Pl. 30).

At some faces where mining continued till the time when it ceased for ever, the method of fire-setting can be observed clearly. The main galleries were at a steep angle down the ore. Shallow, oblong pits, between one and two metres long, are observed in the plunging faces where the last fires had burnt the faces fracturing the underlying and surrounding rocks. Several fires must have been necessary to advance the whole of these broad faces. The characteristic holes formed by the fires suggests the main impact of fire was in the floor. Its impact was reduced on the rocks above or near the flames, this has produced the characteristic arched roof from the workings. Thus, sinking a decline was preferential mode of ore extraction, whilst working up-dip was done by creating and driving forward holes in the floor.

Use of fire-setting for digging of wells in the hard rocks is an age old practice in the country and is still prevalent in the remote areas. 'Vrihatsamhita' (Circa 5th century AD) compiled by 'Varahmihir' a treatise on temple architecture, ground water exploration and astrology; from pre-existing texts recommends the use of fire to split the hard rock for digging of wells or shaft (Kup). It says -

> "If a rock resits splitting, kindle a fire with the fuel from Butea or Ebony and make the rock red hot; sprinkling of water mixed with lime over the red hot rock face will burst the rock instantly".

Similarly, the book recommends various methods for tempering of iron tools used in digging.

Few unusual pestle like hammers and chisels, used by ancient miners have been recovered from ancient mines of Mochia, Zawar (Pl. 31).

Until recently, some parts of India were in the pre-industrial state and it is reasonable to assume that technology used in these areas have changed very little since the early period. In this context, the observations of Capt. Boileau on copper mines of Singhana, District Sikar, Rajasthan, published in 1831 AD, on use of fire-setting for mining are relevant. He says -

> "The mines are torturous and of great extent; at the working faces it was the custom to light fires which cause rocks to split, lamps were used for illumination, which miners carried on their heads. Gads and hammers were used to extract the ore" (Boileau in Ball 1889).

The use of fire-setting for rock breaking must have created major ventilation problem. To solve this small rectangular or circular shafts of 1 to 2 m diameter occasionally upto 3.5 m, as in case of Dariba, connects inclined stopes at various levels to the surface. Because of large vertical shafts, and heating of inside air during fire-setting must have induced a natural air draft to ventilate the mine and the overall air flow perhaps would be reasonable (Pl. 32). However, this would not have been true for small finger-like inclines or stopes in depth. These shafts perhaps were also used for entry to the mine and hoisting of material as evidenced from the presence of wooden ladder and wooden pulley in the ancient shaft at Dariba.

Extensive use of timbers was noted in the mines for roof support, wooden stairways and launder to channel mine water into the sump. The wooden stairways are sufficiently broad enough for two persons to pass without difficulty (Pl. 33). The obvious implication of this is that an almost continuous stream of workers moved both up and down the stairways carrying baskets of ore and debris (gangue) and pitchers of drainage water for dewatering the mine. Remain of both types of containers are quite common in the mines.

There are evidences that the water running down the stopes was channelled by launders hollowed from tree trunks and collected at the bottom of the gallery or stope (Pl. 34). The large clay pots which, perhaps, were used as water carriers, found in the mine have a capacity of about 7 to 10 litres. Thus it can be safely concluded that much of the water was collected at a place for removal with the help of clay pots. Thus several journeys up and down the ladderways must have been necessary each day, in each gallery system, to drain the mines involving sizeable manpower.

An interesting account of mine dewatering at Khetri copper mine, Rajasthan, was given by Capt. J.C. Brooks, in 1864 AD supports the above contention is reproduced below:

> "At Kulhari (present day Kolihan) mine the only means of overcoming the influx of water known to miners is by passing pitchers (gharas) of water from hand to hand through the passage, in one case no less than 27 men were thus employed upwards a month at a cost of Rs. 200.00" (Ball, 1889).

Clay vessels with straight flared sides, resembling small pudding basins or bases of larger vessels trimmed to size were found everywhere in the mines. These are probably ancient lamps. When filled with vegetative oil-soaked cotton seeds, on buring emit sufficient light for several hours. Preliminary tests by Dr. John Evans, North East Polytechnic College of London, have confirmed that the clay walls of the vessels do contain vegetable oils. This seems to confirm their use as lamps (Pl. 35).

Apart from development of major underground mining systems, open cast mines were also developed notably at Dariba and Agucha. So far there has been little radio-carbon dating of these open cast mines with the single exception of the large open cast at Dariba which lies more or less over the deep workings of East Lode. This is about 300 m long by 100 m wide and was water-filled until recently. The recent drought and modern mine drainage have together removed most of this water and revealed that one side i.e. hanging wall is supported for at least 150 m along the strike by a massive timber revetment (Pl. 36). Most of the timber is now buried in silt but at least three stepped benches could be seen, and others may be buried at depth. The revetment is of roughly trimmed uprights hammered into the loose soil, set about approximately every 50 cm. Brush wood and reused timber such as pieces of old ladders were placed behind to hold the soil back. The uprights are held in place by the major beams which run along in front. These in turn are held in place by a system of pegged double beams which lie horizontally back into the soil for about four or five metres, and are themselves pegged together and held at the far end. Each bench is not tied in at all with that above or below it, suggesting the whole revetment developed progressively as the open cast went down. The system is simple in design but incorporates a massive quantity of timber, tentatively identified as teak. This has carbon date of 2140±100 B.P. (Figure 4). The rather poisonous water-logged conditions and the use of the teak explain the generally excellent condition of the revetment. So far only a tiny portion has been recorded, but it is probably the largest timber structure to survive from antiquity anywhere. Both the revetment and the open cast mine itself are at present completely without parallel in the ancient world and are of significance to the history of mining generally.

Beneficiation

Having mined the ore it was then necessary to prepare it for smelting. First the ore was crushed, and this was usually done in the immediate vicinity of the mines.

Provenance	*Material*	*Reference*	*Date BP*	*Cal. BC/AD*
ZAWAR MINES				
Zawar Mala mine				
ZM/LW/85/13	charcoal	BM-2481	modern	
ZM/LW/85/14	charcoal	BM-2482	2150±110	380-45 cal BC
ZM/LW/85/8	charcoal	BM-2483	2180±35	360-280, 255-185 cal BC
Sample from scaffold	wood	BM-2148R	2350±120	755-685, 540-305, 275-200 cal. BC
Launder.	wood	BM-2149R	2140±110	375-40 cal BC
470 MRL, RL 456 MRL	charcoal	PRL 932	2410±100	765-390 cal BC
RL 566 MRL	charcoal	PRL 933	1990±140	175 cal BC-135 cal AD
Mochia				
Plt prop.	Terminalia	BM-2381	2360±50	410-390 cal BC
Balaria				
10 West stope 415 MRL	Terminalia	BM-2338	170±50	1660-1700, 1725-1820, 1835-1865 cal AD, 1920-modern
ZAWAR LEAD SMELTING SLAG HEAPS				
Site 5, Layer 3	charcoal	BM-2484	100±45	1685-1740, 1805-1935 cal AD
Site 14, Layer 3	charcoal	BM-2485	1950±60	20 cal BC-115 cal AD
Site 29, Layer 2	charcoal	BM-2486	200±35	1655-1680, 1745-1800 cal AD, 1940-modern
Site 2, Trench 2	charcoal	BM-2487	1930±80	20 cal BC-135 cal AD
Site 7, Trench 2	charcoal	BM2488	1370±80	610-685 cal AD
ZAWAR ZINC SMELTING RETORT DUMPS				
Ash dump north of Ramanath Temple (adjacent site 5)	charcoal	PRL-935	840±130	1025-1280 cal AD
Site 30, furnace	charcoal (*Artocarpus*)	BM-2223	530±50	1325-1335, 1395-1430 cal AD
Site 5, retort heaps	charcoal	BM-2578	500±50	1405-1440 cal AD
Site 1, trench 1, large retort heaps	charcoal	BM-2222R	240±110	1500-1690, 1730-1815 cal AD, 1925-modern
Site 34, large retort heaps, lower layers	charcoal	BM-2243R	350±130	1430-1660 cal AD
AGUCHA				
Site 1, layer 3 and 4 producing cupel debris	charcoal	BM-2489	2350±40	405-390 cal BC
Mine gallery 30 m below surface	timber	BM-2356	2240±60	390-335, 320-200 cal BC
DARIBA MINES				
Depth 64, layer 436	timber	TF 1117	2245±100	400-190 cal BC
East Lode 4 to 5 m depth. Top of revetted open cast mine	timber	NPL 206a	2140±100	370-45 cal BC
South Lode 100 m depth	timber (*Aogeissus sp*)	NPL 210	3040±150	1490-1485, 1450-1055 cal BC
South Lode 100 m depth	rope	teledyne	2100±280	405 cal BC-215 cal AD
East Lode 263 m depth	bamboo	NPL 209	1790±120	85-390 cal AD
DARIBA SILVER / LEAD SMELTING SLAG HEAP				
Site 9, layer 4	charcoal	BM-2490	2340±60	405-385 cal BC
Site 10, layer 23	charcoal	BM-2491	2170±45	360-285, 255-170 cal BC
Site 7, layer 4	charcoal	BM-2492	1940±70	20 cal BC-125 cal AD

Fig-4

Around the entrances to the mines at all three sites are heaps ranging in size up to several tens of metres diameter made up of rock fragments of rather regular size, typically about 2 cm diameter. The ore was presumably broken up with iron or stone hammers or pestles. At Dariba next to the large open cast are a series of large mortar holes carved into the hard calc-silicate rock (Pl. 37) were observed. Clearly these would have required large iron pestles, but elsewhere small regular depressions abound on every suitable rock surface near the mines were found, where ore has been ground. After crushing the ore was probably separated by hand, picking out mineral from the waste or gangue. The team has found no surviving evidence of washeries or launders where the separation was effected by water, but even up in the hills of Zawar water brought out of the mines themselves could have been used where no other was available. A further possibility, particularly during the dry season was to separate the ore by winnowing.

Metal Production

Silver and lead were the main products at Agucha and Dariba while zinc at Zawar.

SILVER AND LEAD

At Agucha the slags extend to the east of the mine, over an area of about 20 hectares, and to a depth probably varying between one and two metres from surface (Figure 3). Radio-carbon dates from the old workings at Agucha show that the mine was in production during 3rd century BC (Figure 4). These dates together with the presence nearby of a major Mauryan settlement at Agucha village itself strongly suggest this was the main period of working. The heaps are now very eroded and flattened, and it seems there has been little, if any, activity in the last two millennia.

At Dariba the occupation and metallurgical debris lie to the southeast of the mines and extend over about 50 hectares. The debris are now considerably eroded by successive floods and can give the appearance of being just a surface scatter with the occasional tell or heap. However, whenever modern wells have been dug through this surface a deep stratigraphy of occupation debris, can be observed to a depth of upto 6 m. Many of the layers are of slag and refractories. Three radio-carbon dates have been obtained from the material and suggest intensive occupation in the later part of the second millennium BC. Surprisingly, little domestic ceramic material was encountered in trial trenches, but walls eroding out of the heap were of characteristic large Mauryan Empire, and also with the first silver coinage of India.

Thus clearly both at Agucha and at Dariba there was lead-silver production on a prodigious scale principally during the latter part of the first millennium BC. At Zawar lead was smelted on a small scale from about 500 BC possibly as a by-product of zinc mining. Detailed scientific investigation of the process has only just begun and as yet the interpretation is tentative based largely on visual examination of the excavated refractory remains and comparison with similar material from other sites.

There is no evidence to suggest that argentiferous galena was roasted in air prior to smelting. Lead ores are readily reduced and a roasting stage may not have been judged necessary. Fragments of the furnaces and tuyeres abound in the slag heaps at all the three sites. The early furnaces were of clay-lines stone but the more recent seem to have been of coil-built clay. Both the clay coils and linings were very heavily tempered with straw. The impressions left in the clay are typically several centimetres long, and the size and quantity suggest they were added as straw rather than as dung. The coils are typically about 2 or 3 cm thick and the walls had a separate lining of clay both inside and out. Giving a total thickness of about 5 to 7 cm. The furnaces would have been a simple cylinder, possibly tapering slightly towards the top and about 50 to 100 cm in height. Where sufficient curvature survives on fragments the diameter can be estimated at about 30 cm which accords well with the dimensions of the furnace described by Dixon (in Balls 1881). Another furnace, this time for copper smelting illustrated in Ball (1881) shows the probable appearance of the ancient furnace (Figure 5).

Fig. 5: 19th century copper smelting in small shaft furnace (from Ball, 1881)

The tuyeres are relatively thin walled, about 3 to 6 mm typically, and between 3 to 5 cm in diameter. The refractory is of much finer clay than the furnace wall and the straw tempering is generally absent. Impressions in the clay suggest they were made by rolling of clay around a straight stick which was then pulled out. Around the thin walled tuyere a much thicker clay collar was wrapped before inserting into the furnace pointing down at an angle of 30° below the horizontal. As none of the tuyeres were found still set in a furnace, it is impossible to be certain how many tuyeres originally fed each furnace. Examples, such as recorded by Dixon or Ball (1881) and from the recent past in India suggest three was the usual number, and modern experiments with furnaces of this type fed by manually operated bellows found three, set at 120° around the furnace most practical.

Many of the surviving furnace fragments show a considerable degree of vitrification penetrating for several centimetres into the wall suggesting a high temperature maintained for many hours. The smelting of the lead ore probably took place through a working day, with the slag tapped periodically whilst the ingot of argentiferous lead formed at the bottom of the furnace.

The slags found together with the refractory fragments are of the usual fayalitic type, and are not fully molten below 1200° C and in the reaction zone near the tuyeres the likely temperature was between 1300 - 1400° C. The main gangue or waste material with the ore was silica, but the ore did contain considerable quantities of iron and of-course zinc minerals; both of which would act as fluxes to react with the gangue to form the slag. Thus, it may well not have been necessary to add a separate flux generally, but just a little crushed quartz or roasted iron pyrites to maintain the balance where necessary.

Slags from all the three sites have been analysed at the laboratories of HZL and are remarkably low in silver. Typically they contain about 1% or 2% of lead, and between 10 and 100 ppm of silver. In contrast the silver content of contemporary Greek and Roman argentiferous lead slags generally lay between 100 and 200 ppm.

The explanation for the low silver content in the slag at Zawar is probably that the ore never contained very much silver in the first place, but this is not so at the other two sites, and one may conclude that the Indians achieved a very high level of silver recovery from lead. Almost all the concentration of silver was lowered by diluting the freshly smelted lead in the furnace with previously de-silvered lead in order to lower the silver loss to the slag.

The product of the smelt was lead which at Dariba and Agucha at least would have been rich in silver compared to Zawar. The next stage was the separation of the silver from the lead. This was done, as elsewhere, by cupellation. The argentiferous lead was melted in an open crucible or hearth and air blown across the exposed surface at about 900 - 1000^0 C. This oxidized the lead to lead oxide, some of which blew away as litharge to coat and glaze any exposed hot ceramic surfaces, and some remained in the cupel itself. The silver remained unoxidized in the cupel and floated on the surface of the denser litharge. Pliny, graphically described in Book 33, Chapter 31 of the Natural History how the silver lay on the litharge like oil on water in the cupel, completely separated. At both Agucha and Dariba there are enormous quantities of cupellation debris, particularly fragments of refractories. These are distinctive both by being covered in lead glaze, and the body of the ceramic is bright red showing they had been in an oxidizing atmosphere, unlike the black, reduced ceramics from the usual reduction processes associated with smelting. The debris included furnace fragments and very many pot sherds, often with the glaze extending over the break, showing they were used as broken sherds. The cupellation would have been carried out either in an open hearth or clay vessel, but a low wall, screen or hood of clay and pot-sherds would be necessary both to retain the charcoal fire in case of the air blast, and to condense as much of the lethal litharge fumes as possible. The lead glazing is so extensive in some furnace fragments of curved clay walls that long drips have formed showing that some pieces were almost vertical and others horizontal, strongly suggesting a hood or cape over the cupel opposite the tuyere.

Excavations in the area of a well at Agucha produced a remarkable collection of small cupels (Pl. 38) some with the tuyere still attached, tuyere fragments, and small clay pillars with flat bases and concave tops. Almost certainly these are tuyere supports, giving the arrangement seen in Figure -6. These tuyeres have remarkably thin walls, only about 2 to 3 mm, with an internal diameter of about 25 to 30 mm, and they are set right down into the cupels, which are designed to accommodate them. The cupels have walls about 5 mm thick and are heavily tempered with straw. Their capacity is small, 14 mls to the brim, but only 8 to 9 mls to the end of the tuyere, as indicated by the rim of the glaze inside the cupel. Surely these are too small and fine to be the standard vessels in which silver was refined on an industrial scale? These small vessels look more probably like equipment for trial assays, possibly this area contains debris from the mine assay office where the silver content of each batch of ore would have been determined. One large piece of litharge was found. Its shape was a fragment from a plano-convex piece of diameter 6 cm, and a maximum depth of 3 cm i.e. a volume of about 30 to 40 mls. This would seem a more fitting industrial scale and for example would fit perfectly into one of the many cupellation hollows discovered much further west at Sardis in Turkey. It is very possible that cupellation was generally carried out at Dariba and Agucha in larger vessels, or just in hollows scooped into the ground as at Sardis. There is literary evidence for this in India, thus in the 16th century work the Ain-I-Akbari it is stated that a small hollow was scraped in the ground and lined with a mixture of ashes of cow dung and Babul wood.

In more recent times cupels have been made of bone ash which absorbs litharge well, the origins of this practice are not well established. We have found no evidence of the use of bone cupels to absorb the lead oxide at the sites investigated, but the *Arthashastra* written by the Mauryan Chief Minister, Kautilya in the 3rd century BC describes how silver is to be purified with lead mixed with copper sulphate and powdered bone, and also heated in a skull. The description is confusing but bone is clearly associated with the cupellation process. Over a thousand years later the "Rasahridaya" possibly written in the 11th century describes cupels made of goat bone ash and lined with borax.

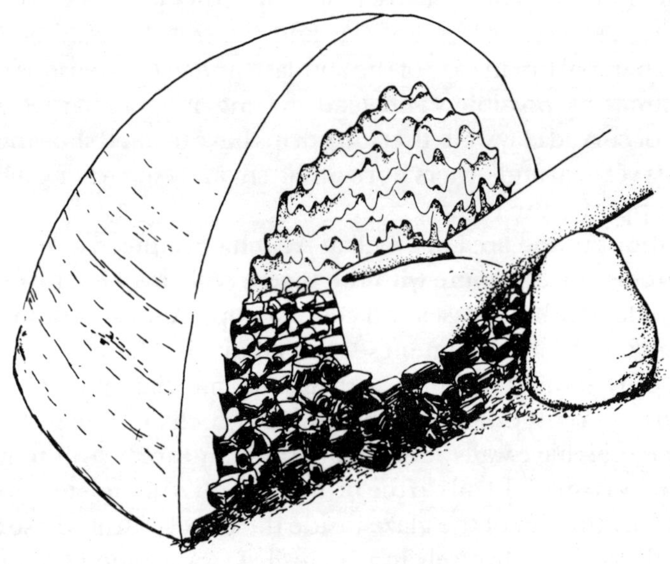

Fig. 6: Artists reconstruction of Cuppelation furnace used at Agucha, 3rd century B.C. (Courtesy B. Craddock)

Zinc

Introduction

Zinc was the last of the common metals to be smelted in early times - the first large scale production probably commenced in India rather less than a thousand years ago. This delay in production of zinc is because of the great difficulty in extracting it from its ores. Not only does it require extreme reducing conditions, but it is produced as a vapour. Thus some sort of system was necessary where the zinc was produced under extreme reducing conditions and the vapour condensed with oxygen rigorously excluded. However, brass, the alloy of copper and zinc was in regular use over 2000 years ago through much of the classical Indian world. This was made directly on an industrial scale by reacting calcined zinc ore with charcoal and finely divided copper in a closed crucible at about 1000° C, known as cementation process. When the Greeks wrote of brass "Oreichalkos" it was as an exotic material usually originating well to the east suggesting the process began somewhere between Greece and India sometime in the 1st millennium BC although the earliest regular series of dated brass are the brass coins from Phrygia and Bithynia in Western Anatolia dating to the beginning of the 1st century BC. Europe, China and Southeast Asia all had access to tin for making bronze. India and the Middle East had at best only very minor sources, and thus it is the area where one would expect brass to have been developed as a substitute. In the west the cementation process became the standard way of producing brass up to the 19th century and metallic zinc was almost unknown until introduced from the east from the 16th century onwards.

In India the situation developed in a very different manner. The few analyses which have been performed, such as on the objects from Taxila, now in Pakistan suggest brass was in use from the 4th century BC, and at least some of the pieces may have been made by adding metallic zinc to copper as their zinc content is analysed to be 34.34% Zn. Some of the contemporary Indian medical and scientific treatises such as the Charaka refer to a metal 'Rasa' to which in addition to forming an alloy with copper, could also be burnt and the ashes used as an eye salve. This can surely only be zinc. There are other scattered references to metallic zinc. However, it is not until about a 1000 years ago that detailed descriptions of the preparation of zinc by distillation appear, and shortly after that there is the earliest field evidence that from Zawar for the industrial production of zinc the team has uncovered.

Lead, Silver and Zinc at Zawar

Before considering the process, it is necessary to digress briefly and discuss the apparent enigma of the major early mines on Zawarmala and Mochia at Zawar. These mines have been comprehensively explored and numerous radio-carbon dates obtained (Figure 4). These show beyond reasonable doubt that major mining operations were carried out during the later part of the 1st millennium BC. The ore extracted from these enormous workings was predominantly sphalerite (ZnS), with small amounts of lead, having low silver content. This is based on the analysis of the remaining ore, but it seems unlikely that the original composition was vastly different. It also seems highly unlikely that the miners were going to such lengths to recover just the small amount of lead present in the deposit. Was then silver after all the metal sought? Although the silver content is now low, it is possible that the ore was originally heavily enriched in silver as in the upper levels at Agucha. The heap of beneficiation debris where the ore was crushed and the mineral picked out are very low in zinc, as well as lead, although it is of-course possible that they could have been reworked more recently, but for which there is no evidence. The slag heaps for the production of lead at Zawar range in date over the last 2000 years and their silver content is quite low, at around 10 ppm which suggests that the smelted lead metal perhaps had a high silver content. It seems probable that such a highly argentiferous lead ore would have been exploited at Zawar. However, excavations of slag heaps at Zawar significantly has not produced a single piece of cupellation debris. Thus there is evidence for zinc mining on a massive scale and lead smelting on a minor scale 2000 years ago without any evidence of contemporary`zinc smelting.

One possibility is that the earlier evidence could be buried beneath the later heaps, but if so it was probably on a minor scale. It is more likely that the sphalerite was taken elsewhere to produce brass directly by the cementation process. Although as noted above, there is some evidence for the existence of metallic zinc in India at this time it seems likely that the bulk of brass would have been produced by the cementation process. The last written reference to that process in India seems to be in the 12th century AD, just as the production of zinc at Zawar was beginning on an industrial scale. Thus it would seem that the mines at Zawar were always worked predominantly for zinc, with lead recovered as a by-product. There is little evidence that silver was ever produced, thus confirming Zawar as the earliest known zinc mine in the world.

These early workings are commensurate in date with Agucha and Dariba, and although no occupation sites have been located at Zawar it would seem likely that these mines were also controlled by the Mauryan Empire, and again after its collapse, production ceased. However, the story at Zawar was to be different from then on. By the 7th century AD the mines were back in production, and by the 12th century AD zinc metal was being produced industrially. Already in the late 14th century production was on an industrial scale and perhaps it is not surprising that the first direct historic reference to Zawar dates from the 1380 AD, when Rana Laksh Singh (Lakha) was credited with founding the mines. Production continued on a major scale for about four centuries before ending during the wars and famine which plagued Rajasthan in the early 19th century, and perhaps, in the face of western competition. Ironically the western technology was almost certainly derived from Zawar.

The distillation technology in use through the centuries was broadly the same with only minor differences in the capacities of the distillation retorts, and is clearly closely linked with the description of zinc smelting given in some of the medieval Indian texts, although Zawar is never mentioned there by name. The first direct reference to zinc production at Zawar in the Ain-I-Akbari written in 1596 AD, which stated that zinc was made uniquely at Jalore which was a dependency of the Suba of Ajmer, now positively identified as being Zawar.

At Zawar there are enormous heaps of debris, metres thick and covering many hectares made up of countless millions of clay retorts. Each retort was used but once and then discarded and the great heaps must have built up quickly, frequently burying the blocks of furnaces in which the process took place. From a detailed examination of the retorts and furnaces linked with the contemporary descriptions it has been possible to reconstruct the process of zinc smelting in some detail.

The Process

The ore at Zawar is principally of sphalerite with some marmatite, pyrite and galena, all sulphides. Before any reduction could take place it was essential to remove both the lead and sulphur. The galena is macroscopically separated from the sphalerite, and could be separated manually, and the remainder of the ore would have been roasted in air to convert sulphides to oxides. Even a few percent of residual sulphides would severely inhibit the reduction process. Thus it is likely that the roasting must have been carefully carried to completion in specially built ovens, but no remains of these have so far been found.

The calcined ore was now ready for reduction in a retort (Pl. 39). The form of the retort was an open ended clay vessel. After filling, an open ended funnel-shaped clay condenser was fitted over the open end and sealed in place with clay. This is clearly described in the 'Rasaratnasamuchchaya' in the 14th century.

> "A crucible of the shape of a brinjal (aubergine or egg plant) to which is attached a tubular and expanding towards its mouth like the flower of the Dhatura (Datura Stramonium) thorn flower which is either 8 or 12 digits in length, is suitable for the extraction of the essence of zinc".

The calcined ore was mixed with a variety of ingredients such as those listed in the "Rasaratnasamuchchaya'.

> "Rasak (Zn ore) is to be powdered with lac, treacle, white mustard, the myrobalans, natron and borax, and the mixture boiled with milk and clarified butter and made into balls. These are to be enclosed in a crucible and strongly heated".

At first reading these instructions seem fanciful and unlikely in the extreme, but where possible they have been verified by examination of the Zawar retorts. The present structure of the charge does suggest an original structure of balls, now sintered together. Such a structure would be necessary for the reaction to proceed satisfactorily. The reaction -

$$ZnO + Co \qquad Zn + CO_2$$

is strongly endothermic, required a large heat input from outside, through the retort walls, and the charge itself. Conduction is quite slow and if it was the main means of heat transference the reaction would have proceeded unequally through the charge. However, with an open structure then the heat transfer was instantaneous by radiation. Also the reaction is largely governed by the easy movement of gases, both the carbon monoxide needed to bring about the reduction, and the zinc vapour to escape. Without an open structure this could not happen, and in a tightly packed retort the reaction would be unable to take place. Our examination of the retort charge combined with the contemporary descriptions given above suggest that the practice was to mix up the calcined zinc ore with an inert carrier (calcined dolomite), various fluxes, (salt has been identified) and charcoal, and made it into balls with sticky organics. These would hold the charge together until the process got underway and sintering commenced. The balls were placed in the open retort, a stick inserted, and the condenser sealed in place. The stick prevented the charge from falling out when the retort was inverted in the furnace (Figure 7).

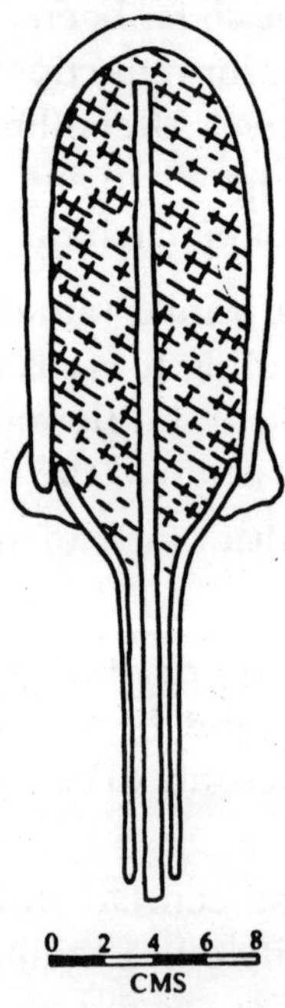

Fig. 7: Section through loaded retort with condenser and stick in place prior to firing

On heating, the organics would char, but the fluxes, including the salt, would promote the sintering of the dolomite preserving the open structure. The carbon monoxide generated from the organics and charcoal would reduce the zinc oxide to zinc vapour which could escape to the condenser down the central passage left by the now burnt stick. It was essential to maintain very reducing conditions as the reaction is easily reversible unless the CO_2 had to be removed or reduced. The proportions of the charge and the temperature were also critical, too much flux or too hot and the charge would melt, but too little flux or too cool and the charge would just fall out of the retort as a dry powder. The sticky organics specified in the contemporary accounts do seem unnecessarily extravagant, more suited to the alchemist's laboratory rather than an industrial process. Cow dung was widely used more recently in India both as a binder and a fuel, and it is likely that this was also the case at Zawar. Thus the large retorts would have contained about a kilogram of calcined ore and dolomite mixed with charcoal and about 1% of salt bound together by cow dung in balls about 1 or 2 cm in diameter.

Turning next to the actual smelting the 'Rasaratnasamuchchaya' once again gives a good description of the furnace or koshthi, "a vessel filled with water is to be placed inside a koshthi apparatus and a perforated cup or saucer placed over it; a crucible charged as above is to be fixed in an inverted position over the saucer and strongly heated by means of the fire of 'jujube' (Zizyphus jujuba) charcoal, the essence which drops into the water should be applied in medicine".

Excavation team had the good fortune to discover several of these furnaces buried largely intact in the retort heaps (Pl. 40). The principles are based on those given above, but the Zawar furnaces are industrial units, and each held 36 retorts in a six by six arrangement, and the furnace were themselves arranged in banks. The particular example described here was situated up on Zawarmala seven furnaces (Pl. 40). Thus no less than 252 retorts were fired simultaneously. The retorts were inverted in the furnace with the condenser necks protruding through the perforated bricks which separate the furnace from the cool chamber beneath (Figure 8). The furnaces have the form of truncated pyramids, and the fire burnt above and around the retorts. Each furnace had four perforated bricks supported on a ledge running round the furnace walls and a central pillar in the cool chamber. Thus perforated bricks separated the furnace from the cool chamber (Pl. 41). Each of the bricks had nine large holes to accommodate the condenser necks and smaller holes for air and ash to pass. The forming zinc vapour condensed and dripped out of the condenser ends in the cool chamber below to be collected in vessels placed beneath.

The 10 or 20 cm length of the condenser must have been very precise. Under ideal conditions zinc condenses at 913° C and solidifies at 419° C, but in practice little condensation takes place before the temperature had dropped below 550° C. However, if the temperature is below 415° C then the vapour condenses as a product known as blue powder which is made up of minute globules of zinc coated with zinc oxide and is completely useless. It was impossible to suppress it totally, and at Zawar too, the condenser necks always have some blue powder present. As mentioned above the zinc vapour had to be kept in an intensely reducing

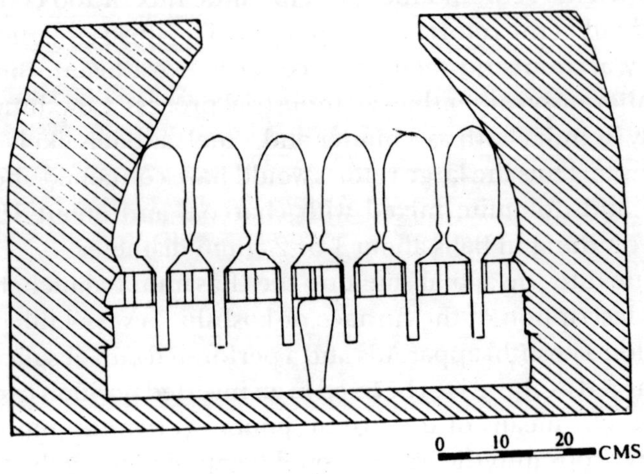

Fig. 8 Idealized section through one of the furnaces or *koshthi* at site 30.

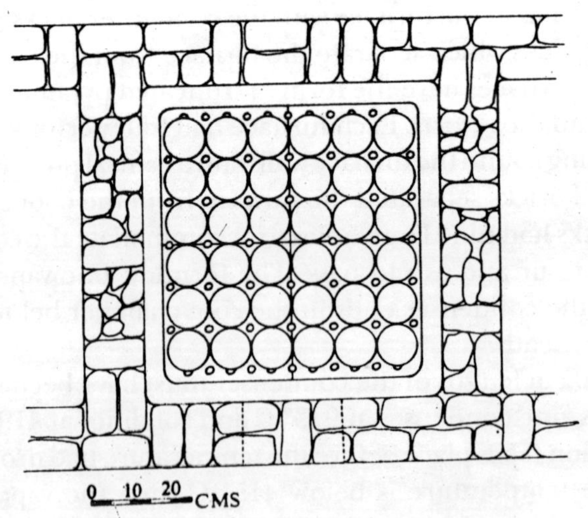

Fig. 9 Idealized horizontal section through a site 30 furnace at perforated plate level.

atmosphere of carbon monoxide, which burnt off at the condenser ends. Some estimate of the conditions in the condenser could be obtained by watching the colour of the flame; whilst the reaction was proceeding satisfactorily the over-pressure of carbon monoxide would burn off with some of the zinc with a highly characteristic blue-green flame. The significance of this flame was noted in Indian, Chinese and European manuals of zinc smelting. Thus the 'Rasachintamani' states that when the flame changes from blue to white, the reaction is over.

Study of the degree and depth of vitrification of the retort and furnace walls by scanning microscope enables some estimate of the temperature and duration of the process to be made. The temperature as calculated from the retorts was in the region of 1150-1250° C which was low compared to the temperature of 1400° C used in the 19th century European retort process. The depth of vitrification on the furnace walls suggest firing times between four and six hours. This accords well with a contemporary account which stated the process ran for three or four hours. Combined SEM and X-ray diffraction study show the conditions to have been extremely reducing with partial oxygen pressures of 10 or less. At the time when the furnaces were in operation these must have been the most extreme conditions ever obtained (Freestone et al., 1985).

DISCUSSION

These discoveries raise broader questions on the nature of the society which had both the resources and the technology to develop these mines and processes, and when such considerable quantity of metal was needed. Also it is necessary to consider the origins and antecedents to the zinc smelting process.

Unfortunately, history as understood in the Europe is largely missing in India, and with the exception of the Mauryan Empire it is difficult to link the mines with any particular group. Organised urban life in north India is supposed to have ceased for almost a 1000 years after the collapse of the Harappan civilization around 1500 BC. Sites such as Ahar near Udaipur are believed to have continued for a few centuries afterwards, but then a veritable dark age descended. Paradoxically, it is at this time that mining seems to start at Dariba, and apparently already before the end of the 2nd millennium BC the mines had penetrated over 100 metres below the water table according to the one NPL date of 3040±150 (Figure 4). To develop mines on the scale of Dariba pre-supposes a good working knowledge of mechanics, pneumatics, hydrostatics and practical geology. A certain competence in mathematics would be necessary, and almost certainly those in charge of the operations were literate. It is difficult to reconcile this with the traditional impression of the cattle-raiding Aryan invaders of the Rig Veda, with their abhorance of urban life and the arts of civilization including reading.

There is also the problem of what the silver was used for as little silver of this period has been found as yet. In the later part of the 1st millennium BC production seems to have greatly accelerated at Dariba and also was mined elsewhere such as Agucha, and also at Ambaji at the south-eastern end of the Aravallis in Gujarat. This new demand for silver must have been created by the introduction of silver coinage in India. The origin and date of the first Indian coinage is controversial, but a 4th century BC date for the widespread use of coins seems likely on present evidence.

During the 4th century BC much of north India became unified under the Mauryans and for a couple of centuries thrived as rarely before or after until the Moghuls in the 16th century. This of-course coincides with majority of radio-carbon dates listed in Figure 4 including the great open cast at Dariba, and by this time deep mining had penetrated to at least 250 metres below the water table. These mining achievements are unparalleled elsewhere. In addition to silver, at Zawar zinc was being mined and in the north of Aravallis copper was being mined on a similar scale at Khetri. Nor was the flowering of mining enterprise confined to Mauryan lands, since in the south of India large scale copper working at Indigandala in Karnataka was in progress and the Hutti Gold Mines, 200 km south-west of Hyderabad were being worked at a depth of 80 metres. The scale and sophistication of this mining activity pre-supposes an engineering and practical scientific basis which is at present only acknowledged to the Hellenic civilization, but which clearly had a much broader base throughout the Middle East and Asia.

Although the mines at Zawar seemed to have shared the fate of most of the big mines in the aftermath of the Mauryan collapse brass making clearly continued in India, as the Hindu copper alloy images dating right back to the Gupta Period (4th century AD) are almost invariably of brass. Given this interest and the existing economic importance of zinc ores in India, it is perhaps not surprising that this is where the technology of its extraction as a metal should have developed.

The remains excavated at Zawar are of the developed industrial process. But we can speculate on the inspiration behind the technology of that process. The only other metal condensed from a vapour during the smelting process was mercury. The conditions for this were much less rigorous since just heating the mercury ore, Cinnbar (HgS) at about 700-800° C would cause it to decompose releasing mercury vapour which could be easily condensed on a cold surface in the atmosphere with no danger of reoxidation. The method employed in early India for the production of mercury seems to have been to smear ore and reducing agents on the inside of a pot inverted over a collecting vessel containing water, and then to heat by means of a fire around the inverted pot. A similar arrangement is shown in Agricola in use in 16th century Europe. This process was known in India as "Tiryakpatana", in Europe as "Distillato Per Descendum" literally, distillation of descending.

To successfully smelt zinc, very different apparatus was necessary, the koshthi, the form of this seems to have been inspired by either the humble pottery kiln, or more likely the furnace used for producing cementation brass, both of which are sometimes divided by a perforated floor. Of-course the principle of operation is totally different in the pottery kiln or cementation furnace. There the fire is in the lower chamber and the flames go through the perforation in the floor of the upper chamber where the pots or crucibles are stacked. In contrast in the koshthi the fire is in the upper chamber and the condenser necks lead the molten zinc through the perforations to the cool chamber below. Thus a combination of the principle of mercury smelting and the form of the pottery kiln seem to have produced the koshthi apparatus for zinc smelting.

The process is described in the Indian literature in increasing detail through the 1st millennium AD, but there is no mention of zinc in the contemporary Arabic or Chinese scientific literature. However, the Iranian 'Abu Dulaf', writing in the 10th century describing zinc oxide and its use in brass making, states that India "it is made from the vapour of tin". All this suggests quite strongly that zinc smelting developed in India. Zawar could not have been the only source of zinc in India as zinc coins dating between 14th and 18th century now in the British Museum, were found by lead isotope analysis not to be of Zawar zinc, or indeed likely to be of zinc from anywhere in the Aravalli hills. Thus other sources probably in Kashmir or in Afganistan were also producing zinc. However, Zawar does seem to have been the major producer as nothing on that scale has been noted elsewhere. The production estimates based on the surviving debris are impressive. A furnace block, could have produced between 20 to 50 kgs. of zinc per day, depending on the furnace type and on the retort capacity, and an overall production through 5 or 6 centuries of the order of 50,000 tonnes is estimated.

CONCLUSIONS

Just as there is a tendency for western archaeologists to be Hellenic or Mediterranean oriented, so there is a tendency amongst other scholars to regard the technical development of the modern world to be a uniquely European phenomena. In particular there is a tendency to regard the science and technology of the rest of the world to have been static in recent centuries and unimportant.

However, studies of technical development occurring in the rest of the world are showing that these ideas of European exclusiveness are far from true. Wagner (1984) for example has suggested that the very advanced Chinese iron and steel industry was far from being static, was still evolving technically in the 17th and 18th centuries. It is also becoming clear from studies such as that presented here that the basis of much of the rapid technical development in Europe was based on the adaption of advanced technologies from the east, as well as on new materials.

Finally, the technical supremacy of Europe and north America that has been until quite recently tacitly assumed to be permanent is no longer tenable. The area of innovation is spreading much wider and the impetus may already move back to the east, or at least to the Pacific Rim. The lesson from all this would seem to be that technical development cannot be satisfactorily studied or contained within any one civilization or period. Only by widening our horizons to include all major regions of the old world will we begin to arrive at a true understanding of the synchronous development of ancient technology.

The importance of Zawar zinc production technology is reflected in the sentiments expressed by Royal Society, London, at "Conversazione" organized on 20th June 1985, which are as follows:

> "Excavation by British Museum, Baroda University, and Hindustan Zinc Limited, at Zawar, Rajasthan, have uncovered blocks of intact furnaces still holding their original charge of 36 retorts per furnace. The technical sophistication and application of scientific principles are unparalleled elsewhere in the late-Medieval period, and the element of standardization and mass production foreshadow the Industrial Revolution. This is the earliest example of high temperature distillation is almost certainly ancestral to all such operations today".

REFERENCES

Allan, J.W., 1979. *Persian Metal Technology* 700-1300 AD, Oxford.

Allchin, R.F., 1962. Upon the Art and Methods of Gold Mining in Ancient India. *Journal of the Economic and Social History of the Orient,* 5, No. 2, 195-211.

Ball, V., 1881. *A Manual of the Geology of India,* Vol. 3, Economic Geology, Calcutta, 1881.

Basu, K., Bortnikov, N.S., Mishra, B., Mookkerjee, A., Mozgovann, Tsepin, A., 1984. Significance of transformation textures in folklores from Rajpura-Dariba polymetallic deposit. *Neues Jahrbuch* 149, 143-161.

Brooke, J.C., 1850. Notes. *Journal of the Asiatic Society of Bengal* 19, 212.

Craddock, P.T., 1983. The copper alloy of Tibet, in W.A. Oddy/W. Zwalf (eds.), *Aspects of Tibetan Metallurgy,* London, 1-32.

Craddock, P.T., 1987. The Early History of Zinc. *Endeavour* 11, No. 4, 183-191.

Craddock, P.T., Burnett, A.M., Preston, K., 1980. Hellenistic copper base coinage and the origins of brass, in W.A. Oddy (ed.), *Scientific Studies in Numismatics,* London, 53-64.

Craddock, P.T., Freestone, I.C., ale, N.H., Meeks, N.D., Rothenberg, B., Tite, M.S., 1985. The investigation of a small heap of silver smelting debris from Rio Tinto, in Craddock/Hughes 199-218.

Craddock, P.T., Huhes, M.J., (eds.) 1985. *Furnaces and smelting technology in antiquity,* London.

Cribb, J., 1985. Dating India's earliest coins in J. Schotsmann/M. Taddei (eds.), *Asian Archaeology,* Napoli, 535-554.

Day, J., Bristol Brass, Newton Abbot, 1973.

E-Tu Zen Sun, Shiou-Chuan Sun., Tien-Kung k'ai-Wa, *Chinese technology in the 17th century,* Philadelphia, 1968.

Freestone, I.C., Craddock, P.T., Gurjar, L.K., Hegde, K.T.M., Paliwal, H.V., Analytical approaches to the interpretation of medieval zinc smelting from Zawar, Rajasthan.

Freestone, I.C., Craddock, P.T., Hegde, K.T.M., Hughes, M.J., Paliwal, H.V., 1985. Zinc production at Zawar, Rajasthan, in Craddock/Hughes, 229-244.

Gandhi, S.M., 1983. Rampura-Agucha lead-zinc deposit. *Mining Magazine,* 315-321.

Habib, I., 1982. *An Atlas of the Mughal Empire,* Delhi.

Hegde, K.T.M., Ericson, H.E., 1985. Ancient Indian copper smelting furnaces, in Craddock/ Hughes, 559-570.

Ingalls, W.R., 1903. *The metallurgy of zinc and cadmium,* New York.

Marshall, J., 1951. *Excavations at Taxilla,* Cambridge.

Merkel, J.F., 1983. Bronze age copper smelting experiments, Doctoral thesis for the University of London.

Paliwal, H.V., Gurjar, L.K., Craddock, P.T., 1986. Zinc and brass in ancient India. *Bulletin of the Canadian Institute of Mining & Metallurgy,* 75-79.

Pearson, G.W., Stuiver, M., 1986. High precision calibration of the radio-carbon time-scale. *Radiocarbon,* 805-839.

Percy, J., 1870. *Lead,* London.

Ramage, A., 1970. Sardis: Pactoculus north. *Bulletin of the American School of Oriental Research* 199, 18-28.

Ray, P., 1956. *History of Chemistry in ancient and medieval India,* Calcutta.

Sankalia, H.D., Deo, S.B., Ansari, Z.D., 1969. *Excavations at Ahar,* Poona.

Schroeder, U., 1981. *Indo-Tibetan bronzes,* Hong Kong.

Tiwari, Kavdia, 1984. Ancient mining activity around Agucha village. *Man and Environment* 8, 81-87.

Tod, J., 1984. Some traditional iron production techniques. *Journal of the Historical Metallurgy Society* 18, No. 2, 95-104.

Willies, L., 1984. 'Ancient lead and zinc mining in Rajasthan'. *World Archaeology* 16, No. 2, 222-233.

Williers, L., 1987. Ancient mining in Rajasthan. *Bulletin of the Peak District Mines Historical Society* 10, No. 2, 81-124.

Zu Li, 1989. Traditional zinc smelting technology in the Guma district of Hezhang county, Guizhou. P.T. Craddock (ed.), *2000 years of zinc and brass,* London.

Metals in Indian Medicine

K.N.P. RAO

Metals and minerals have been used in the Indian systems of medical pharmacopea — like Ayurveda and Sidha systems from their inception. While Ayurveda uses herbal and plant products along with minerals and metals, the Sidha system is said to relie mostly on minerals and metals. This is probably due to the fact that vegetable products are perishable and the active principles would vary due to the differences in the soil, method of growth, climate conditions and such other factors. In any system of medical pharmacopea, it is of fundamental importance that the products should be in a form or condition that they are easily and effectively assimilated by the body fluids and organisms. By years of experimentation and experience in the treatment of diseases, experts have devised meticulously tested and tried out methods of processing different metals, and minerals. The entire system and knowledge is built on Holistic approach and on Religion, Philosophy and to some extent on Science. It is to be appreciated that the lack of sound scientific knowledge, and understanding of the reactions involved, facilities like precise instruments to determine and fix the working conditions like, temperatures, weights, volumes, concentrations, standardisation, much of the results could only be empirical. Further the shroud of secrecy and the knowledge being transmitted to the chosen few has inhibited the development. The few treatises on the subject have been written in Sanskrit, which is not understood by many. Sanskrit language has the drawback, that many of the words and expressions can be interpreted in different ways and meanings.

All the common metals — Gold, Silver, Copper, Tin, Lead, Zinc and Iron also, Mercury, Mica, Sulphur etc. and their alloys have been used in Ayurveda. A metal is converted into 'Bhasma' in common parlance "a burnt-out" product or Ash, before it is administered. The preparation of Bhasma is a very complex process, involving the purification of the metal and selection of a number of ingredients such as herbal juices and plant products like, leaves, flowers, fruits, barks, roots; animal products such as urine, blood, flesh, bones and shells, and minerals like Sulphur, Salts, Mica. The conversion process involves various treatments like grinding, mixing, boiling, filtration, distillation, inceneration, maceration, infusion etc. under exact and very careful handling. The final product 'Bhasma' has altogether new properties. It is so micro-fine that when dropped over water it floats.

There are very few books on Indian medicine describing the scientific usage of minerals and metals. Rasa Ratna Samuchhaya by Vagabhatta (Circa 1300 A.D.) is considered authoritative. I.N.S.A. has published an English translation and the following pages give the salient information contained therein. The notes cover the common metals, their classification and purification before undergoing further treatment for conversion to Bhasma.

Gold

Gold has been categorised into 5 varieties :

1. *Native*—Prakrita - That which is produced by 'Rajoguna' and pervades the entire world. It appears to be unrealistic.
2. *Suhaja*—That which covered Lord Brahma in the form of "Jarayu' and was converted into a divine Mountain "Meru". This also is unrealistic but appears to be placer of alluvial types found on mountain slopes.
3. *Agni Sambhava* — A product of Pyro-metallurgy.
4. *Khani Sambhava* — That which is obtained from underground mines.
5. *Parada Vedha Sanjata* — That which is got from the transputation through Mercury (alchemy).

Purification

(a) Gold foils are enclosed in earthen dishes, and the gold leaves are covered with salt and rouge (red ochre) and heated on a charcoal fire. This process should be repeated several times.

(b) Gold foils are dipped in oil, butter milk, cow's urine acidic liquid (Sap of 'Arka' Calotropis Gigantea Sanskrit - Arka plant) and decoction of Kulatha (Horse gram). To be treated 7 times in each of the above.

Liquifaction

1. Gold can be made to stay as a liquid, if molten gold is poured over calcined bones of frog, borax, fat, saliva of a horse and "Indragopa" (Five fly insect) in equal parts.
2. Powdered "Indra Gopa", macerated with Deva Dali fruit juice, mixed with molten gold keeps it in the liquid state.

Silver

Classification

1. Native - Occurs on the Kailash mountain range.
2. That which is obtained from the mines in Himalayas.
3. Tin when came into contact with Sri Rama's Paduka was converted to silver.

The silver which is compact, heavy, clear, bright, soft, smooth and has colour of 'conch' free from fissures is of superior kind. The variety that turns red, yellow or black on heating, is rough, full of fissures, and light is of inferior quality.

Purification

1. Add Lead and Borax to molten silver in a crucible and immerse the whole in Jyotismati (celastrus paniculatus) oil. Repeat 3 time.
2. Make a lining or raised border of ash and lime on the inner surface of an earthen dish. Keep silver on it, with equal quantity of lead and blow with a blow pipe, until all the lead gets lost.

Liquifaction

Devadali powder should be subjected to bhavana (Maceration) with man's urine seven times. The sprinkling of this powder into molten gold and silver, makes them stay as liquids.

Copper

Copper obtained from Nepal is the best. It has red glowing lustre and good malleability. Other vareities have white, black and red shades, rough surface and a layered (foliated) texture.

Purification

1. Copper foils are melted along with rock salt, and goose-berry juice (acid) or lime. Galrika is added and poured into a vessel having butter milk and diluted dung of buffalow. This to be repeated 7 times.
2. Clean Copper foils are coated with a mixture of rock salt and lime juice. It is then heated strongly and dipped in acid solution. This process is repeated 8 times.
3. Copper foils are boiled in cow's urine, in strong heat for several hours.

Tin

There are two varieties:

1. *Khuraka* — It is white, brilliant, soft, heavy, melts quickly, and devoid of "Tin-cry".
2. *Mishraka* — It is blackish-white in colour.

Purification

1. Molten tin is mixed with Turmeric powder and juice of Lakki leaves (Nivagundi - Vitex Nagundo Linn). This is done three times.
2. Molten metal is poured into butter milk mixed with Punarmava (Bverhavia diffusa) and Kucata (Inux vomica fruits) or in Katafruit decoction.

3. Mishraka tin is exposed to sunshine, and on which is sprinkled Nirgundi swarasa mixed with the powder of Niragundi Mula. This is done 7 times. The same process can be adopted to lead and zinc.

Elaboration of the process

Two litres of Lakki leaves juice are taken in an earthen pot, to which is added 150 gms of Turmeric powder. This is placed on an oven. The mouth of the pot is covered with an earthen lid, having a hole of 1 inch diameter in the centre. About 1-1/2 kg. of molten tin is poured through the orifice. After it has cooled the juice of Lakki leaves is removed and a fresh stock of juice and turmeric powder is placed and the process is repeated 7 times.

Lead

It is a blackish, lumpy mass. It should be heavy, externally black, produce foul odour, and the fractured surface appears shining black.

Purification

Molten Lead is mixed with the powdered Lakki plant root and turmeric and on this is poured Lakki Juice. It is repeated 3 times.

Zinc

It is a darkish solid mass.

Purification

It is done in the same method as for Tin. Molten Zinc mixed with Turmeric powder is poured over the juice of Lakki leaves. It is then cleaned and roasted with the bark of tamarind tree and uttereni. It is further ground to powder and mixed with Talaka and again powdered and heated. It is then mixed with Aloe juice and made into small discs. These pieces are placed inside pulverised bark of Aswatha tree (Ficus Religiosa) and heated until the entire mass becomes soft.

Brass

There are two kinds :

1. *Ritika* — which turns into Copper colour when heated and dipped in Kanjika.
2. *Kakahundi* — which turns black with the same treatment.

Superior brass is heavy, soft, yellow, malleable and brilliant.

Purification

By heating and quenching 5 times in Nirgundi juice containing Turmeric powder.

Bronze (Bell Metal)

Bronze is made by melting 8 parts of Copper and 2 parts of tin. Superior Bronze can produce rich and sonorous sound, soft, brilliant, blackish - white, and turns red on heating.

Purification

Foils or thin sheets are heated in a strong fire and dipped in cow's urine.

Panchaloha

An alloy of Bronze, Copper, Brass, Iron and Lead. It is purified by pouring the molten alloy in Horse Urine.

Iron

There are 3 main types of iron (1) *Munda,* (2) *Tikshana* and (3) *Kanta.* There are again, 3 sub-varieties of Munda, 6 sub-varieties of Tikshana and 5 sub-varieties of Kanta. The qualities of the sub-varieties are :\

Munda

1. Soft, melts quickly, no cracks on the surface and is malleable (Pure Iron).
2. Medium — Not very malleable (contains higher carbon content).
3. Looks black and is brittle (perhaps cast iron).

Tikshana (Steel)

1. Hard, without fibres, breaks on bending and the fractured surface shines (High Carbon Steel).
2. Looks whitish-black, fibrous (tough) and is hard.
3. The structure looks like hair follicles and brittle.
4. Hard, has fine and dense band like structure.
5. That whose edges can be destroyed quickly, has a fibrous structure and obtained from pale earth.
6. Bluish-black in colour, dense, smooth, heavy and bright surface and the sharpened edges do not get distorted when hammered with iron tools.

From the descriptions of the sub-varieties given above it is possible that they refer to the different types of steels obtained from various heat treated, manufactured articles like, tools, swords, files etc.

Kanta (Magnetic Iron) or it may also refer to Load Stone

1. Magnetic — Soft iron and weakly magnetic.
2. That which attracts iron pieces (stronger magnetism). That which kisses or adheres to iron.
3. That which attracts (draws) iron — Stronger magnetism than the second type.

4. That is forcibly attracts iron (stronger than 3).
5. That which gives rise to hair like follicles on coming in contact with iron (articles).

There is also a different type of classification depending on the basis of magnetic strength of the face of the iron (or steel) and direction of the magnetic force which can be in one to five sides and more. Best known (kanta) Magnetic Iron vessel will not allow oil to spread on water kept in it. Asofeteda will loose its pungency when its paste is anointed on Kanta Loha and Lime paste will loose its bitterness. The Kanta Loha vessel does not allow boiling milk to spill over rather makes it attain a cone like shape.

Purification

1. Iron anointed with the blood of rabbit and heated 3 times will loose its impurities.
2. Iron foils — so thin that a needle will pierce through it is cleaned and it is heated strongly and then immersed in a decoction made of 50 gms of Triphala put in 8 times of water and the volume reduced to half. This is to be repeated 7 times, using fresh decoction each time.
3. Iron anointed with common salt is heated and dipped in Triphala decoction.
4. Iron foils are treated and is dipped in a solution made by mixing Tamarind juice or decoction of Triphala powder boiled in 8 times in cow's urine and reduced to 1/4 of the volume.

Liquifaction

The burnt ash of gourd is mixed (macerated) with 21 times the volume of cow's urine. Iron (steel) bits are taken in a crucible and heated with the salt, made as above.

The (burnt) ash of Devadali is mixed with cow's urine and the iron piece is heated strongly with the salt as stated above.

Devadali ash (salt) is mixed with human urine is poured into iron kept in a crucible and heated strongly.

Sulphur is to be treated with the juice of Devadali and treated 7 days converts iron powder into liquified state. The magnetic iron or steel melted and treated with the sulphur mix as above. The iron will remain in the liquid form.

Iron scoria - It is also purified and used for medicinal purposes.

Mercury

Mercury along with sulphur have been used in Ayurveda system of medicine extensively.

'Engaleeka' is the ore-probably, Cinnabai - from which mercury is extracted by the distillation process. The ore is soaked in lime juice or that of margose leaves for 3 hours, dried and ground to powder. It is placed in an earthen pot and another pot is placed over it, in an inverted position so that the mouths fit each other. The joint is sealed tight with red mud and dried in the shade. This set-up is placed over an oven. A wet cloth is placed on the top and cold water is dripped, but the water should not get into the pot. 330 gms of the ore needs 12 hours of firing and another 3 to 6 hours of cooling. Mercury collects on the bottom of the pot placed above as a thin layer or crust. This is collected by wiping with a fine clean cloth carefully. Mercury is washed to get rid of any dust or dirt and is dried before using.

REFERENCES

Rasa Ratna Samuchaya. I.N.S.A. Translation in English.

Rasa Ratna Samuchaya. Dr. Ashok. D. Satpute — Translation in Kannada.

A Scientific Analysis of Iatro-Metallurgical Practices of Ancient Indians

K. NAGESH RAO
AND P.R. RANGESH

Introduction

Metals, the natures fabulous shining meterials, have fascinated the man since the earliest period of human history. From the time of the first knowledge of gold man has traversed a great distance to master the technique of converting the metals to domestic appliances, defence, and disease fighting. Though initially the metals were employed for the first two needs their afficacy in medicine was lately recognised. The ancient Indian civilization which stands pre-eminent in the world for its cultured living was convincingly the first to successfully use metals in medicine. The system of converting the metal into therapeutically effective drugs involved a highly evolved metallurgical process. The present paper undertakes the task of unveiling such scientific aspects of Iatro-Metallurgical practices of ancient India.

Iatro-Metallurgy meaning metallurgical practices as undertaken by physicians has been on records of medicine from the first compendium, *Charaka Samhita* (1st century B.C.) till the birth of an independent discipline Iatro-Chemistry or *Rasashastra* which deals exclusively with such operations. Charaka has advocated the use of iron as a geriatric Medicine (Agnivesh, 1970) and Acharya Sureshwara (Circa 12th century A.D.) has recorded Sushruta, Harita and others as well versed in Iatro-Metallurgy (Dwivedi, 1977).

Rasashastra was developed by alchemists involving many of the alchemical procedures and it took a definite shape of developing a new branch of processing of metals into safer effective medicaments. Leaving aside many whimsical practices, Vagabhata (13th century A.D.) wrote a compilation of earlier observations and his work *Rasaratnasamucchaya* (RSS), still today remains unmarvelled in depicting the state of art of Iatro-Metallurgy of that time. Many of the discussions made in this paper derives reference from this work.

Metals used in Iatro-Chemistry

Metals are called 'Loha' in Sanskrit. The word 'Loha' is derived from the root 'Luh' meaning 'Karsha' - Pulling, drawing, attracting (*RSS,* Ch. 5/1). Hence it is called Loha because of its attractive metallic lustre and thereby drawing the attention of man.

As described in Rasashastra the metals used in medicine are Mercury, Gold, Silver, Copper, Iron, Lead, Tin, Zinc and alloys like Brass and Bronze (Ray, 1956). These metals are employed in medicine only after subjecting them for stringent heat treatment procedures.

The prevention and cure of illness forms the basic aim of all systems of medicine. Ayurvedic system in addition to its herbal drugs had metal based drugs also. Importance of metallic medicine can be seen as, they are the only esteemed remedies. They are useful for ones having low appetite, and are palatable, could be used in small dosage and having high potency and easily assimilated into the system (Dwivedi, 1977). The form with which the metals are used in medicine is called 'Bhasma'.

Extraction of Metals

As the metals are not found in nature in their true form except gold, the need for extraction from their ores or compounds is of paramount importance. Ancient Indians were not lagging behind in the techniques of extraction of metals. Alchemists were the fore runners in experimenting with various ores and oxides and have evolved basic procedures for extraction. It was these people who announced the birth of a new metal zinc and a new technique of distillation procedure for extraction (Ray, 1956). These extracted metals from different ores were called *'Satvas'* (*RSS*, Ch. 5/34) (lit. essence of the ore the form of metal). The alchemists were experts in the preparation (*RSS*, Ch. 10/2) of crucibles called *'Musha'* used for extraction of metals and a complete narration is given on the different shapes, and basic ingredients that go into its manufacture.

General Method

The ore taken is mixed with alkalies, vegetable products, borax, and crushed into fine powder. This is transfered to a crucible and strongly heated with the help of leather blower. The carbon derived from vegetable ingredients reduces the ore into the respective metallic form.

Extraction by Reduction Process

Copper

Chalcopyrite (copper sulphide-*Makshika*) soaked with honey, castoroil, plant products and borax when heated yields the essence in the shape of copper (*RSS*, Ch. 2/89). Similar is the procedure for the extraction of copper from its compounds (copper sulphate - *Tutta/Sasyaka*). (*RSS*, Ch. 2/132).

The reaction can be understood as - various plant materials supply carbon by destructive distillation and when heated in a closed crucible, the carbon and borax reduce the ore or the compound to the metallic form.

Similar principle has been described for the extraction of other metals from their respective ores.

Extraction by Distillation Method

Mercury

The chief ore Cinnabar (mercuric sulphide - *Hingula*) is crushed into a fine powder and rubbed with lime and ginger juice which is applied to the bottom of the sealed retort. Then it is subjected to mild heat. The mercury sublimes over and the vapours are condensed and collected in another vessel through the retort tube (*RSS*, Ch. 3/144). This is basically a *Distillation Process.*

Zinc

Another remarkable extaction process by high temperature distillation was that of zinc. It was to the credit of *Nagarjuna* (8th century A.D.) who was a genius, that Calamina (zinc carbonte - *Rasaka*) was made to yeild zinc metal in a sealed retort in the presence of carbon (Khan). Later *Vagabhata* has given a lucid description of extraction from calamine.

Zinc extraction is indeed a difficult proposition. It oxidises very rapidly and has to be distilled in an air tight reducing atmosphere at about 1100° C. Zinc as a separate metal was unknown in earlier period and it was Indian alchemists who devised the intricate distillation process for extraction of zinc. This itself is their major contribution to metallurgy.

In this process calamine is mixed with turmeric, resin, salts, borax and 1/4th its weight of marking nut and acid juices. This paste is smeared inside tubulated crucible (*Vrintaka Musha*) and dried in sun. The mouth is closed with another crucible inverted over it, and heat applied. When the flame issuing from the molten calamine changes from blue to white, the crucible is caught hold of by means of a pair of tongs and its mouth held downwards and it is thrown on the ground, care being taken not to break its tubular end. Thus the essence of calamine in the form of metal zinc is collected (*RSS*, Ch. 2/158-162).

Alchemists not only isolated the metal zinc, but also processed it into a metallic drug form. Another important aspect of this procedure was the transfer of the laboratory technology into industrial production, as seen in Zawar mines in Rajasthan, making a close link between Science and Technology.

Extraction by Cupellation Method

Bismuth

Apart from the above methods, they were also well versed in the recovery of metals by Cupellation Method. There is a note on the extraction of *Chapala* by Vagabhata (*RSS*, Ch. 8/41-44). There is much uncertainty of identification of *Chapala* but the method of recovery from molten lead and tin suggest that it should be only Bismuth.

This method is similar to modern *Cupellation Process.* Here the Bismuth bearing lead powder is mixed with milky erudation of plants and is melted. When it is molten, part of the lead is converted into litharge (Lead Monoxide). The Oxide of lead is scooped out and this operation of heating and scooping litharge is repeated till bismuth which is not easily oxidised remains at the bottom in liquid form at the end of the operation. The same procedure is given for separation of bismuth from bismuth bearing tin (*RSS*, Ch. 7/41-44).

Heat Treatment of Metals

The ancients were well ahead in the knowledge of varieties of iron and steel and their classification of iron and its property has no parallel in the world. Steel has been classified based on its response to the heat treatment. Bhanu Prakash (1990) has deciphered the classification and has equated with modern nomenclature.

The heat treatment procedures can be viewed under the following headings.

(a) for metals used in internal administration
(b) for metals used in external application (Surgical Procedures)

(a) In case of metals used for internal administration, they are subjected to various procedures like, *Shodhana* (purification) and *Marana* (Killing or reduction). The resultant nature of metal in the form of fine powder is called *Bhasma* and taken for oral administration.

Shodhana: It involves the metallic sheets to be heated to red hot condition and quenched in different liquid media, derived from organic and inorganic sources, in the form of acidic or alkaline nature, for multiple times. In this procedure all the impurities adhering to the metal are dislodged and make the metal soft (*RSS*, Ch. 5/13).

Marana : The above purified metal sheet is reduced to powder with the help of pestle and mortar over a number of days after mixing with equal quantity of Mercuric Sulphide rubbed in the media of juice of specified plant. The resultant paste is made into a round flat form and dried in sun. This dried cake is kept in earthen saucers, sealed with mud layered cloth and heated in a pit furnace fired by cowdung cakes. The dimensions of the pit varies according to the quantity of heat required for the particular treatment of materials. But for processing the metals exclusively, *Gajaputa* alone is used (Govt. of India, 1978).

Accordingly the *Gajaputa* comprises of a pit of 22 inches cube, fired with cowdung cakes and the temperature pattern depicted as given in Fig. 1. The cowdungs are allowed to burn till consumed to ashes and allowed to cool by itself (*Swangasheeta*). Then the saucers are carefully opened to collect the processed metal powder. This procedure of grinding with pestle and mortar and heat treatment in *Gajaputa* has to be repeated till the metallic powder assumes microfineness in which when sprinkled on water should float and not sink (*Varitara*), (*RSS*, Ch. 8/26).

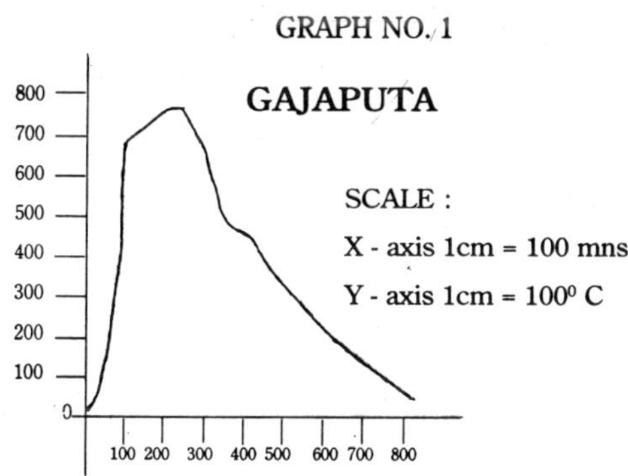

Time in minutes
PITSIZE : 22″, depth 22″
CAPACITY : 26 Kgs (260 Nos) Cowdung

Fig. Temperature Vs Time graph for Gajaputa

This procedure can be described as a thermochemical diffusion process, where the metal is converted to its sulphide and mercury being highly volatile, sublimes over. Sometimes it is converted into its oxide form also. In general, the *marana* procedure is basically converting the metal sheet to a fine powder (Dutt, 1980). Repeated heating and annealing makes the metal soft which aids in the easy grindability and also convert it into a microfine particle size of less than 10 microns, so that the particles can float on water when subjected to test. This processed microfine powder is considered by physicians as a safer form of oral administration of metallic drug (*Bhasma*). This fine metalic powder with its high potency is found fit for easy absorption by the body.

(b) In case of metals used in the form of surgical instruments a high degree of heat treatment is necessary. Ancients were aware of the intricacies involved in the heat treatment of metals. Sushruta, a celebrated surgeon, had developed the procedures for hardening and tempering of surgical instruments. His knowledge of treatment procedure can be compared with the existing modern process of heat treatment. Sushruta had used water, oil and alkaline solution for quenching. The Criteria for the selection of these media for quenching was also given by him. The instruments which are used for cutting bones were quenched in alakaline solution, whereas those used for cutting and drawing muscular tissue were quenched in water and while those which are used for puncturing vessels and cutting ligaments, were quenched in oil. He also suggests that these should be given a sharp edge by honing on slabs of special stones (*Susruta Samhita*).

It is of great significance to note that, instruments quenched in alkaline solution had been used for cutting the bones, since the saline quenched instruments have a least flexibility and attains maximum hardness. While the other two are relatively soft but sufficiently hard, enough to handle soft tissues and vessels.

Further, the subsequent physicians had added dimensions to the above treatment process by introducing carburising of steel, by applying a paste comprising of powders of long pepper, mustard, saussurea, salts and honey, heating in the flame for a long time and quenching in water. This is essentially a Carburising Process (Manunatha, 1990).

Ancients were also aware of various processes of heat treatment such as - Annealing (*Swangasheeta*) and Nornalising (*Bahirsheeta*) (*RSS*, Ch. 8/58).

Melting and Alloying

Knowledge of melting of metals, was of high order. The colour of the flame was used to judge the state of metal in the furnace. During the sequence of melting when the metal emits flames which are of the colour of the substance melted indicate that liquification has commenced (*Bijavarta*) and while being further heated when it emits white flame, means the end of melting process (*Suddhavarta*) (*RSS*, Ch. 8/56-57).

Apart from the above observation, alloying of metals is also defined as combining two metals by heating them to proper temperature to make it into a homogenous solution (*Dwandwana*) (*RSS*, Ch. 8/46).

Brass, Bronze and Varta Loha are the three major alloys mentioned by Vagabhata. Bronze - *'Kamsya'* made in Saurashtra (Gujarat) is of good quality and he mentiones that it is an alloy of 80 parts of copper and 20 parts of tin (*RSS*, Ch. 5/205). While Brass has been described as an alloy of copper and zinc by the later authors the composition has not been specified. This metal is available in two forms as *Reetika* which has less percentage of Zinc and *Kakatundi* has high percentage of zinc which is identified based on reaction to weak acid (*RSS*, Ch. 5/191-192).

Varta Loha is an alloy of Brass, Copper, Bronze, Iron and Lead. (*RSS*, Ch. 5/212). As it contains two alloys in its making, it could probably be due to recycling of old icons and utensils.

CONCLUSION

From the above it is very clear, that ancient Indians had reached the frontiers in metallurgy at a time when instrumentation technology, modern scientific outlook were not in vogue.

They had observed all the basic steps in metallurgy right from extraction, treatment and melting and alloying, which speaks of their ingenuity and skill. They were pioneers in pronouncing the birth of a new metal, zinc much ahead of their counterparts elsewhere in the world. Because of their scientific acumen they were able to classify ancient iron and its properties in a methodical way. Their accomplishments in the field of metallurgy have been overlooked for the reasons that, many of the operations involved unnecessary ingredients and improper understanding of the underlying principles. As seen in this paper, the scientific aspects which could be attributed to their practices, can bring their efforts to the limelight and further development in this field of Iatro-Metallurgy as well as Iatro-Chemistry is very much warranted.

REFERENCES

Agnivesha, 1970. *Charaka Samhita,* Redacted by Charaka, II Part, Sri Kashi Samskrita Granthamala 194, Chowkhamba Sanskrit Series Office, Varanasi, Chikitsa Stana Chap 1.3/15.23

Bhanu Prakash, 1990. Some Aspects of process control of Ancient Iron making, paper read at Conference on Paleometallurgy of Iron and Culture, France.

Dutt, U.C., 1980. *Materia Medica of Hindus,* Chowkhamba Saraswati Bhawan, Varanasi, p. 24.

Dwivedi, Viswanath, 1977. *Lohasarvaswa* (Sureswaracharya), in *Bharatiya Rasashastra,* Shri Sharma Ayurved Mandir, Nagpur, 1977, p. 28.

Govt. of India, 1978. *Ayurvedic Formulary of India,* Ministry of Health and Family Welfare, Part - I, p. 181.

Khan, Iqbal Ghani, *Metallurgy in Medieval India,* p. 79.

Manunatha, S., 1990. *Vriddha - Sharangedhara,* in Experimental evaluation of Shastra Payana using different liquid media, M.D. (Ay.) Thesis, Bangalore University.

Rasaratnasamucchaya (R.R.S.) Vagabhata, Hindi translation (D.A. Kulkarni, Mehar Chand Lachhman Das, Delhi, 1969).

Ray, P.C., 1956. *Rasasnava,* in *History of Chemistry in Ancient and Medieval India,* Indian Chemical Society, Calcutta, p. 138.

Sushruta Samhita, 1976. Sushruta, Kashi Sanskrit Series, 156, Chaukhamba Sanskrit Sansthan, Varanasi, Sutra Sthan B/12-13.

An Historical Review on Ancient Mining with Special Reference to India

R.D. SINGH

Introduction

Mining, one of the oldest industries known to man, ranks as a basic or primary industry of human civilization. The present day civilization and cultural progress of mankind has been influenced to a large extent by mining and metals. From the Stone Age to the present day, mining has made a vital contribution to engineering and growth of civilization, which began with the making of tools and ornaments, and the smelting of ores probably some 9000 years ago (Bryan, 1979).

The French mining historian Simonin (1969) in his classic book *Underground Life of Mines and Miners* writes:

> "The intrepid coal miner, whose advent is of but recent date; the veteran of the mineral world - the metal miner - whose origin dates from the dawn of the history; and the patient seeker after gems"
>
> and then comments:
>
> "everybody has thought, at some time or other, of the important part played by metals in the life of nations. From the most precious to the basest, from the rarest to the commonest, they all play a special part; and material progress is only affected by their means. The history of metal constitutes the true history of inventions and labour".

Utilization of minerals by man began some 1000,000 years ago in Pleistocene period (Sharma and Sinha, 1967). Archaeologists have named this Pleistocene period as Stone Age, and based on the shape and type of implements used by man divide it into three sub-divisions i.e., (i) Old Stone Age (Palaeolithic, Circa 60,000 B.C. - 20,000 B.C.); (ii) Middle Stone Age (Circa 20,000 B.C. - 7,000 B.C.): and (iii) New Stone Age (Circa 7,000 B.C. - 4,000 B.C.).

Mining began some 45,000 years ago with the Palaeolithic man as evidenced by flint implements found along the bones of early man from the Stone Age (Davies, 1990). At first, the early man recovered stone raw materials from the surface excavations but by the beginning of the New Stone Age (Neolithic period some 10,000 years ago) he had progressed to underground mining. However, the oldest underground mine is from the Old Stone Age and is believed to be 40,000 years old. It is a hematite mine at Bomvu Ridge, Swaziland.

The Stone Age was followed by Copper Age, Bronze Age and Iron Age in that order. Gold was probably the first metal discovered followed by discovery of copper and manufacture of bronze for weapons and utensils. Smelting of iron started long after the knowledge of reducing ores of copper, silver, lead and tin.

Ancient Mining in Some Foreign Countries

Copper was mined in Sinai Peninsula before 3000 B.C. by the Egyptians. The miners employed copper chisels and stone picks and hammers to excavate the chambers in sandstone. If the rock was too hard, fire-setting was used. Diodorus (1943) writes about the Egyptian mining as follows:

> "Of those condemned to the disastrous life such as excel in strength of body and pound the shining rock with iron hammers applying not skill but sheer force to the work, and they drive galleries, though not in straight line, out in the direction, taken naturally by the glistening stone; then on account of the winding of these passages, live in darkness and carry round lamps attached to their foreheads. In accordance with the peculiarities of the rock they have to get into all sorts of positions and throw on the floor pieces they detach. And this they do without ceasing, to comply with the cruelty and blows of the overseer. The young children make their way through the galleries into the hollowed portions and throw up with great toil the fragments of broken stone, and bring it outdoor to the ground outside the entrance .... "

The method of mining was a kind of pillar and stall system, the roof being supported by leaving pillars of rock. The roadways were winding, following the ore vein and were very narrow. Ventilation was practically non-existent in the development ends and conditions there must have been very unpleasant (Hinsley, 1956)

Mining was an important industry in ancient Greece. Silver was mined at Laurium, reference to which dates from 500 B.C. It is said that the cultural achievement as well as the fighting strength of Athens were based upon the economic energy derived from the mines. At Laurium there were over 2000 shafts, the deepest of which was 386 ft. (117.65 m). The roadways were 2 to 3 ft (0.6 - 0.9 m) high and followed the vein. Timber props were rarely used. Ventilation was provided by the use of occasional air-shafts or by duplicating the main roadways. Furnaces placed at the bottom of the shaft were also used to maintain the circulation of air.

The leaders of post-Romans were Saxons. Mining in Hungary was revived by Charlemagne. In the eighth century the Spanish gold and silver mines which were closed by Visigoths were reopened by the Arab invaders. The French mines were also revived by the aid of miners from Saxony.

Archaeologists believe that the use of coal began in Bronze Age, some 3000-4000 years ago (*The New Encyclopaedia Britanica*, 1977). But the earliest reference to it in literature dates back to 350 B.C., when Theophrastus, a pupil of Aristotle recorded the use of coal by the smiths of Genoa and Liguria (Bryan, 1979). Probably Chinese mined coal over 2000 years ago. The Romans mined coal from the outcrops in England before 400 A.D. Among the earliest records of coal mining are those contained in the Annals of the Abbey of Klosterrade which indicate that coal wax being mined in the neighbourhood of Ahchen on Belgian - German frontier as early as 1113 A.D. Iron and coal accelerated the pace of industrial growth. The development of mining and metallurgy and use of minerals and metals contained down the ages and the process continues.

Use of Minerals and Metals in India in Ancient Times

The history of mining and use of minerals and metals in India may be traced from about 4000 B.C. onwards (Sharma and Sinha, 1967; Sharma, 1970). In the pre-Harappan period (Circa 4000 - 2000 B.C.) evidence of earliest settlements of agricultural communities in the mountainous and desert areas of the country is found. Copper axes and other articles like copper mirrors have been found which were made by hammering, cutting and rubbing of the copper metals extracted at about 700°-800° C from copper ore probably obtained from Baluchistan.

During the Indus Valley Civilization period (Harappa-Mohenjodaro culture, circa 2500-1500 B.C.) highly skilled industries producing on a mass scale bricks, huge copper pots and jars, glazed pottery wares, and clay bangles flourished. Copper and bronze were used in plenty for domestic utensils, axe-heads, daggers, knives, arrow-heads, sickles, wires and rods and for bangles and finger rings etc. Gold and silver were used for jewellery and silver for ornamental vessels. A large number of minerals, ores and rocks were used in this period. They include quartz and its various varieties, alabaster and steatite, cerrusite, cinnabar, white lead, lollingite, lencopyrites, haematite and slate.

In the post-Harappan period (*Circa* 1500-800 B.C.) refugees migrated from Baluchistan, Harappa and Mohenjodaro and settled in the Gangetic plains of Uttar Pradesh, parts of Bihar, Orissa and Madhya Pradesh. A large number of copper tools belonging to these people have been found. The above period was followed by Vedic Period (*Circa* 1500 B.C. to Pre-Buddhistic era, 600 B.C.) and Ayurvedic period (*Circa* 600 B.C. to 800 A.D.). These periods saw considerable growth in mining and mineral industry.

Rigveda, the oldest composition of the Aryans mentions the use of copper, bronze, gold and silver. *Yajurveda* composed between 1100 B.C. and 1000 B.C. mention gold, silver, copper, lead, tin and iron.

The earliest and most authentic record of information is found in *Arthashastra* composed between 321 B.C. and 296 B.C. by Kautilya (popularly known as

Chanakya). According to Kautilya *Arthashastra* mining and manufacturing industries were the chief source of income. The importance of mines is stressed in the statement that "the treasure depends on mines and that the earth is conquered by means of treasury and the army" (Kangle, 1965). The starting of the new mines and the renewal of old discarded one was an important state activity. The chapter on mining 2.12 of Kautilya's *Arthashastra* is largely devoted to gold and silver ores and disposes of iron ore in a single sutra (2.12.15). While the text devotes two chapters to the manufacture of gold and silver articles (1.18 and 2.14), it merely refers to factories to be started for the manufacture of articles of copper, lead, tin and brass etc. along with steel and iron (2.12.23) and does not contain even a single sutra concerning the process of manufacturing steel for the articles to be made of it.

Metallurgy of copper and its alloys (brass and bronze) was highly developed during the Ayurvedic period. A remarkable use of brass was made in the constructional work of an unfinished Vihara (convent) made near Nalanda (Bihar) in 700 A.D. Taxila (now in Punjab, Pakistan) remains (500 B.C. - 600 A.D.) consist of ornaments, toilet articles, household vessels, surgical and other instruments of this metal and its alloys. A white metal of nickel and copper was in use for coinage and jewellary and objects made of tin and its alloy have also been discovered. A colossal statue of Buddha (24 m high) found near Nalanda is believed to have been constructed during the reign of Purnavaraman in the 7th century A.D. Numerous uses of copper in the early days in the form of coins of Kushan kings like Kanishka and his successor in the 2nd century A.D. and in the following Gupta period are found. In recent excavations near Khairadih in the north-west of border of Ballia district fine types of copper wire of Kushan period have been found. (*Aj* Daily, dated 11.10.1992, p. 8). Large number of iron implements and daggers found during an excavation near Madras are regarded as the products of the 4th century B.C. The famous iron pillar bearing an inscription of 415 A.D. near Delhi by the side of Kutub Minar was constructed in the 4th century A.D. Iron was mined and smelted in India since ancient times and steel making was a highly developed art. Iron does not seem to have been known in the Harappan period but it was known in the Vedic period. Megasthenese (300 B.C.) mentions iron amongst other metals and fine steel had been used in the 4th century B.C. for the manufacture of surgical instruments and for inscribing on stone evidenced by Asoka's inscriptions on stones. Iron produced in ancient India was mostly wrought iron and charcoal was the fuel usually employed.

Description of metals and metallography, especially good quality steel and uses of various salts in medicine have been described in the later *Charaka* and *Susruta Samhitas.* Detailed description of 64 branches of arts and sciences in Vatsayana's *Kamasutra* (5th century A.D.) include knowledge of minerals and mining besides testing for gold and gem stones (Banerjee and Mookerjee, 1984).

India was the chief source of gem stones in the ancient and medieval periods. India was the only supplier of diamonds to the world before the discovery of diamonds in Brazil in the year 1725. Although there is no direct reference to the use of coal in ancient times, archaeological evidence suggests that coal was used in India in ancient times. "Some historians and other experts believe that the evidence of finished high grade copper and steel alloys found in India from 4th century B.C. onwards throughout the Muslim period bears evidence of metallurgical processes undertaken with the help of coal and coke, especially in the eastern coal bearing region of the country" (Banerjee and Mookerjee, 1984).

Ancient Mining in India

From the periods of foreign travellers and historians like Megasthenese (*Circa* 300 B.C.), Pliny (77 A.D.), Fa Hien (5th century), Huen Tsang (7th century), Al-beruni (11th century), Tavernier (1665-69) information on ancient mining and trade of various metals and gem stones, and use of metals and alloys in India in ancient times is found. Techniques of mining copper, gold, lead and zinc, etc. were well developed almost on similar lines as in Egypt, Greece and other foreign countries. Mining of gold in South India, lead-zinc and silver mining in Rajasthan, and copper mining in Bihar and Rajasthan was firmly established in India and the mining standards judged by the standards then prevailing were quite high.

Copper was mined in Singhbhum copper belt (Bihar), Khetri area (Rajasthan), Ambaji and Kumbaria (Gujarat), etc. Mining of copper in Singhbhum copper belt is known from 600 B.C. In Rajasthan copper mining was done at Khetri during the Mauryan period around circa 300 B.C.

In the Singhbhum copper belt the mines descended upto ground water level and almost the entire ore was extracted except in the pillars left for support. Timbering was not an usual practice of ancients, pillars being left to support the hanging walls. The workings were approached by adits driven in the hill side. The adits were 3-8 ft (0.9 - 2.4 m) X 4 ft (1.2 m) in dimensions. Drives were 4 ft (1.2 m) X 6 ft (1.8 m) wide. The slope width was 4 ft (1.2 m) and when timbering was done, the timbers were 5-6 inch (12.7 - 15.2 cm) in diameter. The miners extracted the vein stone by chisel and hammer to a maximum depth of 60 ft (18 m) which is the ground water level. Chips of bamboo were used for lighting. Evidence of same types of work are found in the Khetri area of Rajasthan. The age of these mines is reckoned to be Chalcolithic Age, 1500-1000 B.C. (Bose, 1968). The techniques used for mining were almost the same as used by Egyptians in Sinai Peninsula. Fire setting was used to crack the ore which was extracted with chisel and hammer manually. While at work, the miner seated himself upon his heels with the lamp upon the head, the hammer in the right hand and the 'gad' in the left hand and the small basket upon his knees in which he received all the fragments of ore that were struck off by the chisel. The unfired posture, the lack of fresh air and the excessive heat in which the labourers worked took heavy toll of human lives which were usually extinguished at the age of 30-40 years.

From the very ancient times gold has been produced in India by the inhabitants of the country either by mining or by washing sands for sparsely disseminated alluvial gold (Maclaren, 1905; Munn, 1934, 1936). The actual date when gold mining was started in India is not known. Hutti mine which is probably one of the most ancient mines in the world dates back to Pre-Ashokan period. It is of special note that at Hutti the old workings reached a depth of 640 ft (194.07 m), a depth unparalleled by another truly ancient workings yet discovered in any part of the world. There is clear evidence all over South India that wherever reefs, veins or stringers of quartz occurred carrying anything over 2-3 dwts of gold to the tons, they did not escape the attention of early prospectors. They were systematically explored, prospected and mined. Mostly the depth to which mining was carried out was limited by water. Judged by the standards then prevailing gold mining in India, in ancient times was a highly developed industry. Fire setting was the means for extracting the ore. Timbering of working places was understood. Wind lasses and ropes were employed. Water was the main problem that seems to have finally limited the mining operation in depth.

Lead-Zinc and silver mining has been done in Rajasthan from ancient times (Paliwal *et al.*, 1988,. Willies, 1987). Radiocarbon dating of mine timber and charcoal shows that large scale mining took place during late Chalcolithic period and early Iron Age reaching to a depth of 254 m for lead, silver and zinc. The ancient mines are both underground and opencast. Excavations have been done by 'gads'. The most noticeable feature of the mine is the development of pathways and steps. There are evidences of fire setting near the entrance in the hollows in the roof and at the bottom of workings. Possibly small picks have been used to extract the ore in narrow places. Timber posts have been used to hold up roof flakes and to stabilize the waste on the slopes. Shallow trenches have been dug to carry water inflow, and there are wooden launders for the like purposes. Several pottery lamps are found in the workings. These contain oil residues of vegetable origin. Small jars for carrying water or oil are also found. The earliest workings appear to have been by narrow shafts sunk in the ore bearing loades following apparently irregular deposit. Shafts are common in pairs, often at intervals of only a few metres and have climbing footholds.

Alluvial diamonds were known in India as early as about 800 B.C. The districts of Guntur and Krishna contain many old workings. In North India, diamond mining has been done at Panna in M.P., and continues down to date (Brunton, 1981, Vrendenburg, 1906). For mining diamond the ancients sunk wide shafts upto a depth of 50 ft (15.24 m) from the surface to reach the conglomerate which was worked out by heating with fire for easy extraction. In the alluvial deposits pits upto

30 ft (9.14 m) deep have been sunk. The workings were haphazard in nature and were limited to dry season only. The most famous workings were at Kollur in the great gorge cut by river Krishna. The mines were divided into claims and were worked according to strict rules. The miners worked to a strict religious ritual. Having decided on the place to be worked, they levelled off an area besides it and surrounded this by a 2 ft (60 cm) wall, with holes or gates in it at intervals around the base. The gates were blocked and earth and gravel from the mining area was dumped in the walled enclosure by women and children. When the mining area had been excavated usually to a depth of 14 ft (4.3 m), water seepage prevented further mining. The water was taken in pitchers and tipped into the walled area to bring the earth and gravel to a consistency of soup which it reached after one or two days. The gates were then opened to draw off the water and mud. Sometimes two or three washings were necessary until the hot sun dried off the remaining moisture and left only sand containing diamonds.

CONCLUSION

Use of minerals by man dates back to Pleistocene Period, some 10,00,000 years ago. In India, in the Pre-Vedic Period (Circa 4000 B.C. - 1000 B.C.) mining and use of metals especially gold, silver, copper and gemstones have been mentioned extensively.

Mining methods in India as well as in some foreign countries, e.g. Egypt, Spain, Greece, etc. were almost of the same type. Invariably for all hard rock mining fire setting followed by quenching the fire with water was the main general method for creating a system of cracks in the rock mass/ore body which was subsequently exploited by extracting the ore with chisels. The extracted ore was removed out of the mine by human labour. Ventilation of the working places was solely dependent on natural processes. The ancients, however, appreciated the advantages of separating the in take and return and at some mines shafts were sunk at closer intervals which were interconnected to establish ventilation circuits. Pillars were left to support the slopes and weak walls/backs were supported by timber props. The only inhibiting factor was the water and generally the mining activity ceased at the water table. Mining operation exposed the miners to serious health hazards.

In India mining of gold, copper, lead and zinc and silver and also of diamond was well developed in ancient times and the standards of mining by the standards judged then prevailing in the world were quite high indeed.

ACKNOWLEDGEMENT

The information contained in this paper is based mainly on the published literature in journals, books, etc.

REFERENCES

Banerjee, A. and Mookerjee, A., 1984. "Origin of Mining", in *Coal Mining in India* edited by S.N. Prasad and A. Mookerjee, CEMPDIL, Ranchi, pp. 3-6.

Bose, H.K., 1968. "Mining in Ancient India", *Trans. Min. Geol. Metall. Inst. of India*, Vol. 64, No. 1, pp. 83-89.

Brunton, E., 1981. "Diamonds", N.A.G. Press Ltd., London, p. 23.

Bryan, A., 1979. "Mining Engineering — Stone Age to Computer Age". *The Mining Engineer*, Vol. 123, No. 212, pp. 811-818.

Davies, D.O., 1990. "Mining — An Historial Review". *The Mining Engineer*, Vol. 149, No. 345, pp. 448-451.

Diodorus, S., 1943. Quoted from 'The History of Miner's Diseases' by G. Rosen, New York, Schuman's, p. 13.

Hinsley, F.B., 1956. "An Historical Review of Mining". *Trans. Instn. Min. Engrs.*, Vol. 115, Part 6, pp. 463-471.

Kangle, R.P., 1965. "The Kautilya Arthasastra — Part 3 — A study", University of Bombay, Bombay, p. 182.

Maclaren, J.M., 1905. "Notes on Some Auriferous Tracts in Southern India", Unpublished Report submitted to the Hyderabad (Deccan) Co. Ltd., 32 pages.

Munn, L., 1934. "Economics Dealing specially with the Ancient Gold Mining Activity of the Area with Suggestions for Further Development", *Journal of Hyderabad Geol. Survey*, Vol. 2, Part I.

Munn, L., 1936. "Observations and Notes on the method of Ancient Gold Mining in Southern India with special Reference to the Raichur and Shorapur Districts of Hyderabad State", *Trans. Min. Geol. Inst. Ind.*, Vol. XXX, pp. 103-116.

Paliwal *et. al.*, 1986. "Zinc and Brass in Ancient India", *CIM Bulletin*, Vol. 79, No. 855, pp. 75-78.

Sharma, N.L., 1970. "A Review of the Ancient and Modern Mineral Industry of India". *The New Sketch*, August 17, 1970; Reprint p. 17.

Sharma, N.L. and Sinha, R.K., 1967. "Minerals and metals used in Ancient India". *Mines and Metals*, Vol. 4, No. 4, pp. 8-15.

Simonin, L., 1969. *Underground Life: Mines and Miners.* Chapman and Hall, London, p.522.

——— 1977. *The Encyclopaedia Britanica*, 15th Edition, Helen Hemingway, Benton, London, 1977, Vol. 4, p. 773.

Willies, L., 1978. "Ancient Lead-Zinc-Silver Mining in Rajasthan, India—Interim Report". *Bulletin of the Peak District Mines Historical Society*, Vol. 10, No. 2, pp. 81-123.

Vrendenburg, E.M., 1906. "Geology of the State of Panna Principally with Reference to the Diamond Bearing Deposits". *Rec. Geol. Survey of India*, XXXIII, pp. 261-314.

The Ecology and Archaeology of the Seruwila Copper-Magnetite Deposit, North East Sri Lanka*

SUDHARSHAN SENEVIRATNE

The ecological significance of resource zones sustaining pre Industrial metallurgical operations as a critical factor in the dialectics of social formation and technological change, is seldom taken up for discussion in archaeometallurgical studies in South Asia. The ecosystem of the Formative Proto and Early Historic period in South Asia essentially functioned within an interacting process linking the resident community with the subsistence pattern, technology, resource use and settlement pattern. For the first time we are able to witness during this epoch a coincidence between the technocultural zone and the physical zone. Elsewhere we have pointed out that the momentum of technological evolution associated with such societies largely depended upon certain pre conditions or determining factors leading to a viable utilization of mineral resources, especially metallic ores. The utilization of a particular raw material in the production of luxury, prestige, utilitarian or ritual items was conditioned by the variations in the demand based on factors such as the functional value of the item, the level of material development in society, direct or indirect access to strategic resources, the possession of a suitable technology for resource extraction and production including the existence of an exchange mechanism facilitating the movement of raw material and finished products (Seneviratne 1987: 132-133; 1990).

Technological phases in early Sri Lanka indicate the absence of intervening Neolithic and Chalcolithic periods in the transition from the Mesolithic to the Early Iron Age, the latter being represented by the Megalithic Black and Red Ware culture. Investigations at Megalithic burial sites and Proto Historic habitation levels reveal the occurrence of personal ornaments, weapons and objects of domestic use

---

*An analytical study of the preliminary investigations conducted at Seruwila was presented in a paper entitled "The historical archaeology of the Seruwila copper-magnetite deposit" at the *Seminar on the archaeometallurgy of Sri Lanka* (September 1987. Institute of Fundamental Studies. Kandy. Sri Lanka). Preliminary discussions leading to this Seminar brought into focus the critical role of resource zones and as a consequence an entire session was devoted to metallic resources in Sri Lanka at the Seminar.

made of iron and copper/bronze including industrial remains associated with metallurgical operations (*vide* Deraniyagala 1972; Parker 1884; Raghupathy 1987; Seneviratne 1984; 1987). The commencement of the Proto Historic Iron Age at Anuradhapura according to recent C-14 dates is assignable to B.C. 750 ± 50 (Deraniyagala 1990: 14). Recent studies conducted by us also suggest that during the Early Historic period (B.C. 300) there was a relative increase in the level of utilizing metallic and other mineral ores. The relative improvement in the production technology, a demographic expansion, an extension in the settlement zones, a better regularization of specialized craft production and an intensification of the exchange vortex and the emergence of the early state may be listed as crucial factors contributing towards a wider use of metals and associated mineral resources during the Early Historic period (Seneviratne 1984; 1985; 1986; 1987; 1987a; 1988; 1988a; 1990).

It is therefore apparent that the Iron Age communities had a supply of metallic ore (mainly iron, copper and tin) either from their micro ecological niche or obtained through an inter-zonal exchange mechanism. On the basis of this assumption a synthesised study of the Early Iron Age settlement archaeology and the distribution pattern of mineral zones was carried out in different macro and micro regions of Sri Lanka. This revealed a remarkable correlation between archaeological and textually recorded sites with resource zones (Map 1). The copper-magnetite prospect at Seruwila therefore was a natural selection to the Early Iron Age communities residing in the northern plains over less accessible sources in the central montane region. The latter region, which is also the primary repository for mineral resources, came under intense exploitation only by the Early Historic period and after. The significance of the Seruwila deposit is to be viewed within the above context.

The geo-environmental setting

The discovery of the mineralized zone at Seruwila in modern times was made in 1971 by the Geological Survey Department of Sri Lanka. Seruwila is located south of the Koddiyar Bay housing the large natural port of Trincomalee in north east Sri Lanka. The micro area brought into focus in this study is situated between the Allai tank and the Ulakali lagoon. Its eastern limits are demarcated by the lagoon, beaches and dunes. This whole region along with the Koddiyar Bay submerged during the Holocene transgression which was followed by the filling in due to fluvial depositions. Sedimentation in this region is extremely active to this day and the primary landscape identification is marked by alluvial plains and residual terrain. It is interesting to note that the Allai tank is situated within a vast natural depression, which is about 12 miles in circumference. As late as 1857, small boats travelled a distance of six miles along the escape channel from Koddiyar Bay to reach the Allai

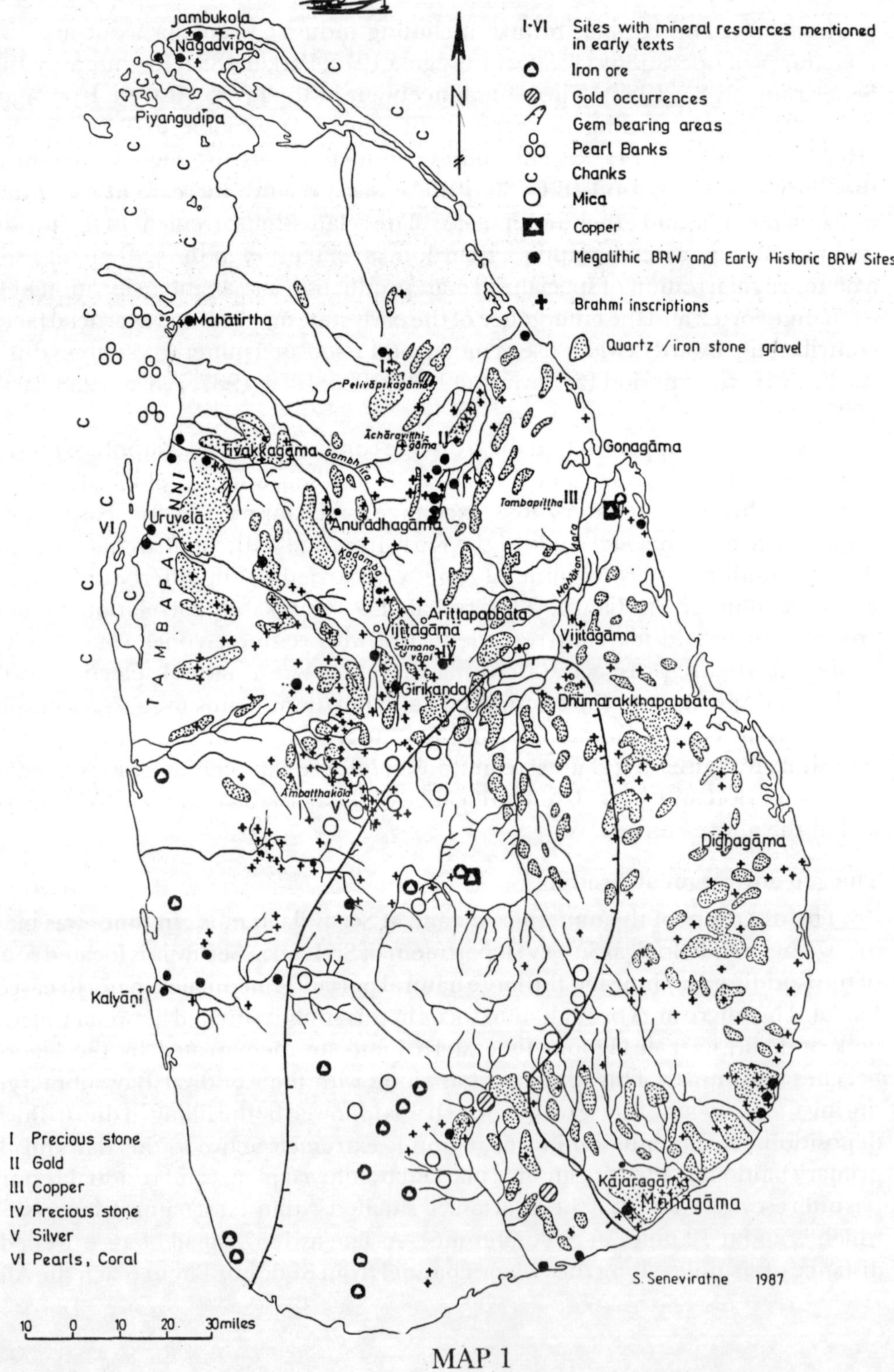
I-VI Sites with mineral resources mentioned in early texts
Iron ore
Gold occurrences
Gem bearing areas
Pearl Banks
Chanks
Mica
Copper
Megalithic BRW and Early Historic BRW Sites
Brahmi inscriptions
Quartz / iron stone gravel
N
Jambukola
Nāgadvīpa
Piyaṅgudīpa
Mahātirtha
Tivakkagāma
Uruvelā
TAMBAPAṆṆI
Anurādhagāma
Ariṭṭapabbata
Vijitagāma
Girikaṇḍa
Dhūmarakkhapabbata
Gonagāma
Tambapiṭṭha III
Dīghagāma
Kalyāṇi
Kājaragāma
Mahāgāma
Ambatthakōla
I Precious stone
II Gold
III Copper
IV Precious stone
V Silver
VI Pearls, Coral
S. Seneviratne 1987
10 0 10 20 30miles

MAP 1

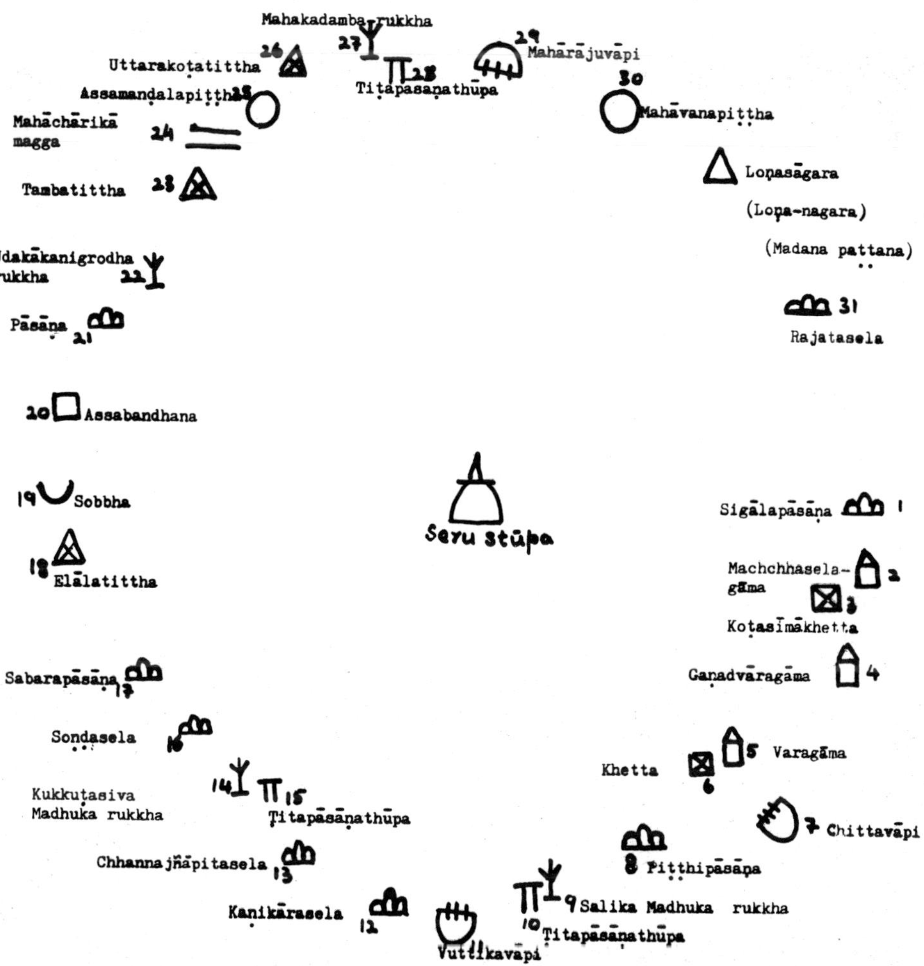

Seruwila historical topography according to the Dhatuvamsa account. Schematic presentation

Sudharshan Seneviratne 1991

PLAN I

tank. It is suggested that the change in the course of the lower Mahaweli river during the Late Historic period may have been caused by the existence of a sesmic focus in this region. This may have consequently had a direct impact on the landscape evolution and the settlement pattern along the lower valley and deltaic region.

The geological formation most obviously had a direct bearing on the landscape evolution and the historical geography of this region. Seruwila has a matrix of high grade metamorphic rocks belonging to the Pre Cambrian age represented by the Highland and the Vijayan series. The Highland series, located to the north west, predominently has charnokites and quartzites while the Vijayan series in the south east has granites, granitic gneisses and hornblende-biotite gneisses as the major rock types. The rocks in the Seruwila area indicate parallel repetition of beds thereby showing isoclinal folding (*vide* Brohier 1935: III 39; Cooray 1967; Dissanayake 1981; Fernando 1971; 1982; Jayawardena 1982).

What is significant to our study is the occurrence of the ore mineralization at the boundary of the Highlands and the Vijayans. The highly crushed and altered nature of the host rocks forming this boundary indicates that the mineralization is at a tectonic trust plane which forms an extensive mineralized belt, approximately 10 miles wide and 250 miles long, extending from Trincomalee to Ambalantota in the south west. The ore mineralization at Seruwila may have taken place during the Upper Cretaceous period. The mineralized zone in the micro area coming under our study is rich in magnetite ore and sulphides (mainly copper) including other minerals such as nickel (pentlandtite rich in cobalt), silver, bismuth, zinc and mica. Future discovery of gold is not ruled out as mica and gold occurrences have been reported from other localities having charnockitic gneisses in association with quartz formation in the Highland series.

Ore bodies containing magnetite and copper sulphides are lenticular in shape and concordant with the dip and strike of the host-rocks. The magnetite-sulphide ore bodies vary in thickness from 3 to 30 ft. and are coarse-grained with magnetite, chalcopyrite, pyrrhotite and pyrite. It is estimated that the massive sulphide-magnetite ores are very rich in iron with values up to 99.5%. Investigations by the Geological Survey Department also indicated the existence of nearly 7 million tonnes of magnetite ore extending nearly 200 ft. below the surface level. The massive ore type also carry a higher content of copper over the disseminated ores. The copper ore deposit at Seruwila is 1:3 and the amount of metallic copper in the Arippu area alone is quantified to about 68,000 tonnes. It has also been observed that the negative anomalies show a high concentration of ore.

The general topography of Seruwila is relatively flat though the ultra-basic rock, which is also the host-rock for mineralization, is highly weathered on the surface and survive as outcrops in the micro landscape. The magnetite outcrops at Seruwila vary in height from 3 to 45 ft. and some of these measure 3 to 15 ft. in length. Secondary copper minerals such as malachite and azurite are exposed on the highly weathered surface of the rock (for details see, Jayawardena 1982; Herath 1975).

Historical geography and topography

Archaeological data and corroborative evidence derived from epigraphical (early Brahmi inscriptions) and textual sources provide us with information relating to the historical geography and topography of ancient Seruwila. In this connection, Pali texts of the Middle Historic period, namely the *Dhātuvaṃsa* and to a lesser extent the *Mahāvaṃsa* (*MV*), carry useful references on the historical topography and mineral resources found in this region.

The historical antiquity of this region extends well into the Early Iron Age. The dolmenoid cist burial site at Kadiraveli, which is only a few miles southeast of Seruwila, is good evidence for the existence of Proto Historic settlements in this region. Surface investigations at abandoned pits yielded scattered Black and Red Ware embedded to the soil. Further to this, in its description about the *sīmā* of the *stūpa* at Seruwila, the *Dhātuvaṃsa* records at least three sites known as *ṭhitapāsāṇa-thūpa* located in the vicinity of the *stūpa* (*DV* 68), which appears to be a reference to funerary monuments of upright stone or dolmenoid cists (Seneviratne 1990 : 147). An echo of this early community movement is found in the *Mahāvaṃsa* (VIII. 12-13; 24-25) which records the existence of early settlements at Gokanna, a site located somewhere along the Koddiyar Bay. Early Brahmi inscriptions at Seruwila clearly point to the existence of a settled agricultural community and several other inscriptions from the region confirm the patronage extended by merchants and chieftains to the Buddhist clergy who depended on the agricultural surplus produced in this area (for inscriptions see Paranavitana 1970: Nos. 382-387; *Epigraphical Notes* 1974: XIII Nos. 5-6). The *Dhātuvaṃsa* description accounts for *nagara*, several *gāma* (village) settlements, *khetta* (fields), *vāpi* (reservoirs) located in the vicinity of the *stūpa* during the reign of Kāvantissa (*DV* 68-69), datable to around mid 1st century B.C.

The micro topography at Seruwila indicates a gradual inclination from the south-east to the north-west, where the mineralized zone with its rocky outcrops marks the landscape in that sector. The location of the eastern sector as the primary habitation nucleus during the Early Iron Age was due to the natural selection based on the elevation of the micro region, the drainage pattern and its proximity to the easily exploitable ore resources. Archaeological investigation revealed an extensive habitation mound nearly 600 ft. to the east of the *stūpa* (Solheim and Deraniyagala 1972: 23-26). This evidence is compatible with the textual description giving the location of habitations, reservoirs and cultivated fields in the same direction. It is

MAP SHOWING THE REGIONAL GEOLOGY AROUND SERUWILA AREA

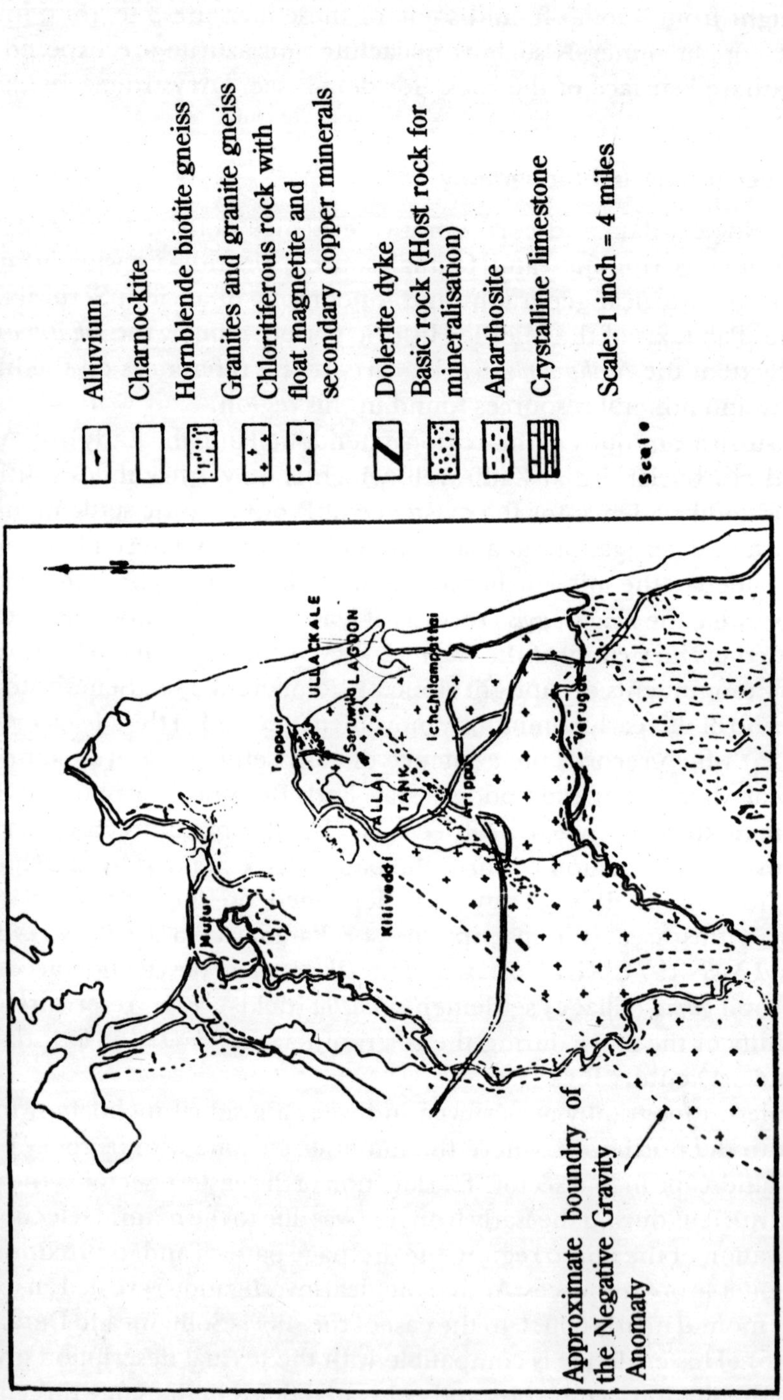

Jayawardena 1982: 141

MAP 2

interesting to note that the *nagara* of Seru is described as a city located in the midst of rocky outcrops or *pabbata* (*DV* 68-69). Similarly, the incorporation of the term *sela* (from *śaila* or stone) within the name of residential villages e.g. Machcha-sela-gāma (*DV* 18, 68), points to a topographic description associated with the settlement.

It is apparent that during the Early Iron Age the Mahaweli river flowed along the relatively flat land located to the west of the resource zone. Aerial reconnaissance clearly show the dry bed of the Mahaweli located in close proximity to the *stūpa*, while the present river flows nearly 10 miles to the south of the *stūpa*. The landscape and human ecology associated with this reverine deltaic region is well documented in the texts. For instance the *Dhātuvaṃsa* (68) description mentions a natural depression or *sobbha* to the west of *stūpa*. The present Allai tank is in fact located in a vast depression associated with the ancient drainage system. Further to this there are at least three *tittha* (from *tīrtha*) or ford/river crossing to the west and north of the *stūpa*. One ford is very specifically mentioned as the Uttarakoṭatittha (*uttara*/north). These most obviously refer to the river crossings of the ancient Mahaweli river. In fact immediately after the crossing named Tambatittha, the text mentions a prominent highway known as Mahāchārika-magga, which was quite possibly the major route connecting this region with the major settlement zone at Anuradhapura to its west.

The topographic description in the *Dhātuvaṃsa*, commences from the east and follows a full circle before arriving at the same point. The final landmark in this description is the Loṇasāgara (lit. salt sea) in the east, which is a reference to the present Ulakali lagoon, the nearest salt water body. The text also refers to a city by the name Loṇanagara and a port or a *paṭṭana* named Madana, where a *stūpa* was constructed at the entrance to the port (*paṭṭanamukhadvāra*) (*DV* 48). The present archaeological site of Ilankaturai is located at the estuary of the Ulakali lagoon and this site has yielded the remains of an ancient *stūpa*, Early Historic Black and Red Ware and early Brahmi inscriptions.

The Archaeometallurgy of Seruwila

Proto and Early Historic vestiges at Seruwila quite clearly indicate that this resource zone was known and inhabited from a very early period in history. It is therefore important to understand the very specific functional value of this particular region to the Early Iron Age community.

First, the very location of the Seruwila resource zone in proximity to the primary habitation zone in the northern plains is of great significance. Seruwila was within easy reach and easily accessible by land, river and sea routes. In contrast the primary mineral deposits in the central montane region had to be reached, recognized, extracted and transported through difficult terrain. In addition, as mineral stones

THE LOWER MAHAWELI SYSTEM & ANCIENT STUPA SITES

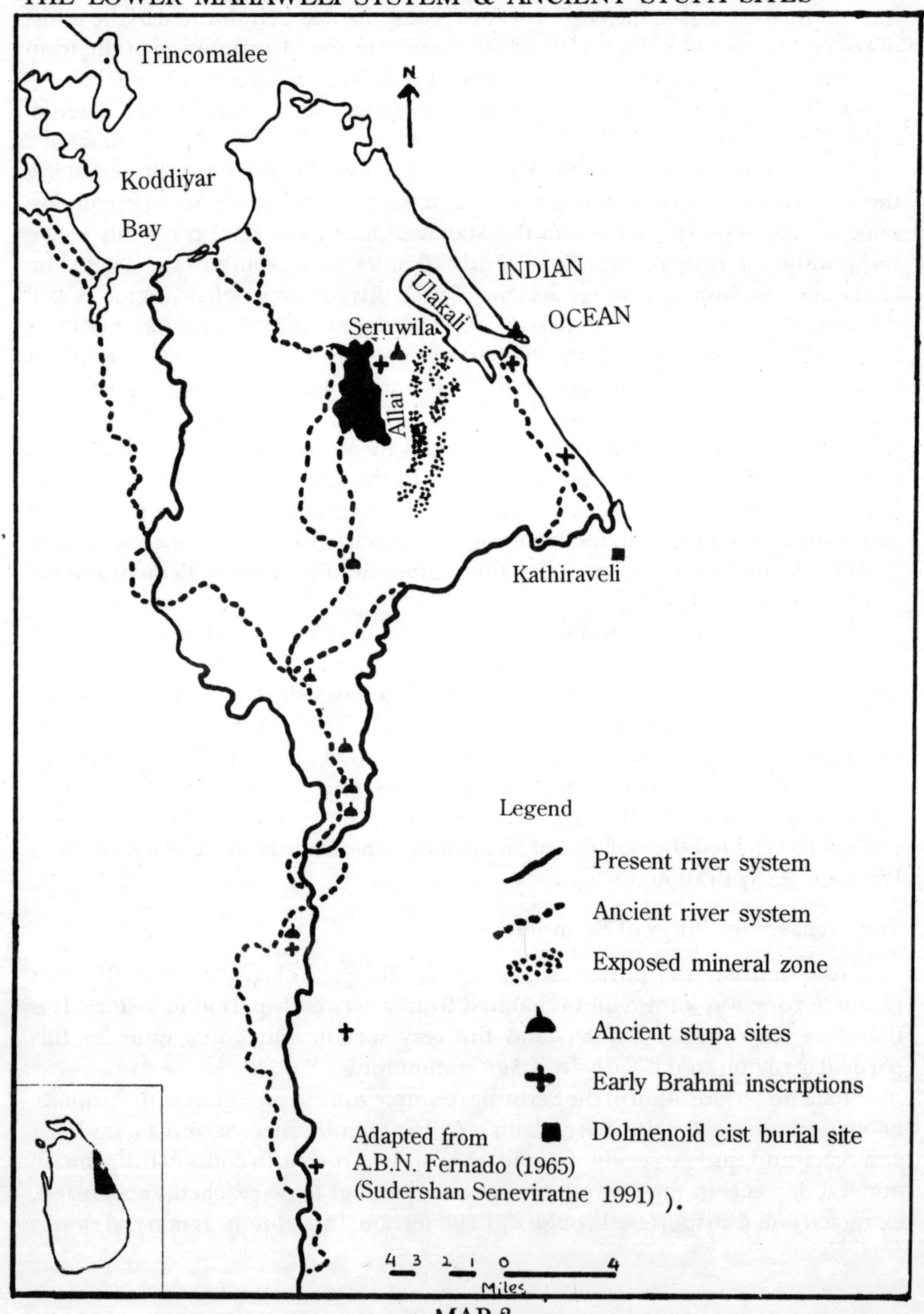

MAP 3

and metallic ores are 'weight-gaining' objects in relation to the distance factor, it was not practical to exploit distantly located sources during the Formative Period.

Second, there is the question about the resource zone itself. The primary ore formation at Seruwila is represented by magnetite and sulphides, mainly copper. Though the native smelters of Tamilnadu worked the magnetic sand that is found in the beds of the surface streams in Coimbatore, the magnetic sand washed down the Tambraparni river or the magnetic ore obtained from the disintegrated mineralized rock in the Selam district, there is little evidence to substantiate the fact that the Early Iron Age folk in Sri Lanka were technologically equipped to work the highly concentrated magnetite formation (99.5%) by reaching a temperature level of 1800° C. to melt the ore. In addition, the existence of sulphide ores such as pyrrhotite and pyrite/marcasite in the Seruwila prospect (Jayawardena 1982: 131-132; Yapa 1982: 25), creats a problem to the smelter as the metal becomes brittle due to the sulpher content and it consequently entails a difficulty in forging the metal (Hodges 1965: 81, 85). Excavations at Anuradhapura yielded iron stone nodules of haematite and limonite in direct association with iron slag within the Proto Historic levels (Deraniyagala 1972: 152, 155). Our studies on the settlement archaeology and resource zones point to a high coincidence between areas with nodular iron stone and Early Iron Age sites (Seneviratne 1985: 137-141). Haematite and limonite provided the Early Iron Age smelter with certain advantages over magnetite. These were less compact ores, silicieous by nature and relatively free from sulphur and phosphorus (Hodges 1965: 81). It is also known that these communities applied the method of roasting the ore to reduce the water content of the limonite ore (see Hegde 1973: 403; Coomaraswamy 1908/1956: 190). Archaeometallurgical studies from the broad region indicates that the Proto Historic smelters in Peninsular India reduced at a temperature around 1200° C and that steel can be obtained by smelting ore containing 50% metal (Gogte 1983: 74; Kularatnam 1979: 222), thus giving limonite and haematite a high utility value over magnetite. The utility value of the Seruwila prospect therefore was less associated as a source of iron ore in antiquity, thereby making the natural selection as a source of copper a very obvious one.

Proto Historic burials and habitation levels in Sri Lanka indicate a somewhat high occurrence of copper-bronze items representing a range of personal ornaments and objects of domestic use.[1] The discovery of copper slag and copper inclusions from crucibles unearthed from the Proto Historic levels at Anuradhapura (Deraniyagala 1972: 145) clearly point to the working of copper during the Early Iron Age. Therefore contrary to the view held so far, raw material in the form of copper and tin was available to the Early Iron Age smelter from sources located within the island. Archaeometallurgical, textual and epigraphical studies indicate that, until the central montane region was effectively penetrated for resources, the

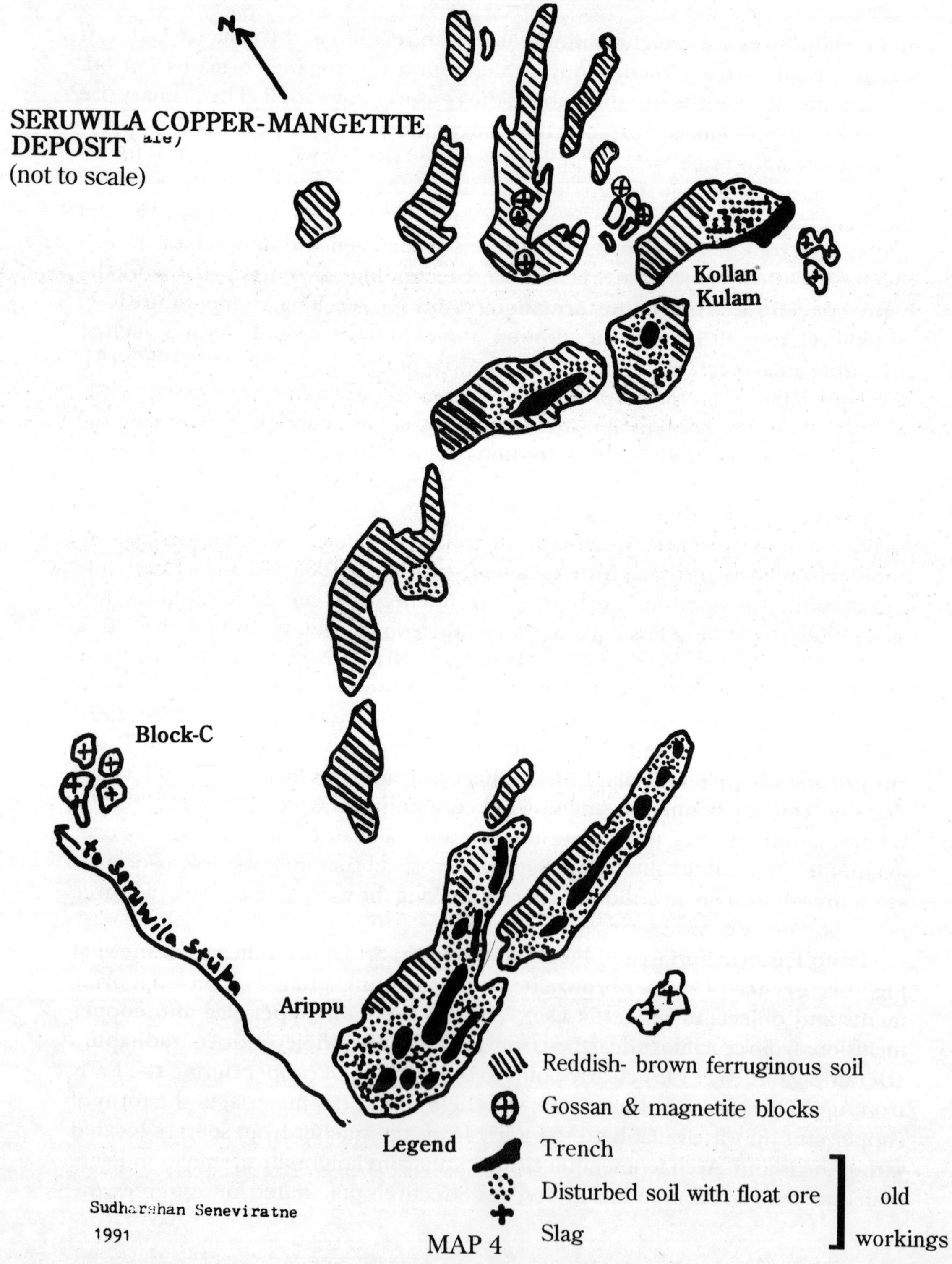
N
SERUWILA COPPER-MANGETITE DEPOSIT
(not to scale)
Kollan Kulam
Block-C
to Seruwila Stūpa
Arippu
Reddish- brown ferruginous soil
Gossan & magnetite blocks
Trench
Disturbed soil with float ore
Slag
old workings
Legend
Sudharshan Seneviratne
1991

MAP 4

prospect at Seruwila was the primary source of copper and related minerals used in the industry by the Early Iron Age coppersmith known as *tabakara* in the Brahmi inscriptions and *tambakāra* in the Pali texts.

Recent archaeometallurgical studies indicated that the iron implements obtained from the Proto Historic context and also from the Early Historic levels carry high iron, copper, chromium and nickel contents and that the copper-nickel combination was a consistant one (Maliyasena and Seneviratne 1987). Further to this, the excavations at Anuradhapura in 1969 yielded copper slag having iron inclusions (which were probably secondary compounds of the ore) and the 1985 excavations yielded ore containing zinc from the Proto Historic levels. Interestingly enough, trace element studies from the Seruwila deposit indicated among other elements, the existence of cobalt, nickel, zinc, chromium and manganese. No other mineral deposit in Sri Lanka represents such a complex mineral formation. Spectrographic analysis of copper objects obtained from the Proto Historic levels at Anuradhapura revealed a high iron content (.84%) in the metal (Maliyasena 1986), which is a feature associated with the Seruwila ore.

In addition to its rich mineral content, convenient location and easy access, the Seruwila deposit may have drawn the attention of Early Iron Age folk due to two other functional reasons. These may be listed as the natural setting of the ore formation and technological factors associated with production techniques.

The magnetite at Seruwila is mainly associated with iron sulphide minerals such as chalcopyrites and to a lesser extent bornite. Though sulphide ores such as bornite contain copper up to 55.5% and chalcopyrites up to 34.6%, these ores were probably not extracted during the Early Iron Age due to problems related to extraction and the working of sulphide ores. In addition, sulphide ores exist at relatively deep levels at Seruwila. On the contrary, the Early Iron Age smelters were more attracted to the secondary copper minerals such as malachite and azurite found on the highly weathered surface of the rocks.

The existence of oxide ores at Seruwila is an interesting feature on several counts. First the carbonates of this group, such as malachite and azurite have colours (blues and greens) that may have been useful to the early smelters to easily identify and locate their raw material (see Hodges 1965: 65). The very formation of the Seruwila deposit in the form of small outcrops provided an additional advantage to extract the ore. In this connection textual evidence may substantiate the above point. In the *Mahāvaṃsa* (XXVIII. 16-17) narration, it is described that during the reign of Duṭṭhagāmaṇī (150 B.C.) copper emerged in an area called Tambapiṭṭha, 7 *yojana* east of Anuradhapura, which is a notice about Seruwila. The term *tamba* in Pali means copper and *piṭṭha/piṭṭhena* means "what stands out, along, over, beside, by way of" (Rhys Davids and Stede 1959:457). The *Dhātuvaṃsa* (68-69) in its topographical description, quite specifically mentions the rock or mineral

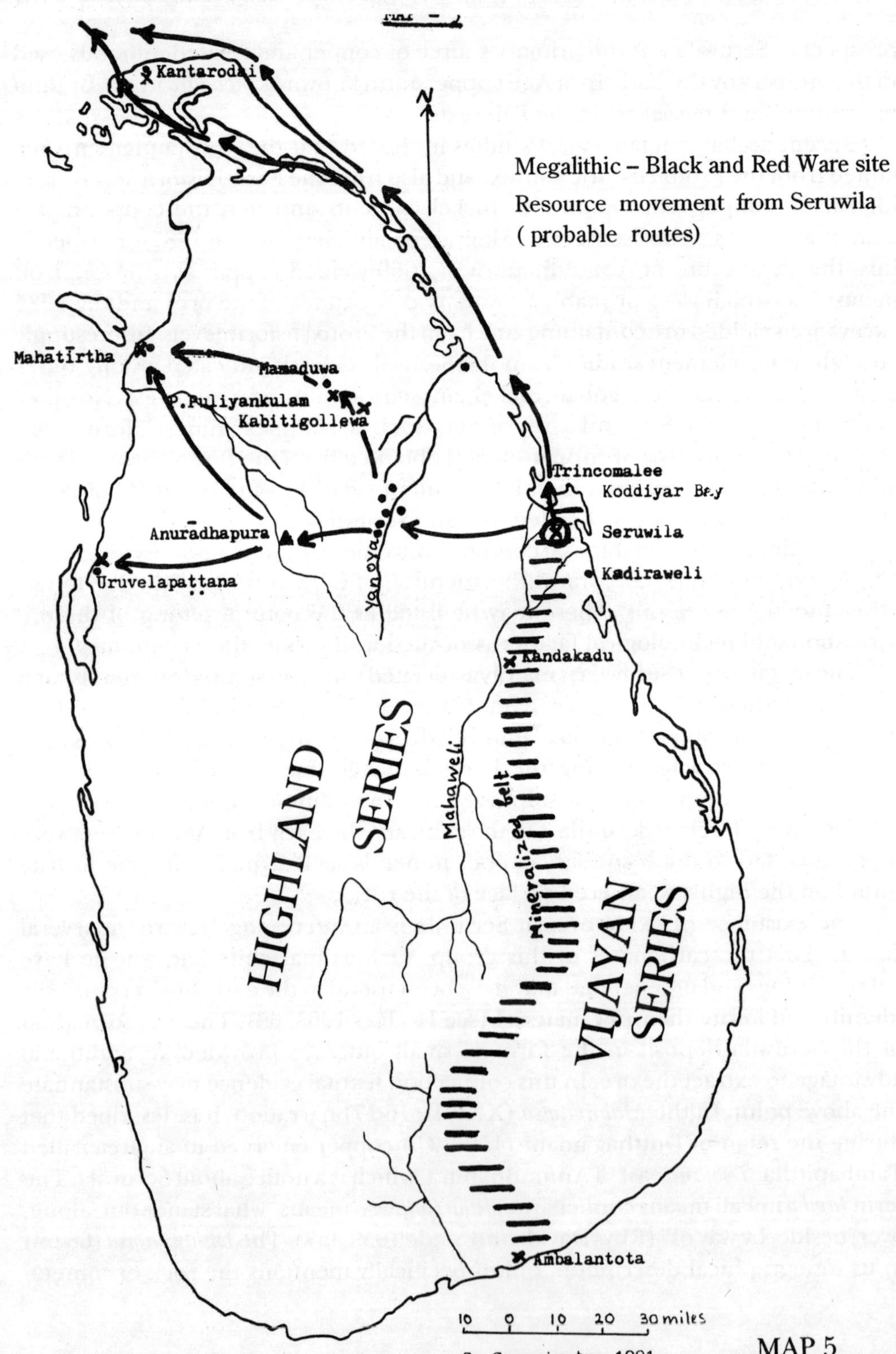
Megalithic – Black and Red Ware site
Resource movement from Seruwila
(probable routes)
N
Kantarodai
Mahātīrtha
Mamaduwa
P.Puliyankulam
Kabitigollewa
Trincomalee
Koddiyar Bay
Seruwila
Kadiraweli
Anurādhapura
Uruvelapattana
Yan oya
Kandakadu
HIGHLAND SERIES
Mahaweli
Mineralized belt
VIJAYAN SERIES
Ambalantota
10 0 10 20 30 miles
S. Seneviratne 1991

MAP 5

outcrops i.e. *piṭṭhipāsāṇa* situated near the fields of the village Varagāma. The same text also speaks of a series of other rock and mineral formations such as Kaṇikārasela, Chhannajñāpitasela, Soṇḍasela and Sabarapāsāṇa before reaching the *sobbha* or the natural depression. Of these, Chhannajñāpitasela may carry the meaning the rock or mineral with a hidden light or glitter. Interestingly enough Sabarapāsāṇa may derive from Śvabhra-pāṣāṇa, where *śvabhra* denotes mica. The occurrence of mica in the mineralized zone has been indicated in a previous section. The extraction of such surface ores may have been done by splitting the surface of the rock and by working narrow tunnels by the gad and hammer.

Surface investigations at Seruwila revealed the existence of abandoned pits with scattered magnetite and secondary copper minerals, indicating a wider use of oxide ores, which were conveniently extracted from the upper and the exposed levels of the deposit. The scattered remains of numerous magnetite minerals may also reveal another feature related to the form of extraction. It is possible that the highly concentrated magnetite may have been 'chipped-off' en bloc, thus separating it from the copper ore. For the Proto Historic smelter who was not technologically equipped to work the magnetite ore, the chipping-off process may have served as a convenient extracting method. In all probability, the copper nuggets or *tambalōhabīja* presented to king Duṭṭhagāmaṇī by the residents of Tambapiṭṭha (*MV* XXVIII.17), may well have been such blocks of oxide ore extracted through the chipping-off process. In addition to this, the smelters and craftsmen utilizing a limited technology found oxide ores to hold additional advantages in the actual operations. Unlike sulphide ores, the oxide ores are considered to be the most simple ores to work as well as profitable due to the copper content of malachite and azurite is as high as 57.3% and 55.1% respectively.

There are also other salient features associated with the deposit at Seruwila and the copper-bronze industry in antiquity. Several field studies report the remains of large heaps of slag, especially in the Seruwila-Arippu area (Solheim and Deraniyagala 1972: 4, 19; Jayawardena 1982: 129; Yapa 1983: 169). It is in fact reported that the slag remains were in such large heaps that it obstructed the construction of the Right Bank Canal of the Allai Scheme in 1957 (Yapa 1983: 169). This clearly points to the fact that at some time in history industrial activity was conducted in the vicinity of the deposit. Surface investigations in the Arippu area, where metallic copper is highly concentrated, yielded beaten copper. The metallic copper found in this locality represented the type of native copper that has been described as a raw material requiring ".... neither smelting or casting and could be worked into small articles by simply hammering ...." (Hodges 1965: 65). This suited a community that operated within a limited technology.

The existence of some amount of iron pyrites in the slag heaps at Seruwila (or for that matter at any Pre Industrial foundry hearth) does not necessarily point to iron smelting. This may have been a result of copper extraction as iron and copper pyrites are known to occur together (Hogdes 1965: 81; also see Buchnan 1982). At Seruwila too magnetite is mainly associated with sulphide minerals. Spectrographically analysed copper objects from the Proto Historic levels at Anuradhapura indicating a relatively high iron content (.84%) may point to another technological feature associated with the functional use of chalcopyrite-magnetite ore at Seruwila. It is known that the presence of iron streaks tends to give additional strength to copper implements. The deliberate extraction of chalcopyrite ore by the Chalcolithic coppersmiths of Ahar in India may have been associated with this technological advantage (Hegde 1969).

It is logical that smelting activity had to be carried out at the source in view of the difficulty involved in transporting large quantities of copper ore from Seruwila. The most convenient semi product in this context is copper ingots (*tambalōhabīja* in the *Mahāvaṃsa*). Large quantities of copper ingots have been reported from the Kantarodai habitation deposit, a settlement that existed from the Proto Historic period. Similarly, the development of a major habitation nucleus in the fertile Yan Oya valley points to its location as a production-distribution zone in relation to two primary resource zones (Seruwila and Kabitigollewa) and the primary habitation zone in the north central province (Seneviratne 1987b).

Copper in the form of ore or ingots may have moved out of the source area through land and water courses. The location of Tambatitta to the west of the *stūpa* points to the existence of a port transporting copper (*tittha* also means a port or place of landng). Small boats carrying ore or ingots may have moved along the old river course of the Mahaweli and traversed the Koddiyar Bay to reach the Bay of Bengal. It is reported that as late as 1857 ".... boats drawing a foot of water were able to come from Koddiyar Bay up the escape channel to within a few hundred yards of the (Allai) tank, a distance of 6 miles" (Brohier 1935: III. 39). Twenty miles south west of Seruwila, at Kandakadu, an early Brahmi inscription records an endowment made by a *parumaka* (clan chieftain) named Tabara Veḷ (a) (Paranavitana 1970: No. 319). *Tabara* appears to be a prefix to the personal name Veḷ (a). *Tabara* derives from *tambakāra* (coppersmith) and it may indicate the involvement of this chieftain in the production of copper, as a master craftsman, a leader of coppersmiths, his authority over an area yielding copper or control over a route connected with the movement of copper. Several early Brahmi inscriptions at Periyapuliyankulam (located near the Mamaduwa burial complex), record endowments made by *tabakara* (coppersmiths) and a *topaśa* or tinsmith (*ibid.* No. 349, 350, 370). The significance of this source at Seruwila as the primary repository supplying the northern region was so well recognized that the *Mahāvaṃsa* author did not hesitate to credit the reign of king Duṭṭhagāmaṇī with the 'discovery' of this metallic source.

The existence of bronze implements within the Proto Historic context in Sri

Lanka points to resource requirement beyond copper. The existence of tinsmiths during the Early Historic period at distantly located sites from the source of copper indicates the importance of that industry. Cassiterite is known to occur within the mineralized belt though some amount of tin may have been obtained from sources such as Kadavur and Ururakkarad in central Tamilnadu.[2] It is not difficult to work tin as it tends to melt at 232° C and additional contraptions were not required to work this metal. As late as 1884 cassiterite-granules were smelted in iron furnaces in Central India (Chakraborti 1979: 62).

Another aspect related to the utilization of mineral resources supplementing the technology of the craftsman was his ability to use antimony and arsenic, in order to give a hardening effect to antimonial and arsenical bronze implements. Several strains of sulphur minerals are found in the mineralized zone at Seruwila and it is quite likely that much of the antimony and arsenic requirements may have been obtained from this source. Terms such as *haritāla* (orpiment or arsenic trisulphide), *manosilā* (minium or red lead) are found in the *Dhātuvaṃsa* and the *Mahāvaṃsa* descriptions.

In this connection, it may be noted that the ability to work copper and use alloys did not require great technological advancements to the metallurgists already familiar with the iron technology and the process of making steel. For instance, if the raw material consisted of native copper small objects could be hammered into shape. The ability to obtain oxide ores with relative ease was an additional advantage. This is with special reference to the time and labour factors involved otherwise if the smelter had to work sulphide ores which carried unwanted matter such as gangue (silica) in the ore. In working the ore too, the early metallurgist had met with little trouble in smelting or melting copper ore as the minimum temperature required for that purpose is 400° C. (Bharadvaj 1973: 391-392), while the reduction of carbonate/oxide ores can be performed at 1100°C (Hodges 1965: 66). In certain cases it can be done even at 700°-800° C and pure copper in fact melts at 1083° C (*ibid.* 97). In addition, though the metallurgy of iron has little in common with copper and bronze technology, a close proximity between these two industries in terms of their physical location was an advantage. The utilization of iron slag as flux in the native method of Indian copper smelting is a case in point. Excavations at Early Iron Age habitation sites revealed that copper and iron workings were situated in the same locality in the Sri Lankan context.

Finally, the significance of the Seruwila copper deposit may be located within a regional context. The near absence of any significant copper sources in South India may be noted here. Though copper is found in the southern Deccan, especially in southern Karnataka and south westren Andhra, it is quite likely that the Seruwila copper deposit may have exported copper to South India. The high occurrence of copper at Adichchanallur and the occurrence of Megalithic sites in

Seruwila historical topography (suggested Locational pattern based on the Dhatuvamsa)

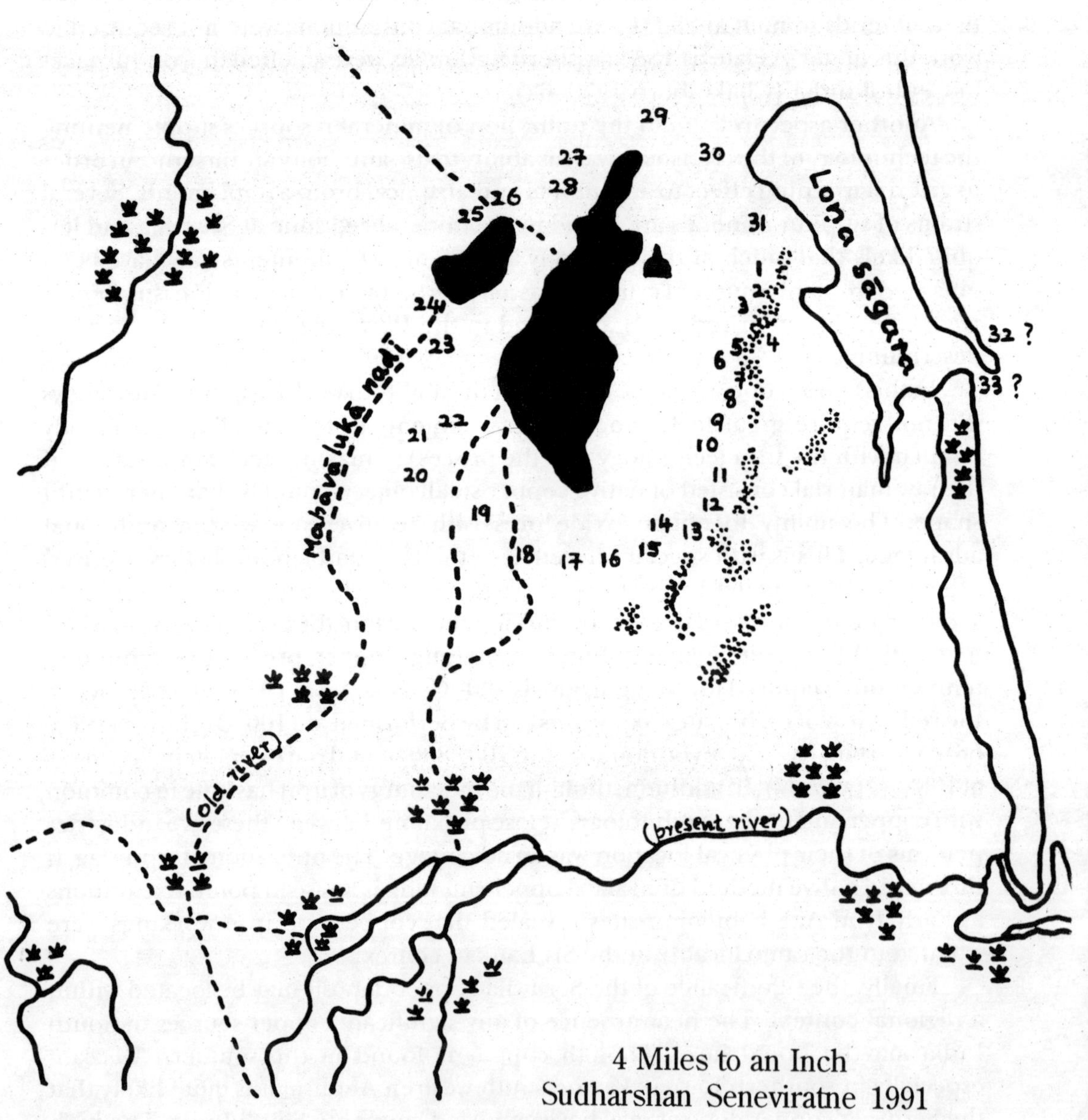

MAP 6

eastern Sri Lanka may have been a consequence of resource extraction at Seruwila during the Proto Historic period (for Early Historic South Indian traders see Seneviratne 1985).

"The Director of mines, being conversant with the science of (metal) veins in the earth and metallurgy, the art of smelting and the art of colouring gems, or having the assistance of experts in these, and fully equipped with workmen skilled in the work and with implements, should inspect an old mine marks of dross, crucibles, coal and ashes, or a new mine, where there are ores in the earth, in rocks or in liquid form, with excessive colour and heaviness and with a strong smell and taste".

"The treasury has its source in the mines; from the treasury the army comes into being. With the treasury and the army the earth is obtained" (*Kauṭilya Arthaśāstra* II. 12.1, 37. Translation Kangle 1972).

TABLE 1
Anuradhapura Citadel (Gedige) 1969 iron analysis

Sample No.	*Layer*	*Weight*	*Fe%*	*Mn%*	*Ni%*	*Cu%*	*Cr%*	*Cu/Ni%*
1	3A	0.0135	50.75	Nil	0.0247	0.0327	Nil	1.32
2	3A	0.0014	40.86	Nil	0.2381	0.42	Nil	1.76
3	3A	0.0135	50.75	Nil	0.0563	0.1589	Nil	2.82
10	3B	0.0113	47.09	Nil	0.0590	0.1431	Nil	2.43
4	4A	0.0049	40.72	Nil	0.0255	0.2101	Nil	8.24
5	4A	0.0346	53.83	Nil	0.5118	0.0425	Nil	0.08
7	4A	0.0017	86.07	Trace	0.4902	3.2871	0.4425	6.71
8	4A	0.0191	53.63	Nil	0.1309	0.0539	Nil	0.41

Maliyasena and Seneviratne 1987.

REFERENCES

Buchnan, Hans-Gert, 1982. *The identification of slags from archaeological sites.* London Institute of Archaeology (Occasional Pub. No. 6)

Bharadwaj, H.C., 1973. 'Aspects of early iron-technology in India'. *Radio-carbon and Indian Archaeology.* Bombay TIFR. Ed. D.P. Agrawal and A. Ghosh. 391-399.

Coomaraswamy, Ananda, 1908. *Mediaeval Sinhalese Art.* New York. Pantheon (1956 reprint).

Brohier, R.L., 1935. *Ancient Irrigation works in Ceylon.* Pt. III. Colombo. Ceylon Government Press.

Cooray, P.G., 1967. *An introduction to the geology of Ceylon.* (Spolia Zeylanica Vol. 31, pt. 1). Colombo. National Museum.

Deraniyagala, S.U., 1972. 'The Citadel of Anuradhapura 1969: Excavations in the Gedige area'. *Ancient Ceylon,* No. 2 : 48-169.

———, 1990. 'The Pre Historic chronology of Sri Lanka: Appendix - I, List of radio-carbon dates from Proto and Early Historic periods'. *Perspectives in Archaeology: Leelananda Prematilleke Fetschrift 1990.* Ed. Sudharshan Seneviratne, Moira Tempoe et. al. Peradeniya. University of Peradeniya. 1-24.

Dhātuvaṃsa Ed. Kamburupitiye Nandaratana. Colombo. The Cultural Publications Ltd. (Published in Sinhala). 1984.

Dissanayake, C.B. and S.W. Navaratne, 1981. 'The occurrence of gold in the gem beds of Sri Lanka', *Economic Geology* Vol. 76. No. 3: 733-738.

Epigraphical Notes. Archaeological Survey. Colombo.

Fernando, A. Dennis N., 1971. 'Aerial survey and archaeology with special reference to the Old course of the Mahaweli'. *Ancient Ceylon.,* No. I.

———, 1982. *The ancient hydraulic civilization of Sri Lanka in relation to its natural resources.* Special Number of the Sri Lanka Branch of the Royal Asciatic. New Series Vol. XXVII.

Gogte, V.D., 1983. 'Iron and copper in Mahurjhari megaliths: chemical metalographic analysis'. *Bulletin. Deccan College Research Institute,* Vol. 42: 74-77.

Hegde, K.T.M., 1969. 'Technical studies in copper artefacts excavated from Ahar'. *Excavation at Ahar.* Ed. H.D. Sankalia et al. Poona. Deccan College. 225-228.

———, 1973. 'Early stages of metallurgy in India'. *Radio-carbon and Indian Archaeology.* Bombay. TIFR. Ed. D.P. Agrawal and A. Ghosh. 401-405.

Herath, J.W., 1975. *Mineral resources of Sri Lanka.* Colombo. Geological Survey Department.

Hodges, Henry, 1965. *Artefacts.* London. John Baker.

Jayawardena, D.E. de S., 1982. 'The geology and techtonic setting of the copper-iron ore prospect at Seruwila. North East Sri Lanka'. *Journal of the National Science Council of Sri Lanka,* 10 (i): 129-142.

———, 1986. 'Records of ancient steel making and copper refining in Sri Lanka'. Paper presented at the *1st National Archaeological Congress.* Colombo. SLFI. (November).

Kularatnam, K., 1979. 'Ceylon building stones'. *Ancient Ceylon,* No. 3 : 211-223.

Mahāvaṃsa, Ed. W. Geiger. London. Pali Text Society. 1958.

Maliyasena, H.B., 1986. 'Metallurgical studies of Sri Lankan artefacts'. Paper presented at the *1st National Archaeological Congress.* Colombo. SLFI. (November).

Maliyasena, H.B. and Sudharshan Seneviratne, 1987. 'Metallurgical and archaeological studies of some iron objects from the 1969 Anuradhapura Citadel (Gedige Area)' Paper presented at the *Seminar on the archaeometallurgy of Sri Lanka.* Kandy. IFS (September).

Paranavitana, S., 1970. *Inscriptions of Ceylon.* Vol. I. Colombo. Archaeological Survey of Ceylon.

Parker, Henry, 1884. *Ancient Ceylon.* London. Luzac & Co.

Raghupathy, Ponnampalam, 1987. *Early Settlements in Jaffna: An archaeological Survey.* Madras. T. Raghupathy.

Rea, A., 1902-03. 'Pre-historic antiquities in Tinnevelly'. *Annual Report of the Archaeological Survey of India.* 111-140.

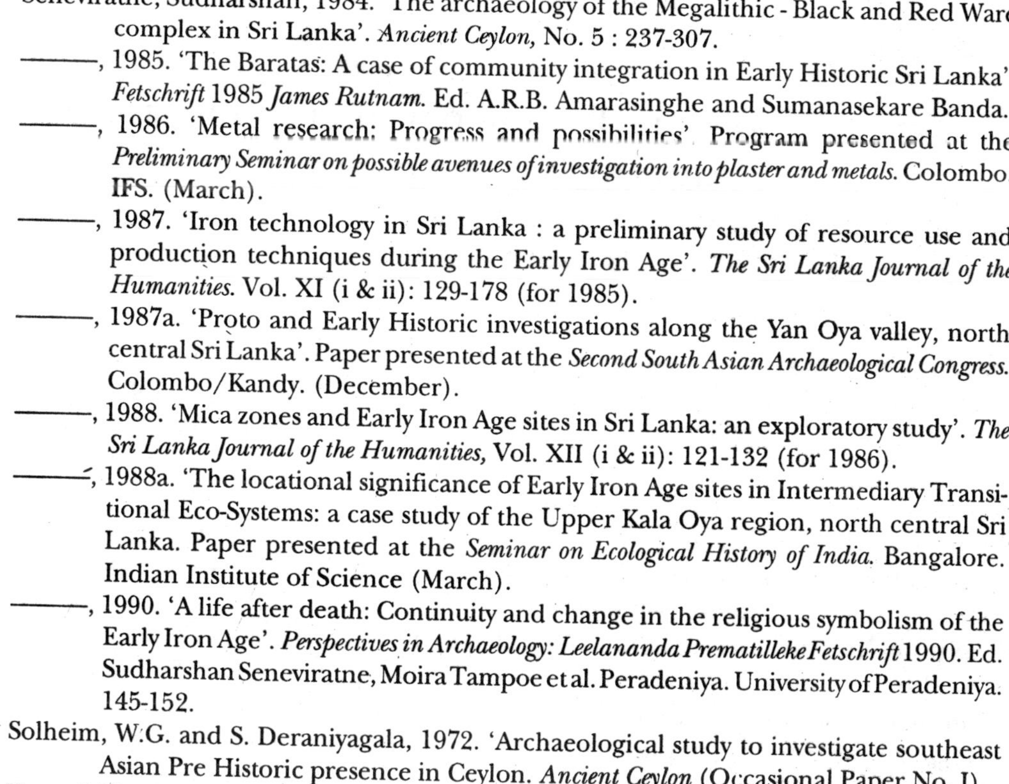

Seneviratne, Sudharshan, 1984. 'The archaeology of the Megalithic - Black and Red Ware complex in Sri Lanka'. *Ancient Ceylon,* No. 5 : 237-307.

———, 1985. 'The Baratas: A case of community integration in Early Historic Sri Lanka'. *Fetschrift* 1985 *James Rutnam.* Ed. A.R.B. Amarasinghe and Sumanasekare Banda.

———, 1986. 'Metal research: Progress and possibilities'. Program presented at the *Preliminary Seminar on possible avenues of investigation into plaster and metals.* Colombo. IFS. (March).

———, 1987. 'Iron technology in Sri Lanka : a preliminary study of resource use and production techniques during the Early Iron Age'. *The Sri Lanka Journal of the Humanities.* Vol. XI (i & ii): 129-178 (for 1985).

———, 1987a. 'Proto and Early Historic investigations along the Yan Oya valley, north central Sri Lanka'. Paper presented at the *Second South Asian Archaeological Congress.* Colombo/Kandy. (December).

———, 1988. 'Mica zones and Early Iron Age sites in Sri Lanka: an exploratory study'. *The Sri Lanka Journal of the Humanities,* Vol. XII (i & ii): 121-132 (for 1986).

———, 1988a. 'The locational significance of Early Iron Age sites in Intermediary Transitional Eco-Systems: a case study of the Upper Kala Oya region, north central Sri Lanka. Paper presented at the *Seminar on Ecological History of India.* Bangalore. Indian Institute of Science (March).

———, 1990. 'A life after death: Continuity and change in the religious symbolism of the Early Iron Age'. *Perspectives in Archaeology: Leelananda Prematilleke Fetschrift* 1990. Ed. Sudharshan Seneviratne, Moira Tampoe et al. Peradeniya. University of Peradeniya. 145-152.

Solheim, W.G. and S. Deraniyagala, 1972. 'Archaeological study to investigate southeast Asian Pre Historic presence in Ceylon. *Ancient Ceylon* (Occasional Paper No. I).

Yapa, S.W., 1982. 'Froth floatation of Seruwila chalcopyrite'. *Transactions. Institute of Engineers, Sri Lanka.* 21-28.

FOOT NOTES

1. The relatively high quantity of copper-bronze items from the urn burials at Adichchanallur is a significant feature. The workmanship of the bronzes is of a very high standard and they are invariably deposited inside the urn and very rarely outside. The primary agricultural products of this period i.e. paddy and millet were placed in bronze dishes and deposited in the urns (Rea 1902-03: 127-130). All this points to a high degree of prestige value attached to copper-bronze objects by the Early Iron Age folk.

2. Carnelian was imported to Sri Lanka as raw material for bead making from South India during the Proto Historic period. This mineral stone is not found in Sri Lanka.

Bronze Tools and Technology in Protohistoric Afghanistan : A Review

V.C. SRIVASTAVA

There is a lot of controversy over the concept of civilization or urbanism and the factors responsible for their evolution (Redfield, 1964, 364-375; Gordon Childe, 1950, 3-17; Adams, 1968; Flannery, 1972, 399-426; cf. Kroeber, Milton Singer, Mandelbaum, Spengler and Toynbee, in Redfield, 1964, Service, 1975) but it may be agreed in a broad way that technology is one of the factors which by its interaction with ecology, human population and social organization ushers in a new society and cultural change. The rise of metal technology had been responsible to a very large extent for what is termed by Gordon Childe as "the Urban Revolution" because the rise of metallurgy led to the development of a new concept of wealth, a more developed social organization, qualitatively a far improved technology of utilitarian and luxury goods and ultimately a somewhat stratified society or urbanism or complex society or civilization (Renfrew, 1972).

In view of this assumption it is desirable to discuss the new technology of metals in Afghanistan in the first instance as there is very little development in the technology of lithic and bone tools in protohistoric Afghanistan. The technology of metal may be viewed as the determinant factor for the change to urbanism while the continuance of the lithic-bone techology may be seen as the factor for the survival of the earlier tradition.

In our present state of knowledge Afghanistan can claim the honour for the use of bronze (tin + copper alloying) technology at the earliest in the world context because there is evidence of the use of bronze technology as early as the late sixth millennium B.C. at Ghar-i-Mar (Aq Kupruk, Balkh province). So far, the earliest occurrence of tin-bronze had been dated to the 4th millennium B.C. (Necropolis A at Susa and Sialk III), though its general occurrence started around 2700 B.C. (Early Dynastic III, Mesopotamia), (Agrawal, 1971). Three fragments of sheet metal with an embossed motif—two of a rectangular rod and third of another sheet fragment have beer reported from Ghar-i-Mar. Two C-14 dates for Ghar-i-Mar assemblage are 5487 B.C. and 5291 B.C. Since these artifacts had been reported in association with the Neolithic and the dates have been unbelievably early Dupree, (1962, 1965) the excavator named it as Chalcolithic but this has been disproved by the chemical analysis (Caley, 1972) which shows that these contain 7% tin with very high percentage of copper and traces of iron and nickel. It is a low-tin bronze-characteristic of the early stages of bronze (Caley, 1972: 45). Conservative quantitative limit for bronze is more than 10% tin in copper. A tin content lower than this may be regarded as accidental but recent studies (Cleuziou and Berthoud, 1982, 15) have taken the view-point (which appears quite reasonable) that "tin values above 0.5" indicate deliberate combination of copper and tin at some point during the process which transformed ores into metallic bronze, though it may not have been totally controlled and its results may not be standard ones. But this must have facilitated the understanding that higher levels of tin would produce a significantly harder metal. If this view-point is accepted then the tin-copper alloy at Ghar-i Mar in the late sixth millennium B.C. may be regarded as the first definite step in the emergence of bronze technology not only in Afghanistan but in the world as such. It is no matter that this has been reported in association with the Neolithic culture because it may be explained on two grounds. Firstly, the introduction of metallurgy often takes a long time before its full socio-economic and cultural effects are manifested and detectable in archaeological record (Shaffer, 1978: 89) and secondly, the assemblage associated with the metal artifacts appears to reflect a specialized adaptation to pastoral nomadism. In Afghanistan specialized agricultural sedentarism and specialised adaptation to pastoral nomadism developed as contemporary and reinforcing adaptations allowing maximum use of resources. (Shaffer, n.d.). This evidence has wide implications for south Asian protohistory (Shaffer, 1984).

There is a gap of about 2000 years after the first appearance of bronze technology in the sixth millennium B.C. It reappears in the middle of the 4th millennium B.C. at Mundigak (Kandahar). Mundigak I2 gives us evidence of the use of metal tools and it goes on in frequency throughout the sequence. The earliest example was a flat blade like instrument with a probable hafting tang of copper or bronze. A simple type of bronze point or punch with a circular cross-section first noticed in I4 and easily available in later periods (so much so that it replaces bone awl/punch in Pd. IV) is the most common tool type. Unfortunately no chemical analysis is available for these earlier metals (bronze or copper). But elemental analysis of one artifact from Pd. Mundigak I5 shows interesting aspect of metallurgical sophistication. It has been an alloy with 1% tin, 0.15% iron and remaining percentage of copper. This is a case of very low tin + copper alloying in comparison to earlier Ghar-i Mar 7% tin alloy. But it cannot be regarded as accidental. It is purposive and shows a stage in the process of metallurgical sophistication in view of studies of Cleuziou and Berthoud (1982).

Period II3 provides the first example of a true projectile point which becomes frequent in subsequent periods. In Period III a tanged lozenge-shaped point was introduced and in Period IV a tanged oval shaped point was introduced. Period IV has given us a rare large lancehead and knife. Period II3 has given us luxury type metal artifacts—two pins—one with a double volute and the other had a flattened and perforated end. Period III continues this tradition with the presence of pins with flat ends and twisted shafts. Two socket hole axes and one adze from period III6 are examples of important metal artifacts. Fortunately they have been subjected to chemical analysis which shows important developments in metallurgy during this period (3000 B.C.). They had a tin content of about 5% — the highest recorded at Mundigak. It approaches the composition usually associated with bronze, though still lower than Ghar-i Mar tin content. Similar tin content occurs in almost contemporary metal objects such as from Susa B (5.3%) but they are still rarer in a period characterized by the use of arsenical copper. But it shows that alloying technique with controlled process and standardised results had been known and developed at the early 3rd millennium B.C. in Afghanistan. It is also significant to point out that other artifacts from the same period (III) had tin (1%) composition as of Period I low tin Bronze. The inference that can be drawn for this differential composition for different type of tools is that they realized the functional variability of tools and gave a higher tin content to tools such as axes and adze requiring weight, force and hardness while give lower tin content to artifacts such as pins or points or knives which need sharpness, neither weight nor hardness. Thus they appear to have attained the sophistication and specialization by utilizing differential characteristics of varying alloying compositions (Shaffer, 1978: 144). Period IV shows further development in the art of metal objects. Period IV provides us with an example of punch hafted into a bone handle, an oval shaped tanged point, large lancehead and a knife. As regards luxury artifacts, Period IV is characterized by the presence of the greatest variety and number of luxury types such as concave disc (mirror), double volute, lozenges and broad flat headed pins, handle for discs and buckle. At least in two cases smelted iron decorative buttons were found in Period IV. Small curved knives or sickles and chisels are frequently found in Period III-IV while a few barbed hooks are also reported from Period IV. Period IV pins have slightly increased tin content (1.5%) and traces of lead (1.6%) and other elements are found. The increase in luxury types in Period IV shows the development towards urbanisation. Period V continues the tradition of utilitarian and luxury metal objects.

Said Qala (Kandahar) has given us bronze objects similar to those of Mundigak III in the latest protohistoric periods - II-IV. (2300-2100 B.C.). Utilitarian artifacts included sickles, blade fragments, a lanceolate or lozenge tanged point, and a point or punch, with circular cross-section. 'Luxury' objects include pins of many types - double volute head; flattened with forked end and a simple forked end with twisted end and bronze handle fragment with a rounded and perforated distal end. No chemical analysis of the tools is available.

There is evidence of the use of copper tools at Deh Morasi (Kandahar) in periods II a-c-IV (Dupree, 1963: 98). The "shrine complex" in IIa has given us two fragments of a hollow tube and a handle fragment while II b-c has given us many fragments of simple pins while Period IV has given evidence of a single fragment of a compartmented seal. Unfortunately no chemical analysis is available. On the analogy of Mundigak metal objects it is quite likely that here also it may have been bronze. Dupree has termed even Mundigak culture as the Chalcolithic but it has proved to be the Bronze Culture. C-14 date for II-b (MASCA corrected) is 3200 B.C. (Dales 1973: 159).

Hilmand-Seistan sites have given us large quantities of cuprous slag (Fairservis, 1961) on the surface and graves with metal artifacts at Garden Rig. It demonstrate that copper was locally available and exploited.

Bronze objects from protohistoric northern sites including Dashli (Jauzjan) include utilitarian artifacts* of various types of axes, daggers, knives, sickles, razors, slashers, awls and piercers. Axes are of five types—(1) ground stone, (2) shaft hole, (3) adzes, (4) hewing adzes, and (5) socketed axes.

Daggers are of two categories:

(1) Rhombic with a straight haft sometimes curved at the end;
(2) Leaf-shaped often with a twisted haft—only recognized in Bactria.

The knives are single bladed and slightly curved. The sickles are toothed, slightly curved, projecting hafts are curve-ended. The razors are straight bladed with slightly bent hafts. Only one example of slasher has been reported. It is a massive tool with a sharp blade and little curved haft.

The awls and piercers are of two types:

1. Round double edged;
2. Square in cross section.

Luxury items are mostly mirrors of different types, pins and miniature vessels. The mirrors are round either without handles (Type-I) or with a projecting handles (Type-II). The mirrors with handles may be subdivided into two types - (1) straight handles; (2) female-figured handle which is rare.

Such mirrors have been reported from Baluchistan (the Mekli Burial ground) and this led Stuart Piggott to suggest that they may have come from Baluchistan but in view of its discovery in Southern Baluchistan (the Sapalli Burial ground) Sarianidi (1977) has suggested that perhaps a common source existed. This suggestion gets some support from the presence of similar mirrors among Luristan

* The above account is based mainly on V.I. Sarianidi, *Drevnie zemledeltsy Afganistana*, Moscow, 1977.

Bronze objects.. Various types of pins have also been reported from Northern Afghanistan. These types are as follows:

1. Conical pyramidal headed
2. Spade headed
3. Spiral headed
4. Bispiral headed
5. Rosettee headed
6. Ribbed headed
7. Hemispheric headed
8. Figure headed

The figure headed pins provide unique example. One depicts the fronts of two seated rams and the other shows a human headed bearded ox—very similar in style to the designs on vessels from Fullol hoard in Northern Afghanistan and Ur tombs (Mesopotamia).

Three types of miniature vessels have been reported:

1. Bottles with a rounded body and long neck from graves and many have long pins inside.
2. Cylindrical vessels so far known only in Bactria.
3. Jar shaped vessel with a carinated bottom—a rare variety.

Besides pins and miniature vessels two types of bracelets have also been found as follows:

1. Unclosed end; and
2. Unclosed but overlapping ends.

The temporal rings are either plain with unclosed ends or coiled ones.

The Soviet study of the metal industry is based upon three-types of analysis:

1. Spectroscopic research by the laboratory of Scientific Research Methods by E.N. Chernykh.
2. A metallographic investigation by N.N. Terekhova.
3. A chemical analysis of copper ore.

These studies show that Afghanistan belongs entirely to the Central Asian Zone and constitutes a single metallurgical province. According to Sarianidi (1977) within Afghanistan there existed two metallurgical centres - one in the South and the other in the North. The North Afghan metallurgical centre constitutes a part of the Irano-Turkmenistan metallurgical province (Sarianidi, 1977).

Shortughai (Takhar) has given us only fragmentary evidence of copper. The copper objects are not numerous nor very typical. They include a small number of pins, fragments and a discoidal mirror. More significant is the discovery of clay crucibles with traces of copper inside. They attest to the existence of small scale metallurgy since there is no slag nor evidence of smelting of ores. Some pieces of gold and lead from the same layer (the earlier) complete the picture of metallurgy. A discoidal gold bead perforated along its diameter has also been discovered (Francfort, 1984: 303). Analysis of 27 metal objects has been reported as follows: (Francfort, 1989: 75, 208 Table A)

TABLE A
Analysis of Metal

Inventory	*No. Lab.*	*Period*	*Objects*	*Metal*
SHAF 79 PO 151	Berlin	A/5/1	Shaft	Copper Oxide
SHAC 79 PO 125	Berlin	A/5/1	Shaft	Copper Oxide
SHAC 79 PO 116	Berlin	A/5/1	Shaft	Copper Oxide
SHAF 79 PO 158	Berlin	A/6/1	Fragment	Lead Oxide
SHAF 77 PO 41	Berlin	A/6/II	Fragment	Lead Oxide
SHAH 79 PO 37	Berlin	A/3/II	Fragment	Lead Oxide
SHA 78 PO 8	CEA 3129	A/2/II	Shaft	
SHA 78 PO 2	CEA 3131	A/2/II-IV	Shaft	Copper-Tin not negligible
SHBB 78 PO 94	3124	B/3/III	Mirror	Copper-Tin not negligible
SHB 76 PO 32	3126	B/3/III	Shaft	Copper - + Lead
SHB 76 PO 70	3127	B/3/III	Shaft	Copper-Tin not negligible + Lead+Arsenic
SHA 78 PO 95	3128	B/3/III	Point Shaft	Copper + Lead
SHB 76 PO 38	3134	B/3/III	Emmanchee	Copper
SHBB 78 PO 64	3135	B/2A/III	Spatula	Bronze-Tin + Arsenic
SHBB 76 PO 4	3136	B/4/III	Shaft	Copper-Tin not negligible + Arsenic
SHBC 78 PO 69	3138	B/3/III	Fragment	
SHBC 78 PO 19	3123	B/2R/IV	Shaft	
SHBC 78 PO 52	3125	B/2R/IV	Shaft	Copper - Silver
SHBB 78 PO 84	3130	B/2R/IV	Mirror	Copper
SHBB 78 PO 121	3133	B/2R/IV	Dagger	Copper
SHBB 78 PO 32	3137	B/1/IV	Spatula	Copper + Arsenic
SHBB 77 Surface	3132	B/III-IV	Mirror	Copper + Arsenic
SHA 76	Berlin	A/2/II-IV	Fragment	Lead
SHA 79, 40	Berlin		Coloured pebble	Cinabre
SHBC 78 PO 75	Berlin	B/1/IV	Mineral	Pure Iron
SHAF 79 PO 89	Berlin	A/1/II-IV	Slag	Iron Slag
SHA 78 PO 93	Berlin	A/6/1	Slag	Iron Slag

It shows that 10 objects have copper oxide while three have copper with lead oxide while others have got negligible tin with lead and/or arsenic, copper and arsenic and copper and silver. Iron minerals have also been reported for the transitional period from the copper to the early Iron age. Broadly speaking, the time-bracket for Shortughai II-IV may be fixed as 3000-1000 B.C. Since the percentage of tin/arsenic/lead is not known it is not possible to say whether it may be branded as true Bronze or not but it cannot be denied that the technique of alloying was well known at this site during the protohistoric period.

The question of source of copper and tin has been a matter of controversy (Cleuziou and Berthoud, 1982). It has been suggested that Zarafshan valley may have been the source for copper (Francfort, 1989: 208) used at Shortughai. Recent explorations (Cleuziou and Berthoud, 1982) and studies (Besenval, 1988 in Gardin (ed.): 229-236) have tried to locate sources of copper and tin in Afghanistan itself. It has discovered tin and nearby copper mines (worked in ancient times) at Misgaran (south-west of Herat) with 600 p.p.m. (0.06) tin values. The Russians have reported about tin deposits in Uzbekistan in Central Asia exploited from the mid 2nd millennium B.C. All these evidences lead us to agree to the suggestion that Afghanistan is a good potential source of tin used not only in Afghanistan but in south-east Asia in the 3rd millennium B.C. The discovery of Misgaran with higher tin values and nearby anciently worked copper mines provide almost a definitive evidence for the source of tin + copper in Afghanistan. This geological evidence is in harmony with the evidence (of course later) of Strabo (XV 2.10) who says that Drangiana (modern Seistan) has plenty of tin.

The above review shows that the use of metal tools and artifacts begin as early as the late sixth millennium B.C. and after a gap of about 2000 years reappears in the middle of the 4th millennium B.C. at Mundigak and continues at other sites in 3rd and 2nd millennium B.C. such as Said Qala and Deh Morasi in Southern Afghanistan and Dashli and Shortughai in Northern Afghanistan. It is quite probable that Afghanistan specially Herat region and northern region had sources of tin and copper in protohistoric times which provided these metals not only to Afghanistan but to the entire south-west Asia. It is also quite probable that northern Afghanistan belonged to a common metallurgical zone termed as the Irano-Turkmenistan zone. Alloying technique and bronze technology came to be developed here at the earliest in the whole region and has been responsible in association with other variables for the ushering of the urbanisation.

REFERENCES

Agrawal, D.P., 1971. *The Copper Bronze Age of India*, New Delhi.

Childe V. Gordon, 1942. *What Happened in History*, Penguin Books.

Caley, E.R., 1972. Chemical Examination of Metal Artifacts from Afghanistan, in *Prehistoric Research in Afghanistan* 1959-66; ed. L. Dupree, pp. 44-50, *Transactions of the American Philosphical Society*, n.s. Vol. 62, p. 4, Philadelphia.

Cleuziou S. and Berthoud Th., 1982. Early Tin in the Near East, A Reassessment in the light of new evidence from western Afghanistan, *Expedition*, Vol. 25, I, pp. 14-19.

Dales, G.F., 1973. Archaeological and Radio-Carbon Chronologies for protohistoric South Asia, in *South Asian Archaeology*, ed. N. Hammond, pp. 157-169, London.

Dupree, L., 1963. Deh Morasi Ghundai: A Chalcolithic site in South-Central Afghanistan, *Anthropological Papers of the American Museum of Natural History*, Vol. 50, Pt. 2, New York.

Fairservis, W.A. Jr., 1961. Archaeological Studies in the Seistan Basin of South-Western Afghanistan and Eastern Iran, *Anthropological Papers of the American Museum of Natural History*, Vol. 48, Pt. I, New York.

Francfort, H.P., 1984. The Harappan Settlement of Shortughai (N.E. Afghanistan) and some Aspects of Central Asia in the Bronze Age in Lal B.B. and Gupta S.P., (ed.) *Frontiers of the Indus Civilization*, New Delhi, pp. 301-310.

Francfort, H.P., 1989. *Fouilles De Shortughai Recherches Sur L'Asie Centrale Proto-historique*, 2 Vols., Paris.

Gardin, J.C. (ed.), 1988. *L'Asie Centrale et ses rapports avec les civilization orientales des origines a l'Age du Fer*, Paris.

Piggott, S., 1950. *Prehistoric India*, Harmondsworth, Penguin Books.

Renfrew, C., 1972. *The Emergance of Civilization*, London.

Redfield, R., 1962. *Civilization as Things Thought about Human Nature and the Study of History*, pp. 364-75.

Service, E.R., 1975. *Origins of the State and Civilization, the Process of Cultural Evolution*, New York.

Shaffer, Jim, G., 1978. *Pastoral Nomads* in, n.d. *Prehistory*, unpublished.

———, 1984. *Prehistoric Baluchistan*, Delhi.

———, 1984. Bronze Age Iron from Afghanistan, Its implication for South Asian Protohistory in Kennedy, K.N., and Possehl, G.L. (eds.), *Studies in the Archaeology and Palaeo-Anthropology of South Asia*, Delhi, pp. 41-62.

Probable Sources of Silver in Ancient India: A Historico-Scientific Approach

BHAGIRATHI PRASAD AND NISAR AHMAD

India has an impressive heritage in the production and use of ferrous and non-ferrous metals. The remains of ancient mines and smelting furnaces located in the vicinity of today's mining centres as well as the evidence of common occurrences of slags and few metal-working places presented by the unearthed ancient sites reflect man's earliest endeavour to recover and utilise these metals in the process of the cultural development. Similarly the most precious metal gold (Au) was also in greater demand as it played an important role in the socio-economic life because of its use for beautification as the mark of advance culture and its intrinsic value, and the testimony related to its mining is known.

Notably silver (Ag) was also in demand like gold. It served as a convenient medium of exchange due to its value second to gold, and thereby it gave stability to the currency and price. Silver was almost used throughout as the basic metal for the Indian currency after the appearance of the metallic money, presently termed punchmarked coins in the 6th century B.C. owing to its transaction feasibility, and remarkably, in the Nanda-Mauryan period, the minting of silver coins outnumbered the later silver coinages which were issued by Indo-Greeks, Indo-Scythians, Indo-Parthians, Western Kshatrapas, Guptas and others. However, a question which has been raised frequently is whether the silver ore was extracted indigenously or the metal was an import. Here, therefore, the nature and form of silver deposits known in India and its neighbourhood and the question of its source in ancient India are being discussed.

Silver ores occur as veins, replacement-deposits, contact-metamorphic deposits or as alluvial. Its presence in nature is by no means frequent and well known. The most common ore of gold or native gold often containing some of its portion is found in scattered grains, scales or plates in rocks of various types and ages, in veins, lodes, or placers or nuggets, grains, specks of dust in appreciable quantity. Silver is found as sulphides, sulpho-salts, arsenide, antimonide, and chloride, and also associated with ores of lead, zinc, copper and other metals.

Thus the silver deposits are of three categories - deposits with silver as a major component; gold deposits with silver as a co-element; and argentiferous deposits with silver as by-product or co-product. The argentiferous deposits are the most important primany ores as the largest amount, 90% of the world's silver is disseminated from them (Smirnov, 1983: 246). To present the clear picture of the co-elements of the silver sources in short, they are tabulated here alongwith their specific gravity :

Sl. No.	*Ore-mineral*	*Metal type*	*Sp. gravity*
1.	Native Silver	Pure silver	10.5
2.	Native gold	Gold-silver	18.1 - 11.1
3.	Sulphides	Argentite (Silver glance) Ag_2S	7.19 - 7.36
4.	Complex sulphides	Stephanite, Ag_5SbS_4	6.26
		Pyrargyrite, Ag_3SbS_3	5.7 - 5.9
		Proustite, Ag_3AsS_3	5.55 - 5.64
		Preieslebenite $(Pb, Ag)_8Sb_5S_{12}$	6.00 - 6.5
		Polybasite$AgCu_{16}(SbAs)_2S_{11}$	6.00 - 6.2
5.	Telluride	Hessite, Ag_2Te	8.4 (Sectile sp. gravity)
6.	Chloride	Cerargyrite, AgC1 (Horn silver)	5.0 - 8.0

Silver is extracted both from open-pit and underground mines. It is also taken out through a floatation process which results in its recovery during lead, copper and zinc smelting. In India (Krishnaswamy, 1988: 429), the occurrence of native silver is rare. It is only reported to be present about 3 km. to the east of Chargaon (Himachal Pradesh) in pegmatic bodies but its mining has yet to be worked out. The gold deposits formed in South also yield silver but the quantity of the latter recovered from them is very little. However, it appears more in secondary sources, associated with galena.

About 70-75 kg. silver is produced annually as the byproduct at Hutti gold mines in Karnataka (*Indian Minerals Yearbook*, 1990: 853) and to a much smaller extent at Ramagiri gold field (Anantpur district, Andhra Pradesh). Before 60-65 kg silver was also recovered every year from the gold refining at Kolar gold field situated in Karnataka. The basal conglomerates near Kalasapura (Chickmangalur district, Andhra Pradesh) are said to contain silver and gold in addition to copper and uranium.

Silver is obtained in considerable quantity from lead and zinc ores of various Indian localities of India. It is recovered at Vizagapatnam (Andhra Pradesh), Tundo (Singhbhum, Bihar) and Zawar (Udaipur, Rajasthan); the plants of these places belong to Hindustan Zinc Ltd. The amount of silver production from three mines are about 17,000, 11,000 and 1000 kg per year respectively. A small quantity (1100-1700 kg per year) of silver is also found by Hindustan Copper Ltd. during electrolytic refining of copper at its Ghatasila smelter in Bihar.

In Sargipalli (Distt. Sundargarh, Orissa), silver is found associated with lead, copper and zinc. Here the extraction about 500-600 gm of silver from a tonne of galena concentrate is estimated. It is also mentioned that the Sargipalli deposit bears 4.74 million tonnes of lead ore over a strike length of 1,400 meters at a vertical depth of 120 metres below the surface, hence the *in situ* silver is perhaps 237 gm without allowing for losses. The presence of silver is also brought to light in the multi-metal deposit near Mamandpur (Tamilnadu). About 120 gm silver per tonne is recovereable from this deposit. In the Sindhuvallitalur belt it exists along with platinum and iridum.

Thus presently, using all the technological advancement, about 40,000 kg silver is produced (*Indian Mineral Yearbook* 1992: 112). Out of that, about 83% output accrues from lead and zinc and again of this, only the recovery of silver at Vizagapatnam is more in quantum. The extraction of silver from it in ancient time can be only thought; the silver removed from copper and zinc at that time was beyond question. Most of the silver is used in the form of bars, grains, sheets, strips and flakes.

In the south-east neighbourhood of our country, in north Myanmar (Burma), Bowdwin mines (Shan States) of argentiferous galena are of special economic importance (Brown *et al.*: 137 and Wadia : 479). Here the slag heaps around the mines hint at the extraction of silver alone to meet its large demand leaving the lead thereat, it may to be pointed out that inspite of being primarily a lead-zinc-copper sulphide ore, it was named 'Bowlwin' (silver mines in Burmese language). This proves that the main purpose of its mining was silver extraction. In the north-west neighbourhood, there are mines in Armenia, Iran and Afaganistan (Marshall, 1931: 677) which were also in operation in ancient time.

As most of the native gold possesses silver, its appearance, according to the quantity of the latter, range from yellow, which is mostly gold (at one end of the scale containing various composition of electrum) to yellowish-white and toxic white coloured (because of mostly silver at the other). The two metals gold and silver are so closely related in the free form that the old Egyptian name for the silver is white gold. In *Taittirīya* (I.5.1.2) and *Kāthaka* (X. 4) *Saṁhitās* and *Śatapatha Brāhmaṇa* (XXII. 4.4.7; XIII. 4.2.10; XIV. 1.3.4), it is called *rajata-hiraṇya.*

Gold and silver were collected as early as 3800 B.C. They were smelted into usable lumps in the combined state. Further, notably silver is also found as a chloride ore which is smelted by the simple process. The deposite of this nature are known to have existed at various places in the ancient world. But the most plentiful sources of silver, almost from the beginning of its history, has been galena since the lead ore generally contains its small proportion.

In the history of the metallurgy of silver, besides gathering of native silver which is quite unimportant in relation to the general development of the metal, three stages may be recognised :

First stage covers the use of the natural white electrum alloy that passed for silver in antiquity but this in reality contained a small portion of gold and this kind of silver has been found in relative profusion in the ancient remains but notably it was always in smaller quantities than gold. Some electrum form of silver occurs in Egypt such as those at Ur dated to the middle and late pre-dynastic periods. It was apparently rarer than either gold, or copper or lead during the times of the pyramid-building dynasties. Similar materials has been found at Troy I and Mohenjo-daro. It was relatively common in Mesopotamia during the late Al-ubaid period and has been discovered at Lagash, Uruk III and in the contemporary royal graves at Ur. A few references of *rajata-hiraṇya* in some of the early Indian texts also hint at its rare use around 1000 B.C. We know that Kolar and Hutti mines contain silver in association of gold. But the use of these sources by the Vedic Aryans is doubted. Therefore, seemingly, the white gold was an import.

The second stage in the history of silver began with the production of the metal as a byproduct of smelting lead galena. In its bare essentials, the winning of silver in this process had no great difficulty. A piece of galena when dropped into fire got reduced to lead; if however, this lead is heated long enough, it is oxidised into a powdery ash leaving behind a drop of virtually pure silver. It is most probable that the presence of silver in galena was the result of this technological innovation.

Since pure silver could be obtained by the ancient people only by using a cupellation process, this method must have been successfuly developed at least a century or so, before 2500 B.C., since object made of very pure silver had appeared by that date; they have been found at Ur, in Egypt and Troy. Remarkably, in India, the ancient relics pertaining to the time brackets of 1500 and 700 B.C. do not yield silver objects and its rare references in the literature ascribed before the said last datum line alludes to the unawareness of the technology related to the extraction of the metal as byproduct during this period. And, the mention of a silver *nishka* in *Panchaviṃṣa Brāhmaṇa* (*SB*, XVII. 1.4) as one of the ornaments of the outlandish Vrātya suggests the use of silver more with alien people; and thereby it also indicates its import.

The third stage in the development of silver metallurgy was the invention of a method for separating the silver from the gold in electrum alloys, this discovery was not, however made until somewhat later than the middle of the second millennium. We find objects made of (*rajata*) silver in the later Vedic texts. And, the phrase *suvarna-rajata* mentioned therein (*SB*, XII. 8.3.11), if taken as one term, may indicate to the use of this kind of silver during this time. However, we do not have evidence at our disposal to infer the adoption of the technology referred to above during this period. So the *rajata* (silver) for the objects or ornaments again was an import.

No doubt Kautilya speaks of the occurrences of silver at Tuttha (Tatta), (*AS*, 2.12.6) Gauda, Kambu and Chakravāla (*Arthaśāstra*, 2.13.14). The silver ores, according to him (*AS*, 2.12.6), are 'of the colour of a Conch-shell or Camphor or crystal or butter or a dove or a pigeon, or *Vimalaka* (gem) or the peacok's neck, (or) of the colour of *sasyaka* (gem) or *gomedaka* (gem) or jaggery or unrefined sugar, (or) of the colour of the flower of *Kovidāra*, or lotus or *pātali*-flower or *kalalya* flower or the flower of flax or of linseed, (and) containing lead, containing antimony, smelting like raw flesh, when broken (either) black with a white shimmer (or) white with a black shimmer, or all variegated with lines and spots, soft, (which) when being smelted do not split and yield plenty of foam and smoke'. Thus the silver ores contained this metal as the major component or co-element.

The above author also hints (*AS*, 2.12.13) at the presence of silver in the lead ore. It is said that this is iron-black or of the colour of the dove or yellow pigment or studded with lines (and) smelting like raw flesh. Thus it has the smell of raw flesh like silver ores and its white lines are the parts of silver. In this deposit silver is minor-element. The silver which is now produced in India is from the argentiferous ores as the byproduct. Silver ores are uncommon. It is only reported from Himachal Pradesh; and of the argentiferous deposits, those at Vizagapatnam are of the cognizance nature. However, the fact of silver extraction from this deposit in the remote past has yet to be established.

The places of silver occurrences referred to above have to be looked in the light of reality that existed in ancient time. Gauḍa is identified with Kāmarūpa (Kangle: 111) but it is unwarranted (Ahmad: 95). It was certainly in Bengal and this region is devoid of silver deposits (Law, 1967: 258f). So the silver of Gauḍa (Bengal) was the imported silver from the mines of Burma. Tuttha can be equated with Tatta (Sind, Pakistan) and the silver of this place presumably was the imported silver from the Southern Iran.

Kautilya also says that silver which is white, smooth and soft is best and one that bursts is bad. He also mentions about the purification of silver with one fourth part of lead (*AS*, 2.13.11-13). He also describes that the silver which is clear, lustrous and of the colour of curds is pure (*AS*, 2.13.14). These statements thereby suggest the presence and use of pure and native silver.

Here mention may be made of bags of gold dust for the presentation, evidently they might have also been used for the voluminous purchases. Further the use of the great objects or ingots for the gifts is known to us since the Vedic period. Notably, in India the rate of the ratio of gold to silver was 10 to 1. As their rate of exchange in Persia was 13 to 1, a very fair profit was earned by the Persians by exchanging the gold in return of silver. It may also be noted here that during the first century A.D. exchange between indigenous and imported foreign currency of the western coast of India, a profitable undertaking is referred to in the Periplus of the Erytharean sea. (McCrindle: 123). Probably the Roman silver coins had only intrinsic value in the Indian context.

It is interesting enough to read the reference, given in the Periplus, of the import of silver into India from outside through the West Indian ports. (Mc Crindle : 109). Since the Periplus is generally dated to the first century A.D., it is held that this phenomenon did not characterise the earlier period (Shastri : 46). It is also said that the silver punch-marked coins of the Saurashtra region were minted out of the Zawar mines in Udaipur (Shastri : 47). Not withstanding, it is yet to be substantiated by the scientific investigations; notably, these coins are only a few in number. The possibility of the use of the foreign silver for them can not be ruled out in view of the reference of the export of this metal in the later period in this region.

No doubt, the chemical analysis of some specimens of the silver punch-marked coins reveal admixture of lead and this indicates that the source of silver for these coins was argentiferous mines. It is suggested that Andhra, Asmaka and Mulaka janapadas might have got their supply of silver from Cuddapah and Kurnool (Shastri : 47). However, we remember that the coins of the Asmaka and Mulaka janapadas are scarce and those of the Andhra principality are open to question in respect of their attribution. However, the use of the indigenous silver for the manufacturing of some of these punch-marked coins can not be ruled out altogether.

Indeed it is suggested that the janapadas in Uttar Pradesh, Bihar and Bengal, for coin minting of silver coins used the silver obtained from the argentiferous lead deposits of Bhagalpur, Manbhum, Monghyr and Singhbhum in Bihar, and Rewa, Drug, Hoshangabad and Jabalpur in Madhya Pradesh (Shastri : 47). But this hypothesis can not be adhered.

In view of this fact the presumption by some scholars that prior to the advent of the Indo-Greeks, silver was coming into India from Balkh (Bactria) (Narain : 165) seems convincing. However, it is possible that the Yaudheyas and the Kunindas made effort to avail and use the local silver to have bi-metal currency system. But as they utilized this source alone the scarcity of silver hampered successful execution of the system. Thus the silver used in our country was mostly imported from the mines of Burma, Afganistan and Iran.

REFERENCES

Ahmad, N., Economy of Ancient Assam: A Review in Historical Perspective in *Essays on North-East India* presented in memory of Prof. V. Venkata Rao, (ed.) Milton S. Sangma, p. 95.

Arthaśāstra, 2.13.14; 2.12.6; 2.12.13; 2.13.11-13; 2.13.14.

Brown, J.C. and Dey, A.K., *India's Mineral Wealth,* pp. 127 ff.

Indian Minerals Yearbook, Vol. I (1992), p. 112; Vol. II (1990), p. 853.

Kangle, R.P., *The Kautilya Arthaśāstra,* Part II, p. 111 fn.

Kāthaka Saṁhitā, X. 4.
Krishnaswamy, S., 1988. *India's Mineral Resources*, p. 429.
Law, B.C., 1967. *Historical Geography of Ancient India*, pp. 258 f.
Marshall, J., 1931. *Mohenjodaro and Indus Valley Civilization*, II, p. 677; *Taxila*, II, p. 607.
Mc Crindle, J.W., *The Commerce and Navigation of the Eryathraean Sea*, pp. 109, 123.
Narain, A.K. *et al.*, *Seminar Papers of Local coins of Northern India*, p. 165.
Śatapatha Brāhmaṇa, XXII.4.4.7; XIII.4.2.10; XIV.1.3.4; XVII.1-4; XII 8.3.11.
Shastri, A.N. (Ed.), *Coins and Early Indian Economy*, pp. 46-47.
Smirnov, V. (ed.), 1983. *Studies of Mineral Deposits*, p. 246.
Taittiriya Saṁhitā, I. 5.1.2.
Wadia, D.N., *Geology in India*, p. 479.

Mineral Resources: A Possible Archaeometallurgical Correlation

AMIT TRIPATHI AND VIBHA TRIPATHI

India is known for its rich and large iron ore reserves since the antiquity. Today the country is one of the leaders in iron ore production and export in the world. The total iron ore (of all types) reserves of the world, according to a Mineral Commodity Summary, stand at over 263,000 mill. m.t. India's contribution to this is 17,565 mill. m.t. which stand after USSR, Brazil, Canada and USA (Table 1). According to 1982 data, India was ranked sixth in world regarding the production of iron ore (Table 2) with an estimated production of 40,902,000 m.t. The bulk of this huge iron ore production comes from the states of Goa, Madhya Pradesh, Orissa, Bihar, Karnataka, Maharashtra and Andhra Pradesh (in decreasing order) contributing to 99.96% of the total iron ore output (Table 3).

This large iron ore reserve is being worked for thousands of years in India, going back to the beginning of the first millennium B.C. or earlier still when it was used for weapons, like arrowheads, spearheads, points etc. by Painted Grey Ware and Black and Red Ware cultures.

This paper attempts to focus on the location and nature of iron ore deposits and associated rocks and then to examine whether there is any correlation between the ancient settlement patterns - archaeological or ethnological - and the probable resource areas exploited by them.

State and Districtwise distribution of Iron Ore deposits and associated rock types

An attempt has been made here to give a brief geographical and geological distribution of the iron ore resources and associated rocks. Other possible usages for these rocks and their effect on the smelting process have also been highlighted.

RAJASTHAN

Jaipur

1. *Morija Area*: Massive hematite ore, some of which occur as hematite schist are found in association with arkose and quartzites of Alwar Group of Delhi Super Group. The ore contains:

TABLE 1

Total iron ore resources (estimated) of seven most important iron ore producing countries

Country	*Reserve (in million long tonnes)*
USSR	1,09,000
Brazil	34,000
Canada	26,000
USA	25,000
India	17,565 (in million metric tonnes)
Australia	17,000
China	6,000

Source : Krishnaswamy and Sinha, 1988.

TABLE 2

Total world production (7,83,000 thousand metric tonnes) and leading world producers of iron ore, iron concentrates and iron ore agglomerates (1982)

Country	*Production (in thousand metric tonnes)*
USSR	2,44,410
Australia	87,694
USA	36,002
Brazil	99,457
China	70,000
India	40,902
Canada	34,496
France	19,390
Liberia	18,165
Sweden	16,138
Venezuela	11,680
Others	1,24,486

Source : Krishnaswamy and Sinha, 1988.

TABLE 3

Leading Indian producers of iron ore lump and fines (1984)

State	*Production in m.t.*
Goa	1,31,42,000
Madhya Pradesh	90,48,000
Bihar	66,75,000
Orissa	64,66,000
Karnataka	51,65,000
Maharashtra	13,12,000
Andhra Pradesh	1,23,000
Rajasthan	12,000

Source : Krishnaswamy and Sinha, 1988.

Iron = 54% to 68%
Silica = 1% to 19%

The reserves in this area have been computed to be about 1.38 mill. m.t. down to a depth of 15 m. while total reserves are expected to be much more. Much of the ore, by present day standard, is of blast furnace grade and some of it is even of steel melting shop grade.

Associated rocks in Morija

The ore is found associated with quartzite and arkose rocks. These rocks, on account of their high quartz content are both hard and compact. The former being much harder on account of its higher quartz content and lesser matrix as compared to the latter which has comparatively larger amount of feldspar in the matrix and is hence less hard.

2. *Rampura Area*: The ore occurs as lenses of micaceous hematite associated with Quartzite, Micaceous Quartzites and Calciphyres of Delhi Super Group (lower part of Ajabgarh Group). The ore contains:

Iron = 65% to 67%
Silica = 1.5% to 2.5%

This iron ore deposit although comparatively smaller than Morija deposit, yields very good grade iron ore, much of which is of S.M.S. grade, after proper beneficiation.

Associated Rocks in Rampura

An important rock associated with deposit is the calciphyre. Calciphyre rocks are rich in $CaCO_3$. Limestones, that are presently used as flux in iron extraction process, owe their fluxing property to $CaCO_3$ which is their principle component. These calciphyraes too, could have caused a better iron output if mixed with the ore used, knowingly or unknowingly.

PWG Sites in Jaipur: Agtari, Bhairat, Bhagwanpura, Gondi, Jodhpura, Viratnagar, etc.

N.B.P. Sites: Bairat, Gondi.

2. Jhunjhunu

1. *Taonda Area*: Iron ores, mostly hamatite is found associated with calc-gneisses and para-gneisses of Ajabgarh Group of Delhi Super Group. One ore body in this area carries 'kidney ore' analysing iron = 67% and phosphorus = 0.06%. But the average ore analyses:

Fe = 43.12% to 69.79%
SiO_2 = 0.01% to 35.36%
P = 0.01% to 0.34%

The reserve iron ore in Taonda area are estimated at 0.25 million m.t.

Distribution of Sites and Iron ore deposits

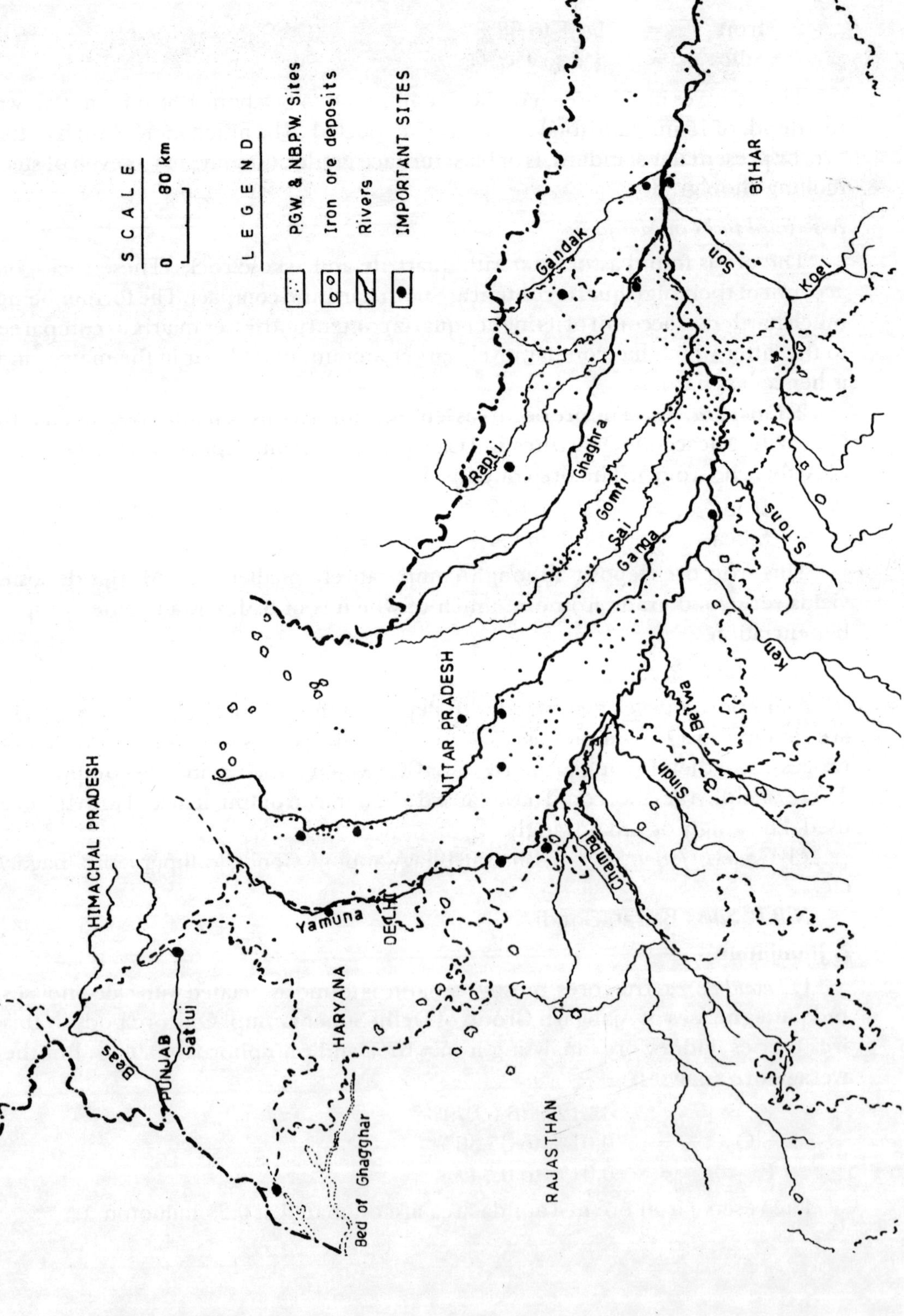

SCALE
0 80 km

LEGEND
P.G.W. & N.B.P.W. Sites
Iron ore deposits
Rivers
IMPORTANT SITES

HIMACHAL PRADESH
PUNJAB
Beas
Satluj
HARYANA
Bed of Ghagghar
Yamuna
DELHI
RAJASTHAN
UTTAR PRADESH
Chambal
Sindh
Betwa
Ken
Ganga
Sai
Gomti
Ghaghra
Rapti
Gandak
S.Tons
Son
Koel
BIHAR

MAP 1

2. *Jamalpur and Siror Area*: Iron ores are found associated with calc-gneisses and para-gneisses of Ajabgarh group of Delhi Super Group. But ore bodies are smaller and reserves much less as compared to Taonda area.

Associated rocks in Jhunjhunu

The Jhunjhunu district is well-known for its copper deposits (Khetri copper belt) at the contact between Ajabgarh Group and underlying Alwar Group. This copper deposit is associated with several other sulphide minerals (besides chalcopyrite, azurite, malachite and covellite) like galena, phyrhotite, pyrite, etc. and is believed to be of a secondary origin formed by a process known as supergene sulphide enrichment. The iron deposits on the other hand occur as primary deposits associated with the host rocks and have a wider geographical distribution. The iron deposits are generally associated with calc-gneisses and at places these rocks are overlain by limestones. The calc-gneisses and para-gneisses are both tough rocks with a high mica, quartz and feldspar content. The calc-gneisses have a large amount of calcium carbonate which is a natural fluxing agent in the iron extraction process.

3. Sikar

Lenticular ores of magnetite and some hematite are found occurring as lenticular bodies of fine grained, massive, steel grey to black ore bodies. This ore might, at places contain high amount of iron, but the occurrences are too small to be worked profitably.

4. Alwar

Magnetite deposits occur in an extensive manner. Some occurrences of the extension of hematitic calc-gneiss and para-gneiss bands from adjoining Jhunjhunu have also been reported. No deposit is large enough to be exploited economically.

PGW Sites : Bajwa etc.

5. Udaipur

Minor occurrences of iron ores.

NBP & BRW Sites : Ahar.

Total iron ore reserves of Rajasthan (estimated) are 12.78 million m.t.

HARYANA

1. Mahendragarh

1. *Karota Area* : In this area the Haryana State Geological Department has estimated a reserve of 1.5 mill. m.t. down to a depth of 45 m with iron content = 46.39% to 69.51%. Of this, 1 mill. m.t. are proved reserves.

2. *Bhanaota-Dhancholi Area* : Magnetite and hematite occurs lensoid bodies around Bhanaota-Dhancholi. The reserves are estimated to be about 1.5 mill. m.t. upto a depth of 75 m. with iron = 53.2% to 67.5% but the ore is high in phosphorus.

3. *Antri-Beharipur Hills*: Some magnetite bands have also been recorded around Antri-Beharipur hills. Here the ores contain about 57.4% iron.

2. Gurgaon

Small occurrences of iron are found in parts of Gurgaon district.

PGW Sites: Asitpur, Autha, Baghala, Bhadsa, Dhankot, Gohana, Malab, Palwal, etc.

NBP: Autha

3. Hissar

Small magnetite occurrences have been reported from some parts of the district.

PGW Sites: Lahoronwali, Theri, Nakora, Raniya, Humayun Khera, Kariwali-1 (Dharampura), Kariwali - 2 (Najjal Ther) etc.

HIMACHAL PRADESH

1. Simla Hills

Iron ore is found associated with carbonaceous and garnetiferous phyllites and slates, quartzites and highly crushed dolomites ($MgCO_3$ $CaCO_3$) of Jutogh Formation. The ore is medium to high grade with an average iron concentration = 60%.

Associated rocks in Simla Hills

The iron ore in Simla is associated with meta-pelitic rocks and dolomite. The phyllites are highly micaceous rocks also containing quartz and feldspar. These are soft and weak rocks that have a tendency to split along the foliations. Slates too are weak and brittle rocks. Quartzites, on the other hand may be quite tough and hard. The dolomite (a rock with $CaCO_3$ $MgCO_3$ as primary constituent) may probably have some fluxing properties as it has $CaCO_3$ as a component and $MgCO_3$ too which is a chemically similar compound.

2. Mandi

1. *Chichot Subdivision hills of Thanag*: Magnetite and hematite occur in the Chichot subdivision in the hills of Thanag. The occurrence associated here as bands in quartzites and sandstones (probably of Murree Formation). The area contains some high grade iron ores.

2. *Kohar Khas Koti-Kohar ridge*: Magnetite and hematite occur extensively as bands in schists and quartzites but with a low concentration.

Associated rocks in Mandi

In Mandi district iron ores are found associated with quartzites and sandstones. Quartzite, a metamorphic rock is very hard and tough as described earlier.

Sandstone, the unmetamorphised equivalent of quartzite is comparatively softer on account of its greater intergranular spaces and higher matrix content.

3. Sirmur

In Sirmur district, dessiminations of magnetite are found as magnetite schists and magnetite quartzites. Deposits near Bihar and Kohad are said to contain good quality iron ore. The rocks containing this deposit belong to the Boileauganj Quartzite Formation.

Associated rocks of Sirmur

The magnetite schist rock is a soft, moderately tough and heavy rock. It has a tendency to part along the foliation planes that are defined by the preferred orientation of flaky minerals mainly mica.

4. Kulu
5. Kangra } Minor occurrences of iron ore
6. Lahaul Spiti

UTTAR PRADESH

1. Sonbhadra

1. *Ningha* : About 2.5 km. north west of Ningha, iron ore in form of hematite is found. This is associated with metamorphic supracrustals of Penninsular Gneissic Complex.

2. *Bargawan-Barhi* : In Bargawan-Barhi area, about five million m.t. of low grade iron ores with less than 35% iron content are estimated from the metamorphic rocks of Penninsular gneissic complex.

3. *Kunwar* : Minor occurrences of hematite has been reported from Kunwar.

NBP Sites : Barkad, Bavan-Jharia, Baraila, Kamhart, Kusi, Mahauwan, Siddhi, Ubadhi, Alamgirpur, Allahpur.

2. Jhansi and Lalitpur

1. *Sonari* : About 30,000 m.t. of soft hematite ore occurs in pockets in the Sonari area in Jhansi district associated with the supracrustals of Bundelkhand gneissic complex.

2. *Maraura Paragana* : At Salda, in Maraura paragana, a considerable amount of iron is reported to have been smelted from hematite (G. Watt, 1980), while a hard steely kind was made in Pura.

NBP Sites : Chhote-Belnga, Erachi.

3. Banda

According to observations made more than a century ago, iron was extensively worked in Kaltangarh Paragana at several points, specially at Gobarhai. Mines of iron ore were also reported at Deori and Khirani.

4. Nainital

1. *Ranibag-Bhimtal-Khairna-Section*: Deep maroon or purple, iron rich slates and siltstones with quartzarinite and basic volcanics are observed from Solani to Talla Bhimtal-Sat Tal. Quartzites, phyllites, basic volcanics and lenticular bands of hematite and magnetite are reported from Ratighat to Khairna near the Ramgarh thrust (Valdiya, 1980). These rocks belong to the Nagthat Formation.

2. *Lusginani*: Iron bands are observed near Lusginani that contain iron = 15% to 62%.

3. *Kala Agar*: Minor occurrences of iron.

PGW Sites: Thakurdwara, Kashipur, etc.

NBP Sites: Kashipur

Associates rocks in Nainital

The rocks with which the iron ores are associated here are highly diverse regarding their physical property. Slate and siltstones are both brittle and very soft rocks rich in clay minerals while quartzarinite and volcanics are very hard and compact rocks. Quartzarinite is a rock containing more than 95% of quartz and contains little matrix. It is hence very hard and tough rock. Basic volcanics are solidified lava flows with low quartz and are consequently moderately hard and compact, although they may contain vessicles of trapped gas within them. These are fine grained heterocrystalline rocks showing a characteristic vesicular structure.

5. Chamoli

1. *Dewal-Jhalia Section, Pindari Valley, W. of Syalkot near Jhalia*: In this area, slates with nodules of hematite occur alternating with white and green marbles of Deoban Formation.

2. *Between Salani Dharinkhet, Mahendru Malla and Lamchula Gad*: Highly ferrugeneous sandstones can be observed alternating with green slates. Iron or ferrugeneous concretions also occur as nodules in maroon slates belonging to the Rautgara formation (Valdiya, 1980) underlying the Deban Formation.

3. *Bhukinda*
4. *Ali Magnetite bearing quartzite* – Minor occurrences of Iron ore
4. *Pokhri Agar*

Associated rocks in Chamoli

In Chamoli the iron ores occur in low grade metamorphosed calc-silicate rocks. Marbles are found in this area, which are a soft but tough and compact form of $CaCO_3$ that are formed by metamorphism of limestone. These rocks can act as a good fluxing agent. Slates are soft and weak rocks whereas sandstones are hard and moderately compact. The ores found in this area are generally of poor quality and are found at great heights (around 1000-2000 m above M.S.L.).

6. Almora

Kasi Valley Subakot-Lachhmanour Area and Garrat Area: Ferrugenous ironstained slates with purple sandstones occur in this area alongwith slates showing ferrugenous concretions. The rocks belong to Rautgara Formation (Valdiya 1980). The ore is of very low grade.

7. Pithoragarh

1. *Chandak Area* : In Chandak area of Pithoragarh district beds of limonite ($2Fe_2O_3 . 3H_2O$) and siderite ($FeCO_3$) are found associated with slates, phyllites and green marbles (Kanalichhina Satgarh track and Chhera - Bans sections) of Deoban Formation. The limonite and siderite have physical and chemical properties much different than more commonly used hematite and magnetite.

2. *Gangolihat Area* : Limonitic and sideritic slates and phyllites alternating rythemically with pink and white marble, marls and some limestones (Typical in south of Dashainthan and South of Chanvala) are seen outcropping in an extension of Deoban Formation.

3. *Thuli Gad Valley*: Southeast of Pithoragarh on the west bank of R. Kali (India-Nepal border). Ferrugenous concretions with low grade ore in purple-slates associated with siltstone and fine grained greywacke are seen to occur associated with subordinate calc slate ($CaCO_3$ rich) of Rautgara Formation.

Associated rocks in Pithoragarh

Pithoragarh district lies west of River Kali which defines the political boundary of India and Nepal. Here, iron ore is dominantly in form of limonite and siderite associated with rocks of Deoban Formation and underlying Rautagara Formation. The associated rocks are slate, phyllites, marbles and greywacke and siltstones with inferior marls, limestone and calc-slates.

Slates are low grade argillaceous metasediments that are soft, brittle and weak. A variety of slate containing high amount of $CaCO_3$ is called Calc slate. Phyllites are metamorphosed form of slates. These are soft, moderately compact rocks that contain abundance of fine grained flaky minerals. Greywacke contains 15% to 75% of matrix and has moderately large intergranular spaces. Greywackes are composed primarily of quartz with some feldspar and lithic fragments. These rocks are of variable hardness and are generally moderately hard to moderately soft and compact. Marbles are $CaCO_3$ rich soft, compact and tough rocks that acts as natural fluxing agent. Siltstones is a very soft and friable rock, consisting mainly of clay minerals.

8. Tehri

1. *Jalkur Valley Area Paturi Section* : In this section abundant dolomitic limestone characterized by chert laminae and high ferrugenous content is seen to occur. The rocks belong to the Deoban Formation.

2. *Between Paturi and Dosari Sirka* : Here quartzites alternating with ferrugenous concretions, exceedingly ferrugenous medium grained purple slates with pink quartzites and bank of volcanics belonging to Rautgara Formation (Valdiya, 1980) are seen to outcrop.

Associated rocks in Tehri

The iron bodies are seen in association with dolomitic limestones. Quartzites, slates and volcanics. The dolomitic limestones are fine grained, soft rocks containing majority of $CaCO_3$ which has been partly replaced by $MgCO_3$. The rock is a good fluxing agent. Quartzites are light coloured very hard and compact rocks primarily composed of quartz. Slate is a low grade metamorphic rock that is soft and has a tendency to part along planes known as slaty cleavage. Volcanics are solidified lava flows with some tuffaceous material that may be vessicular. These rocks are moderately hard and compact.

DISCUSSION AND CORRELATIONS

The above deposits had several archaeological settlements in their vicinity indicating thereby that they were being tapped by man for a long time. It may not be out of place to suggest that man settled there primarily for this very reason i.e. procuring the mineral wealth. Even before metals were used, the ores were used for procuring pigments and various stones for fashioning tools and implements of daily household needs. Quartzite, sandstone etc. were used for making tools, saddle and quern and subsequently as building material. With advent of metal, minerals and rocks including the semi-precious stones for beads and ornaments were all recovered from such areas. Naturally the preference must have been there for such deposits which were of complex nature, yielding a variety of raw material at one place. The emphasis, it appears was more on easy to smelt ores even though the yield was of a lower order. Such deposits could easily be found in complex mineral zones. It is quite understandable in view of the elementary level of technology at earlier stages of advent of iron. The richer iron ores, which demand better skill and higher smelting temperature, therefore are more difficult to smelt, were sought only at the second stage of growth of metallurgy. Such deposits were tapped primarily for ore and secondarily for other material.

This observation is more accurately applicable to PGW & BRW cultural period and continues upto early levels of NBP. Rajasthan, Punjab, Haryana and Western U.P. are the states favoured by settlers of PGW culture. This region, as can be verified from the above data, falls in a zone with medium sized iron ore deposits. To take Jaipur District for an example, we have plotted Agtari, Bhairal, Bhagwanpura, Godi, Jodhpura, Bairat, etc. on archaeological map with PGW-NBP ware cultures. The iron ores in this district have 54% to 68% containing 1% to 19% silica. The ore in this area is found to be associated with $CaCO_3$ rich rocks viz. limestones and calciphyres that happen to act as fluxing agents in iron smelting. Table 4 indicates presence of ore body in calcareous metasediments in Delhi Supergroup thus making the iron ore of Rampura in Jaipur District easier to smelt at earlier stage of iron working. This might be one of the significant reasons of early occurrence of smelted iron in this area. Jodhpura has been excavated and reveals iron from Painted Grey Ware level on a regular basis while bits of iron are reported from pre PGW deposit of Black and Red Ware culture at Noh. Evidence of smelting (of iron) is forthcoming from PGW level, two furnaces alongwith provision for bellows and forges have been reported from early levels of PGW at Jodhpura. Spearheads, arrowheads, spearhead with a leaf shaped point, axe with a broad cutting edge and a socketed tang have been found from PGW level. This region was a significant centre during Mahabharat period as part of Matsyadesh located about 33 km. from Viratnagar. (The site of Noh in Dist. Bharatpur was part of Sursena). More than fifty sites have been located on the bank of River Yamuna in the vicinity. The dried up course of the river has now been reused for irrigation. Several ancient sites are located on this course. For the spread of culture and transportation of raw material River Yamuna must have been vital. The economy was that of a flourishing, affluent village centre. A direct correlation, thus may be shown between ores, other mineral deposits, transportation mechanism and spread and consolidation of cultures.

South Bihar is one of the richest mineralized zones of the world today. Incidently, Bihar is also the centre of imperial power of ancient India. This area grew as a second phase of cultural upsurgence. The neolithic-chalcolithic cultural milieu is more impoverished compared to PGW culture but it is succeeded by NBP ware deposit. The latter is not only more sophisticated and early in date in this region but is also called the epicentre of this culture (Sinha, 1969). The role of mineral wealth present in this region may not be over emphasised in growth pattern of culture in this region. It is another matter why South Bihar the actual centre of these deposits was not centre of culture growth. This is well in keeping with the concepts of theories of urbanization.

Christaller's centre place theory is aptly applicable in the present case. Though, the concept formulated from present day observations, if generalized it can very well explain the distribution of ancient settlements. The central place theory remains unsurpassed as a model of the service activities of man. Christaller was not very much concerned about what went about in towns, but he was very interested in finding out the general principles which determine the number, size and distribution of human settlements.

In formulating his centre place concept, Christaller began with the goods having the greatest range. Such goods, he postulated, would be produced at the centre of the system of centres. Goods of the next range would be produced at the main centre and at several other centres, at the corners for the markets of the main centre. The next class of goods would be produced in several third order centres located at the gravity points of the triangle formed by these centres as well as in the two earlier classes of centres (Fig. 1). This ordering of centres, would continue until the goods of the lowest range are taken into account.

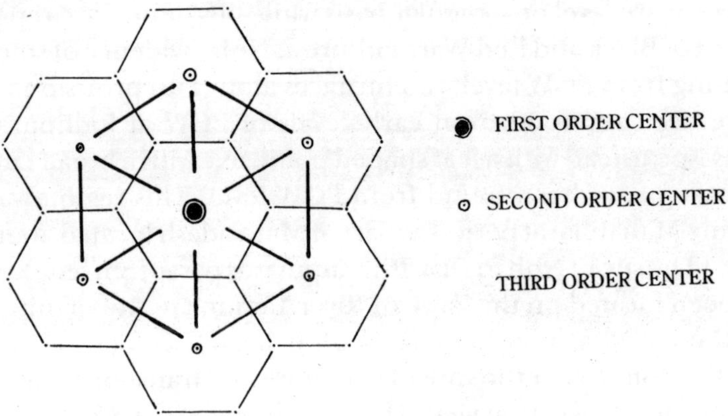

Fig. 1 : Christaller's Center Place Theory

This theory assumes a homogeneous plain with an even distribution of natural resources, population, consumer preference and production technique for each and every product. Human activities are assumed to be space utilizing.

The implications of this theory in the present context is of great significance. If we assume the agricultural products to be the first range goods, the most essential requirement for the development of first order centres are rich and fertile soil, abundant water resources and a warm sub-tropical to temperate climate. If we further consider mineral products to be second range of goods then the location of the first order centres will be fixed by its proximity to mineralized belts (in addition to above mentioned requirements of fertile soil etc.). Further, there should be a good transport connection like navigable rivers etc. to the mineralized belts which would be the location of second order centres.

The above theory explains very well the location of Magadh and its ascent to the largest power centre of ancient India. This region has one of the richest soils and abundant water supply essential for the first range of goods. Further, this area is adjacent to the confluence of Son and Ganga. The Son an easily navigable river, flows through Shahdol, Satna, Rewa, Sidhi, Sonbhadra, Rohtas and Daltonganj. Its tributary Koel flows through Daltonganj, Loharduga and Gumla. These areas constitute one of the richest mineralized zones of India. The minerals could be worked at smaller centres located near the ore deposits or carried through the rivers to the first order centre. No wonder Magadh rose to the status of a 'super

TABLE 4

Stratigraphy of Rajasthan

Orogeny	*Age (in Ma.)*	*Unit and Rock Type*		
	430	Jalore Granite		
	750-505	Malani Volcanics (Acid and calc-alkaline)		
	850±50	Erinpura granite suite in central axial zone		
Delhi orogeny	1700-1500	Granites from Bairat		
2000-900 M.A.	1950-1910	Nepheline Syenite of Kishangarh		
		Delhi Super group	Ajabgarh group	Argillaceous and Calcareous meta sediments
			Alwar group	Mainly Arenaceous
		Unconformity		
Aravalli orogeny	2000	Aravalli Super group		Dharwal and Amet Granites
	2500-2000			Upper Aravalli group
				Lower Aravalli group
		Unconformity		
B.G.C. Orogeny	2580-2950	Berch granite and gneiss		
	2500	Untala and Gingla granites		
	3000-3500	Banded gneissic complex with enclaves of metasedimentary and meta-igneous rocks.		

TABLE 5

Property/Mineral Chemical composition	*Hematite* Fe_2O_3	*Magnetite* Fe_3O_4	*Limonite* $2Fe_2O_3.\ 3H_2O$	*Siderite* $FeCo_3$
Colour	Steal gray to iron black	Iron black	Brownish black to Ochre yellow	Brownish red, brown ash grey, white etc.
Streak	Cherry red*	Black	Yellowish brown	White
Lustre	Metallic to dull	Metallic	Sub-metallic to dull	Vitreous or pearly
Structure & form	Massive to granular	Crystalline to Massive	Amorous botroidal*	Crystalline or massive
Cleavage	Absent	Imperfect	Absent	Perfect
Fracture	Uneven to sub-cocoidal	Sub-concoidal	Uneven	Uneven brittle
Hardness (on Moh's scale)	5.5 -6.5*	5.5 -6.5*	5.0 -5.5*	3.5 -4.5*
Specific gravity	4.9 -5.3	5.18	3.6 -4.0	3.7 -3.9
Others			Strong magnetic	

* Distinguishing properties

power'. Son Valley is full of small habitations right from prehistoric period. This region is also very rich ethnologically. Large number of Pre-industrial ironworking groups including the Agarias inhabit this area. This must have been a potential resource zone for bigger centres developing around this zone. The other Mahajanapadas of 6th and 5th century B.C. are also located in virtually similar settings. Kashi getting its supply from Son Valley and hills of Sonbhadra and Sidhi, Kosal (with Kaushambi as the major town) from ores of Banda which is just about 40-50 km westward, upstream of Yamuna. The Sursena (Mathura) from its own neighbouring deposits of Alwar, Gurgaon and Agra. Avanti janapada could derive its raw material from rich deposits around Udaipur.

It may be pointed out that every mineral deposit which was rich in more than one kind of resource was favoured. A group of persons working in an area with raw materials available for tool making, semi-precious stones as well as metal works were safer and economically on 'sounder ground'. Suitable stones for household appliances and subsequently for construction of monumental buildings and other kinds of structures were carried from these areas to larger towns through major rivers.

The other observation in connection with resource zones is that, at least at earlier stages of the technological growth, specific variety of ores were selected for working, as they were easy to work, even though they yielded lesser amount of metal.

The above data reinforces the vital role played by ecology in human settlement pattern. Minerals and other resources were a deciding factor in growth and prosperity of a centre. Although in India such studies have not been undertaken at a micro level but if seriously worked out, it is bound to throw light on those areas of settlement system which have remained in darkness so far.

ACKNOWLEDGEMENTS

We are thankful to C.S.I.R. for funding the research work.

REFERENCES

Agrawal, R.C. and Vijay Kumar, 1976. The Problem of PGW and Iron in North-east Rajasthan. *Mahabharat: Myth and Reality*, ed. S.P. Gupta and K.S. Ramachandran, Agam Kala Prakashan, New Delhi, pp. 241-244.

Bery, Mason and Dietrich, 1982. *Minerology*, 2nd ed. C.B.S. Publ., pp. 561.

Brown, J.C. and Dey, A.K., 1955. *India's economic wealth*, 3rd ed., Oxford Univ. Press, p. 761.

Bhattacharya, S., 1972. Iron smelters and indigenous iron and steel industry of India; from stagnation to Atrephy. *Aspects of Indian Culture and Society*, ed. S. Sinha. Publ. Ind. Anthopological Soc., pp. 133-151.

Gansser, A., 1964. *Geology of the Himalaya.* Interscience Publ., p. 289.

Hughes, T.W.H., 1873. Notes on some of the Iron Deposites of Chanda, Central Province. *Rec. G.S.I.*, Vol. VI, pp. 77-81.

Joshi, S.D., 1970. *History of metal founding on the Indian subcontinent since Ancient times*, Pub. S.S. Joshi, Ranchi, p. 165.

Krishnaswamy, S. and Sinha, R.K., 1988. *India's mineral Resources*, 3rd ed. Oxford and IBH Pub. Co. Ltd., p. 613.

Roy, T.N., 1986. *A study of northern black polished ware culture, an Iron Age culture of India.* Ramanand Vidya Bhavan, New Delhi, p. 195.

Sen, A.K., 1984. *Laboratory Manual of Geology*, Modern Book Agency, Calcutta, p. 208.

Sen, A.K. and Gupta, P.K., 1981. *A Handbook of Economic Geology*, Modern Book Agency, Calcutta, pp. 259.

Sharma, N.L. and Ram, K.S.V., 1972. *Introduction to India's Economic Minerals*, 2nd ed., Dhanbad Publ., Dhanbad, p. 411.

Sinha, K.K., 1969. The NBP Ware - Fresh hypothesis in the light of Sravasti Evidence. *Potteries in Ancient India*, (ed.) B.P. Sinha, Patna, pp. 174-184.

Tripathi, V., 1976. *Painted Grey Ware: An Iron Age Culture of Northern India*, Concept Publ. Co., Delhi, p. 150.

Valdiya, K.S., 1980. *Geology of the Kumaun Lesser Himalaya*, W.I.H.G. Dehradun, p. 291.

Wadia, D.N., 1966. *Minerals of India*, National Book Trust, Delhi, p. 198.

Watt, G., 1990. *A Dictionary of India's Economic products*, Vol. IV, Cosmos Publ. Delhi, pp. 503-511.

Section II
Ethno-Archaeological Evidence on Archaeometallurgy

Iron Working in Ancient India—A Case Study

VIBHA TRIPATHI

Little attention has so far been paid to the reconstruction of ancient iron working in India. In view of the limited archaeometallurgical researches in the country, it is quite understandable. Excavations rarely reveal furnaces, much less smiths' workshops which can lead to a proper and full understanding of ancient iron working. An attempt has, nevertheless to be made to collect the scattered data and put the pieces together like jigsaw puzzle for understanding technology as practised in ancient India. The excavations conducted at Khairadih, district Ballia in Uttar Pradesh by B.H.U. revealed several furnaces and even workshops of smiths. It is aimed here to study the prevalent technique of iron making at different cultural levels in greatest possible details with special reference to Khairadih excavations. The archaeological data would be substantiated with ethnological evidence to see whether it corroborates with the deductions. Only through such an approach a worthwhile reconstruction of the furnaces and the process and mode of production of iron can possibly be understood. To fulfil this aim, first a brief survey of known types of furnaces found in diverse parts of the world, as worked out by archaeometallurgists would be presented. This would be followed by archaeological evidence on iron technology in India. In the end ethnological data on the subject will be discussed.

Iron Working Furnaces

Archaeometallurgists all over the world have tried to work out the design and shape of furnaces in use in the antiquity. Fragmentary remains make the task of reconstruction pretty ardous. Beno Rothenberg (1983) rightly pointed out, "The uppermost parts of the furnace, including the working level of the tuyere(s), have not yet been found in situ in any furnace mainly because these rather fragile parts protrude above ground and would be the first to be destroyed or disintegrate. This, alongwith the destructive technique of working necessitating the breaking of the superstructure of the furnace towards the end of the operation makes it virtually impossible to find an undamaged furnace in excavations. However, a careful excavation provides sufficient clues for a reasonable reconstruction of a furnace/ forge". Archaeometallurgical researches, e.g. archaeological remains with laboratory experiments have successfully traced the furnace designs of the past. Several kinds of furnaces are discernible in different parts of the world which may be classified as below. (Fig. 1)

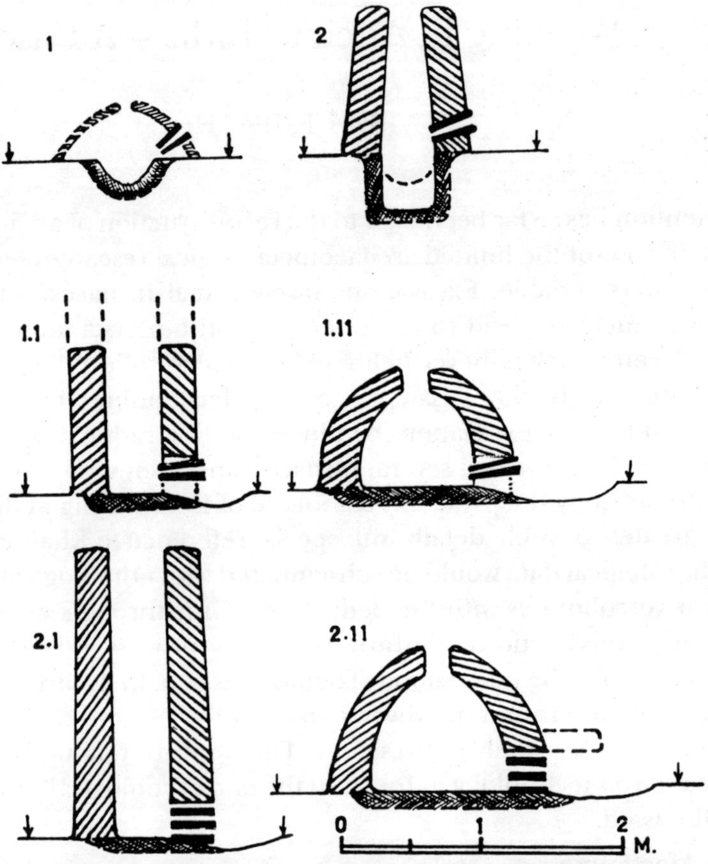

Fig. 1: Proposed shapes of early iron smelting furnaces

Bowl Furnace

This is the most elementary type of a furnace, shaped like a bowl—a simple circular hole into the ground into which bellows were attached for draught through tuyeres. Coghlan (1956) was the first to point out to this type of a furnace. The diameter of the bowl was usually 25-30 cm and was lined with clay. Tylecote (1980) conducted experiment in a fascimily of this furnace and believed it to be the earliest kind of furnace in use, with probably dome shaped superstructure of clay. Unless covered the over-oxidisation would hinder production of metallic iron, resulting only into fused mass of slag and metal (Tylecote, 1965). These furnaces are reminiscent of copper smelting and lead to the assumption that iron smelting was not invented by advanced copper smelters but by more primitive copper smelters (Tylecote, 1980).

These furnaces were capable of producing temperature upto 1150°C - 1200°C. The bowl furnace was charged with broken pieces of ore and charcoal mixed together. Generally such furnaces had no outlet for slag, which accumulated at the bottom of the furnace. This further reduced the capacity of the furnace, making only very small quantity metal to be produced in one charge-the resultant bloom was very small, which rested above the slag. Towards the end, the top of the structure had to be broken to retrieve the bloom which had to undergo repeated heating and beating to extricate the trapped slag producing a homogeneous bloom of iron.

Domed Furnace

This was a slightly more evolved form of bowl furnace. In these structures a slag tap hole was provided along with slag pit for tapping of the liquid slag. Its design helps in maintaining a reducing condition which is quite important for iron smelting. In such furnaces the charge is protected by the clay dome creating the right atmosphere in the reaction zone. Generally an opening was provided at the top (aperture) to facilitate addition of ore and charcoal, whenever necessary as the smelting proceedes. The bloom can, in these furnaces be larger because the slag could run out of the furnace leaving space for bloom. Cleere (1972) observes that in Europe, in certain cases a tunnel was provided for slag to run into. (Fig. 1: 1.11, 2.11)

For forging activity, which invariably followed smelting, no reducing condition was required. Only an open hearth, preferably with provision for bellows was required. A pile of charcoal could be ignited with the help of bellows, pushed down and held in place with a stone. It would easily attain a temperature upto 1200° C. Most of the smithy with wrought iron could be undertaken by cold work — hammering and annealing at 700° C. Many a times open pits, located in excavations could be reminiscent of forges than furnaces.

Shaft Furnace

It was the most evolved type of furnace, developed at a comparatively later stage. Not much innovations have actually been possible to bring about in this design. Shaft furnaces have been described by several metallurgists like, Coghlan (1956) etc. Although these are described as small furnaces with limited capacity, varying from 25 to 30 cm in diameter, but are said to be very effective. It has continued to be in operation, of course, with minor variation, in different regions still among pre-industrial societies.

Alternate layers of ore and charcoal are charged into the furnace upto the top of the opening. Additional charge could always be slid into in between the operation through this opening. There is a provision for tapping of slag through a hole at the base, into a channel or a pit provided for the purpose. (Fig. 1: 1.1, 2.1) The slag in such cases shows clearly the running out character. In these cases the slag is more likely to be found in the shape of drippings than single block (as in the slag pit furnace, below). For removal of the bloom, either the superstructure has to be broken or there was a temporary blocking of the opening at the base that could be broken and the bloom removed from there, at the conclusion of the working as in connection with early furnaces in Denmark.

Slag pit furnace is a variant of the above shaft furnace. The only difference is in the provision for separation of slag. In the actual slag pit shaft furnace, the slag gets deposited at the base in the form of a big block. The bloom rests on the top of the slag heap. There is no provision for tapping of slag out of the furnace.

Archaeological Evidence

In India, iron working has been traced in Chalcolithic milieu from sites of Eastern and Central India, Rajasthan etc. At Pandu Rajar Dhibi (Sudhin De and Chattopadhyay, 1989) iron slags and ash pits identified as iron furnaces have been reported alongwith restricted number of iron objects from a Chalcolithic level (Period II) dated by C-14 from 1045 ± 55BC. to 920 BC. On the basis of examination of slags, it has been deduced that smelting was done there at a rather low temperature, below 1100° C. There is no clearcut evidence of use of tuyeres. Some of the samples belonging to post-Chalcolithic level (Period III) have a high percentage of fayalite (iron silicate) rendering the object brittleness and naturally indicating an elementary knowledge of iron smelting. Seeing the early context of occurrence, it is quite understandable, the technique, it seems had just been learnt.

At Jodhpura in district Jaipur, Rajasthan, pits associated with slags and ashes have been noticed at Painted Grey Ware level (Agrawal and Vijay Kumar, 1976). The details of the working or shape of the furnace etc. have not been discussed by the excavators. But more recently some light was thrown on the evidence of smelting furnace at Noh by Lal again at PGW level (at the Seminar on New Archaeology and India held by ICHR in New Delhi in October, 1988). According to Lal the furnace with a side hole into which the nozzle of bellows could have been inserted was found at Noh.

Atranjikhera, an important site of upper Ganga basin, having a rich collection of iron right from the earliest levels of Painted Grey Ware culture yielded a pear-shaped fire pit from the upper phase of PGW culture. In words of the excavator (Gaur, 1983) in some cases the openings for introducing nozzles of the bellows for fire could be clearly seen. Inside these pits were found tapering clay-lumps and finished iron tools. A pair of tongs which might have been used by the blacksmiths lie outside the pit alongwith iron slags and also charred animal bones. The significance of charred bones in context of a furnace is not very clear and leads Gaur to assume that some of these pits might have been community hearths. Nevertheless the occurrence of slags nearby, with finished iron objects and smiths' tools is significant indeed.

At Ujjain a square furnace has been reported to have been found (Banerjee, 1968: 178-79) alongwith a forge, iron slags, ore and a crystalline material identified as calcite and aragonite as well as lime which is said to be used as flux here. But the evidence remains not very clear nor is conclusive. Jakhera, the site excavated by Sahi (personal communication) has yielded excellent evidence of iron smelting in the form of semicircular chunks of slag and cinders. Though Sahi calls them ingots but even on a superfluous examination they look to be the residue of slag pit furnace which must have been abandoned in situ by the iron-workers.

Excavations at Dhatwa conducted by Hegde (1991) has provided us with a model for understanding ancient iron working. The reconstructed model is an example of domed furnace described above.

Megalithic culture of South India and Deccan were very rich in iron and local smelting has been attested to from several sites. Most valuable evidence has come from excavations at Naikund in the Vidarbha region of Maharashtra. It has been dated on the basis of C-14 dates to 700-400 B.C. (690±110 B.C.). A careful excavation with the help of resistivity meter led to the discovery of an iron smelting furnace (Deo and Jamkhedkar, 1982). The furnace at Naikund was a small one with a diameter of 30 cm and height of 25 cm. It was constructed with circular clay bricks (Pl. XIII, 1982, *op. cit.*). The walls were tapering towards the opening at the top. A hole was provided at the bottom for tapping slag. Two tuyeres of 16 cm length and 2.5 and 3.6 cm, thickness in heavily vitrified condition were recovered. Nearly 40 kg of slag and cinder were found in the trench, a corroded iron rod was also noticed in the vicinity. Gogte (1982) has reconstructed the furnace and its model has been prepared. A large number of finished and well-prepared iron objects were found in excavations at Naikund. This excavation provides us with a model to work on.

At Khairadih three furnaces have been found in a row (see Pls. 42a and b). Though heavily damaged and possibly destroyed while working by the ancient smelters themselves but it could be reconstructed. Underground pits with flat base were excavated, they were about 39 cm deep and had diameter of 20 cm. The clay lining of this pit is thinner than the furnace top. At this level a superstructure was raised above ground, the shape appears to be akin to the shaft furnace from Czekoslovakia described above (Voss, 1989; Plainer, 1989). Clay mixed with straw and sand was used for the furnace walls. Although no tuyere could be found but bamboo, possibly plastered with clay was used. A thick burnt bamboo has been found *in situ* just at the level of the top of underground pit. The slag ran through a channel provided for it at a lower level. A close observation of slags reveal that they flowed out of the furnace in a liquified form and fell as drippings at a lower level,

A large amount of slag (over 30 kg) was picked up nearby leaving the rest *in situ*, indicating a fairly heavy working in the area. Being partially underground the furnaces must have been more efficient than the Naikund furnace which was completely above ground level. Such an inference is possible to draw on the basis of ethnological evidence in Orissa (discussed below, Srinivasan, 1991).

The entire northern side of the site was occupied by the artisan or metal workers. In another trench (CC '6), the adjacent trench similar evidence has come forth. Significantly, traces of glass making and iron smelting have been located in the same area. A mud floor with earthen lamps and other household accessories have been found alongwith pestle, quern (a four legged oblong platform and a straight, short handled rounded pestle) of sand stone, used probably for breaking the ore, a crucible etc. were found in association with lots of burning activity, whitish ash, slags and finished iron objects and glass beads etc. The exact shape and size of the furnaces could not be traced in this trench despite the best efforts. It appears to be a multiple furnace area with working in several pyrotechnologies. The associated pottery belongs to somewhat later stage of Northern Black Polished ware culture. Most of the objects are of wrought iron but some of the iron objects analysed at B.H.U. (Department of Metallurgical Engineering) show a uniform pattern of pearlite cementite. Thus, speaks of high standard of smithy at Khairadih. Carburization has also been attested to.

The types of furnaces described above were in operation at different cultural levels in India. In the chalcolithic Bengal, at the time of introduction of iron technology simple bowl furnaces without tuyeres appear to have been used (Pandu Rajar Dhibi). The yield in these furnaces is generally very small and the quality poor as shown above. On the basis of analysis of slags and finished products it is clear that this kind of technical know how will lead to a very limited use of iron and thus will have little impact on the cultural set up. Nevertheless, introduction of iron was made, howsoever limited its use. An improvisation on this simple bowl was introduction of bellows and tuyeres with a dome shaped superstructure to which additional charge of ore and charcoal could be added as the smelting progressed. At Atranjikhera and Noh oval pits with provision for tuyeres for bellows could be classified under this category. Additional and a more careful retrieval of data on metal working may confirm this. The comparatively large number of iron tools repertoire and their good quality could only be possible in this kind of operation. The slag in the form of drippings collected at those sites may suggest that provision for tapping slag was made in these furnaces. As compared to the most elementary bowl furnaces, comparatively a larger bloom was possible to attain here.

The shaft furnace is one of the most evolved kind. It appears at a slightly later stage of technological growth. At Khairadih this has been found, as described above, only at late NBP culture stage. In fact the number of iron objects shows a marked improvement from early NBP to late NBP level at the site. For a better understanding the archaeological evidence may be well substantiated with ethnological evidence available in India.

Ethnological Evidence

Due to largely prevalent destructive method of iron working and also the problem of survival of superstructures of furnaces, ethnological evidence on iron working is a great asset to proper understanding of smelting—forging processes being traditionally followed. Till the capture of market by British pig iron in the beginning of 20th century, iron was being locally worked at small level in almost every part of India particularly near the ore deposits. The quality of iron thus produced was very high and does not require any elaborations. A variety of furnaces are known to have been in operation in each region. (A detailed study of the ethnological evidence on smelting of iron has been made elsewhere by the author see Tripathi, 1994, 1995a, 1995b). Certain furnaces used by pre-industrial societies are well comparable with the furnace at Khairadih.

Vikash Bharati, a regional centre for appropriate rural technology traces 'Asur Birjia' tribes in Garha Harup village on Netarhat plateau (near Bishunpur on Lohardaga-Netarhat Road) in Chhotanagpur Division of Bihar. Recently, RDCIS, SAIL, Ranchi has not only studied their working process (Agrawal *et al.*, 1991) but also tried to get it restarted and make it economically viable for the iron workers. It is noteworthy, however, that they could not suggest any improvisation in their basic working.

There, a vertical shaft furnace is used. Lump of iron ore and charcoal of sal wood are the raw material. Twin operated bellows, held down by a stone are used for air blast. The same opening at the base was for inserting the tuyere and was sealed with clay and charcoal mixed together. At the conclusion of the smelting the bloom was removed through this hole. Two shallow pits were provided on both sides of the front opening for catching the flowing slag.

The temperature in the tuyere region was found to be as high as 1500° C but was found to be lower on the inside of the shaft region. Sponge iron of 96% purity is attained in this process as a result of reheating and beating of the bloom. The carbon content varies from 0.01 to 0.04%. Similar furnaces were used in Bastar region with only one basic difference that a platform was added at the opening through which additional charge could be slid into the furnace occasionally.

Similar Furnace with slight variation was used in Orissa in Koraput region (near Baipariguda in Jaypore subdivision of Koraput) and Sambalpur. It was also in practise, though on a smaller scale in several other districts like Kalahandi, Bolangir, Sundergarh, Keonjhar and Dhenkanal. There are some minor variations in designs of furnace, viz., the Sambalpur furnace was built above the ground and that at Koraput underground and was claimed to be more effective (Srinivasan, 1991). In some furnaces near Rourkela limestone was used as flux, indeed a significant evidence.

Some analysis was conducted at Rourkela Steel Plant and discovered 0.3% carbon, though in a non-uniform pattern. Some carburization was also prevalent for imparting hardness (Srinivasan, above).

The process of iron working may thus be understood fully with the help of experimental metallurgy based on the pre-industrial working. It has been conducted by the team of metallurgists at SAIL, Ranchi and Rourkela. Actual production of iron has been succesfully attempted at both the centres above. While it has thrown light on the most intricate details of the technique of iron working at these places, they have highlighted the traditional Indian technology too, preserved and practised by these ethnological groups. They represent the knowledge acquired through generations and thus may be true representative of traditional Indian iron working.

Interestingly, it may be observed that most of these societies live in close proximity of ore deposits. This automatically removes the necessity of long and tedious transportation of the mass of ore from deposit area to the working place. No wonder a somewhat similar situation could be demonstrated by an indepth study of resource zone in connection with specific cultural centres of ancient times.

The other observation which may be made on the basis of an evolution in furnace design—from simple to more complex—is that the technology was developed indigenously on foundation of the existing copper technology. What forced the advent of technology, may be investigated further. It is clear, however, that since the newly acquired metal could be produced only in a very limited quantity because of the small furnaces capable of low production, iron was put to use for very special purposes, like hunting (a favourite sport of the elite) and warfare for winning new lands. However, it paves the way for further innovations in the field of technology.

ACKNOWLEDGEMENT

The author is thankful to Late Prof. K.K. Sinha and Dr. B.P. Singh who gave her an opportunity to work at the site and to study the metal objects procured from Khairadih.

REFERENCES

Agrawal, B.B. and Prasad, K.K. *et al.*, 1991. Evaluation of earlier iron making processes and their relevance in Indian contest. Paper presented at National Seminar on Indian Archaeometallurgy, B.H.U. October, 1991.

Agrawal, R.C. and Vijay Kumar, 1976. The Problem of PGW and iron in northern Rajasthan. In *Mahabharat-Myth or Reality* ed. S.P. Gupta and K.S. Ramachandran, New Delhi, pp. 241-244.

Banerjee, N.R., 1965. *The Iron Age in India.* New Delhi, p. 178.

Cleere, H.F., 1972. The classification of early iron smelting furnaces. *The Antiquaries Journal,* Vol. LIII, Pt. 1, pp. 8-23.

Coghlan, H.H., 1956. Prehistoric and early iron in the old world. *Pitt Revers Museum Occasional Paper in Technology,* No. 8, p. 86.

Deo, S.B. and Jamkhedkar, A.P., 1982. *Naikund Excavations,* Bombay.

De, Sudhin and Chattopadhyay, P.K., 1988. Iron from Pandu Rajar Dhibi: Archaeometallurgical Studies. *Steel India* 12 (1), pp. 33-41.

Gaur, R.C., 1983. *Excavations at Atranjikhera,* New Delhi, p. 128.

Gogte, V.D., 1982. Report on Iron working and furnaces, in Deo, S.B. and Jamkhedkar, A.P. (above), pp. 11, 56-59, 52-55.

Hegde, K.T.M., 1991. *An Introduction to Ancient Indian Metellurgy,* G.S.I., Bangalore.

Pleiner, R., 1980. Early Iron Metallurgy in Europe in *Coming of the Age of Iron,* ed. Wertine & Muhley, Yale University Press, pp. 375-416.

Pleiner, R.E. Pieta, 1989. Iron making and working on the territory of Czeckslovakia until Middle Ages (Recent Results). *Archaeometallurgy of Iron* 1967-1987, Prague, pp. 93-167.

Rothenberg, Beno, 1983. Copper smelting furnaces in the Arabah, Israel—The Archaeological Evidence, in *Furnace and Smelting Technology in Antiquity,* ed. Paul Craddock and M.J. Hughes, British Museum Occasional Papers No. 48, British Museum, London, pp. 123-50.

Srinivasan, N.R., 1991. Revival of rural iron making in Orissa, Paper presented at National Seminar on Indian Archaeometallurgy, October, 2-4, B.H.U. Present volume.

Tripathi, A. and Tripathi, V., 1994. Iron working in Ancient India—An Ethno-archaeological study, *From Sumer to Maluhha,* Wisconsin Archaeological Report, Vol., 3, (ed.) J.M. Kenoyer, Wisconsin, U.S.A., pp. 241-252.

Tripathi, V., 1995a. Understanding Technology: The Ethno-Archaeological Model. Paper presented at Jt. Session of IAS and ISQS held at Gorakhpur, Nov. 1995 (in Press).

Tripathi, V., 1995b. Ethno-Archaeological observations on growth of Iron Technology. Paper presented at 2nd Cong. on Traditional Sciences and Technologies of India, Anna University. Chennai, Dec. 1995.

Tylecote, R., 1980. Furnaces Crucibles and Slags, in *Coming of the Age of Iron,* ed. Wertine and Muhley (*Op. cit.*), pp. 183-228.

Voss, B., 1989. Iron furnaces in Denmark, *Archaeometallurgy of Iron* (1967-87), ed. R. Pleion, Prague, pp. 151-157.

Evaluation of Earlier Ironmaking Processes and their Relevance in Indian Context

B.B. AGRAWAL, K.K. PRASAD, G.I.S. CHAUHAN, P.K. CHAUDHURI AND S.M. AERON

ABSTRACT

The progressive use of metal has been one of the yardsticks in the history of development of human civilization. This is one of the efforts by human beings which has been under continuous development centuries after centuries. With time, the technique of metal extraction has been changing, finally culminating into the development of mammoth Blast Furnace Technology of Ironmaking. Both economic and ecological challenges of the present day warrants a quick look at the evolution of metal and metallurgy in search of a newer appropriate technology. The paper tries to recapitulate the evolution and development of earlier ironmaking processes throughout the world in general and a special status which India had achieved in this field. The relevance of these earlier ironmaking in modern India has also been discussed.

Earlier Ironmaking Art

There has been a lot of debate regarding the first discovery and usuage of metals by human beings. Though it is a common belief that *copper* was the first metal extracted by man from its ore, it has also been stated that *lead* was the first metal extracted about 9000 years ago (Crysko 1980). Archaeological studies showing the metallurgical developments in a time span gives the age for the first extraction of lead as 9000-7000 B.C., Copper as 3800 B.C. and Iron implements from extracted metal as 2500 B.C.

Among the iron extraction techniques, sponge ironmaking is probably the earliest process (Agrawal, *et al.*, 1988). Early experiences of men to extract lead and copper from their ore might have educated them to try accidentally found iron ore. To give a shape to this hard but brittle ore, they might have resorted to heat it (at a temperature of around 800° C which they were able to achieve for the extraction of copper/lead) and then hammer it to give the required shape. Their gradual experience to heat it and work on it must have given the shape but unknowingly they were carving the (early) metallurgy of ironmaking. Heat source (most probably from wood/charcoal) must have produced reducing conditions and hammering of hot iron ore was pushing out the (partly loose) earthy material (slag) out of it, and interestingly, they were making almost pure iron, what we call wrought iron today. Fig. 1 shows, the sketch of furnace which ancient man might have used to derive metallic iron from ore.

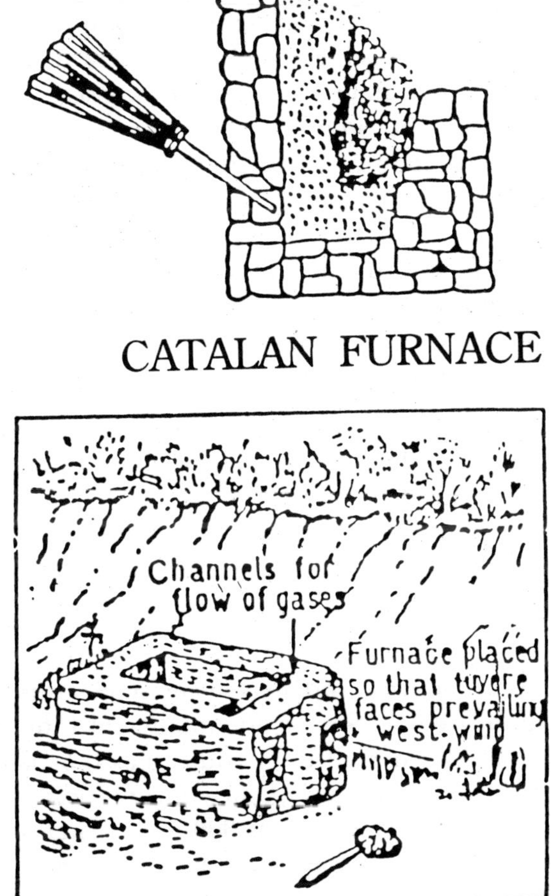

Fig. 1: Sketch of a furnace such as ancient man might have used to derive metallic iron from ore

Excavations have thrown some light on the type of furnaces the primitive men must have used in different parts of the world. The prominent accounts about the existence of such ironmaking techniques in India and Srilanka has been compiled by Prakash (1983), Prakash and Tripathi (1986) and Coomaraswami (1908), in Great Britain by Tylecote *et al.,* (1971) and in Central Africa by Noten and Raymackers (1988). In all these places, though the detailed technique and the furnace used for iron extraction varied from place to place, but there are some common features of the processes practised. In all these processes, there was a (very) low shaft furnace and fuel appears to have been mainly charcoal, the combustion products forming the reducing agent. In the early stages of development of this process, the slag hardly left the furnace and a mixture of metal, charcoal and slag were left behind. With the gradual improvement in the process technique and their ability to achieve higher temperature resulted in a process which produced iron in solid state and a free flowing slag (at least a small part of total slag generated) similar in composition to that of fayalite, $2FeO.SiO_2$. The free running temperatures of slag of about this composition fixes the minimum temperature of operation at about 1200° C. Later on, this technique of semi-smelting had improved sufficiently to allow most of the slag to be tapped from the furnace leaving behind the iron in solid form (Tylecote *et al.*, 1971). The process became more or less obsolete when the blast furnace was introduced during the 16th century.

Earlier Ironmaking in India

India has been one of the most important iron and steel making centers of the ancient world and for most of the recorded history, the only source of "Damascus" steel. The antiquity, the number and the volume of old ore working and slag heaps give ample testimony to this, viz., the great iron pillar near Delhi which weighs ten tonnes and is about fifteen hundred years old (Fig. 2), huge girders of iron at Puri, the ornamental gates of Somnalli, the 24 feet wrought iron gun at Nurvar, etc.

Quality Steelmaking in Early India

The age and the excellence of Indian iron and steel industry can be traced with respect to world famous Wootz Steel made in ancient India. It is generally held (Needham, 1980) that all through the middle ages the Damascus blade of the Muslims were made from Wootz Steel exported from India; and that the origin of the process may date to times as early as 300 B.C. (Yater, 1983). This steel was regarded as being superior to European manufactured steel even up to the middle of the last century and unsuccessful attempts were made to produce it in Europe. As has been mentioned earlier that earlier iron-making processes practiced were more or less same throughout the world producing lumpy sintered sponge iron but it was the steel-making technique which put India on the top among the earlier quality steelmakers.

It is worth mentioning the details (Verhoeven, 1987) of one such quality steel-

Fig. 2: Great Iron Pillar near Delhi

making process. The steel was made in small clay crucibles (hence called "crucible process" also). Each crucible could produce around 500 gms of steel. The crucible charge consisted of sintered sponge iron (made beforehand), wood chips and some fresh leaves. The crucible was closed with a tight fitting clay seal which in some accounts contained a small central hole while in others it was not present. Many such crucibles were then arranged in an array, covered with charcoal and then were heated by combustion with air supplied from tuyeres pumped by two to four bellows. The time of heating ranged from 4-20 hours and it was continued until the whole charge inside melted. There is uncertainty in literature as to whether the wootz metal had a cast carbon composition of around 1.6% or a composition in the range of 2.5 to 3.5% which was then reduced to 1.6% by decarburization. It was this steel which after certain thermomechanical treatment had been used to produce world famous Damascus swords by Muslim in the middle east. This steel was of such a high quality that repeated attempts by Europeans to duplicate the process had failed till the middle of last century, and the then Indian steel was a regular item of export from India to many parts of the world.

From the archaeological excavations and the studies during and after British period in India, evidence have been found that indigenous iron and steelmaking was practised in a big way in the eastern and the southern part of the country. The process of the steelmaking varied from one place to other and a good record of this has been compiled by various authors (Coomaraswamy., 1908; Prakash and Igaki; Bronson, 1986). The Indian iron and steelmaking industry flourished till 18th and 19th centuries. Later indigenous iron and steel industry started to decline and ceased to develop further due to various socio-technical reasons.

Characterization of an Earlier Ironmaking still being practised in Chhotanagpur Area, Bihar

Tribals of Chhotanagpur Division of Bihar State have been found practising an earlier ironmaking technique. Evidence of same was located on and near Netarhat Plateau. A few old artisans were located who had practised the art till almost 50 years earlier. Vikash Bharati, a regional centre for appropriate rural technology, has located the "Asur Birjia" tribals in Garha Harup Village on Netarhat Plateau (near Bishunpur on Lohardaga-Netarhat Road) who were practising an ancient process of ironmaking to make implements for their use. Recently RDCIS, SAIL, Ranchi made an attempt to study and characterise the process. Suggestions were made to improve the process without altering the basic technology and methodology followed by the tribals. These have been described in detail in an earlier paper (Prasad, K.K., 1990).

Here a vertical shaft type furnace is used. Lump iron ore and charcoal from sal wood are the only raw materials. Feet operated bellows are used for air blast. Furnace produces sintered lumpy sponge iron (some liquid slag which is generated during operation is tapped beforehand). Sintered lumpy sponge iron is repeatedly heated in a separate furnace and beaten to give the shape of final implement. For lase of comprehension, some of the diagrams of the paper are being reproduced here (Fig. 3 a, b, c & d). A few photographs taken subsequently, which give a clearer idea of the process (Pls. 43, 44, 45 and 46). Pl. 43 is a photograph of an excavated furnace, which was presumably operated for iron production over fifty years ago and the site is on plains near Bishunpur not far from the region where the climb for Netarhat Plateau starts.

Studies by RDCIS, SAIL

RDCIS, SAIL, Ranchi initiated investigations by temperature measurement and analysing the iron ore, charcoal, tuyere refractory, clays, sponge and metal. These samples have been subjected to chemical analysis, optical microscope and scanning electron microscope examinations as well as other physical tests.

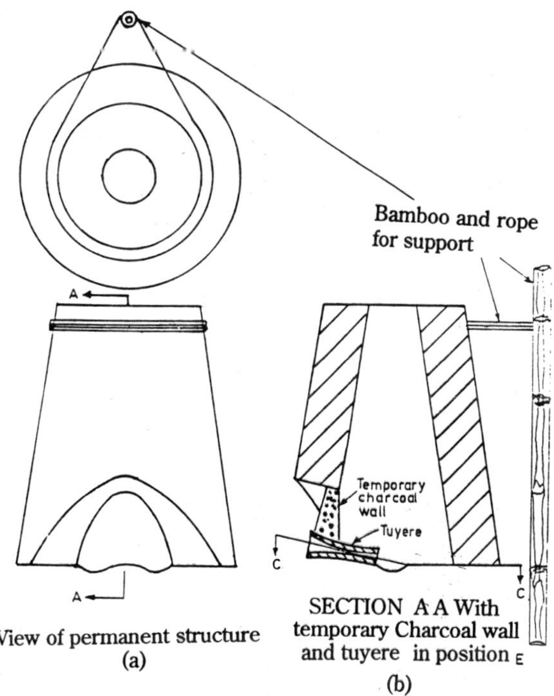

Fig. 3 Furnace for extracting iron being used by Tribals of Chotanagpur. Scale - 1:10

Studies indicate that this process operates at a temperature midway between the direct reduction and liquid ironmaking process temperature. The temperature attained in the tuyere region is as high as 1500^0 C. The total iron content of sintered sponge iron is only 55%. The final product obtained after reheating and beating contains around 96% iron, the balance being FeO and gangue. The carbon content is in the range of 0.01 to 0.04% and sulphur 0.007 to 0.013%. The inclusion level is found to be as high as 10% on average. Some scientists have described it as "Wrought iron-slag composite". More details are available in earlier paper (Prakash *et al.,* 1990).

Techno Social Viability of Earlier Ironmaking Practices in Modern India

The very fact that the process had died down and could be revived only through special efforts and incentives given by Vikash Bharati shows that the process does not work out to be viable and cannot be used as a means of livelihood by the tribals.

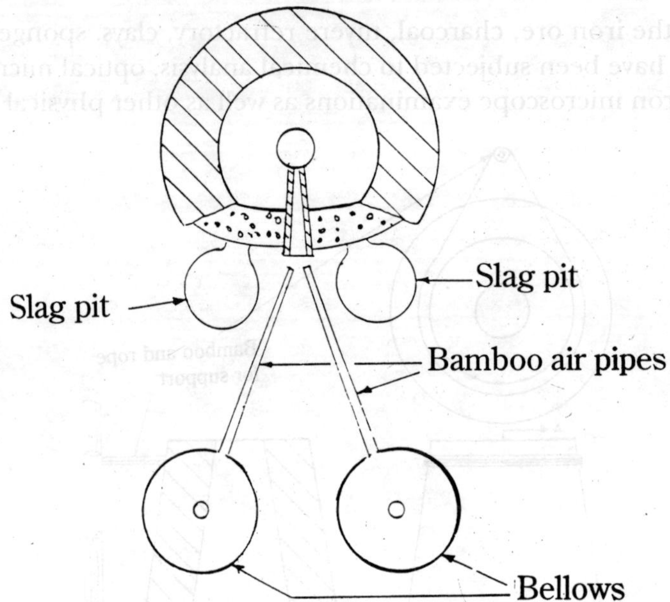

Fig 3 (C) Section C-C with foot bellows in position

Scale – 1:10

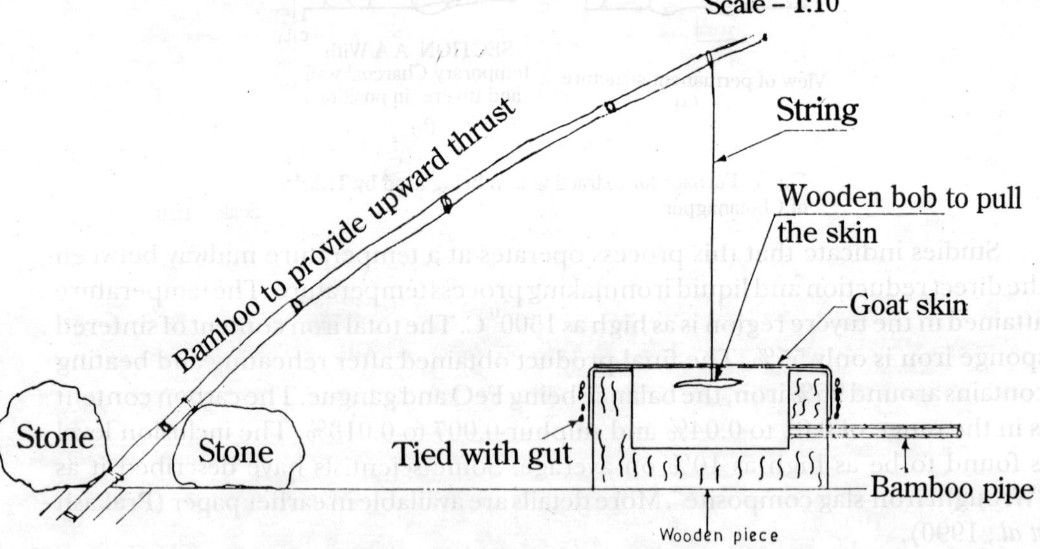

Fig 3 (d) Foot operated bellow

SCALE - 1:5

This is mainly due to large consumption of raw materials (iron ore and charcoal), low yield and inconsistency in the quality of iron. Further these processes cannot compete with the modern Iron and Steel Industry and due to the same reason, these earlier ironmaking processes are becoming extinct. But there are exceptions too. Media reports have indicated that in Bastar region of Madhya Pradesh similar furnaces are being worked for ironmaking except for the difference that a chute like structure is added to the top of the furnace where a part of the additional charge material is kept and it slides down automatically into the furnace as the burden level in the furnace goes down.

It was not clear whether this method of Bastar region is being used as a means of livelihood. But a recent discovery by Vikash Bharati in the same region (but which obviously is not the same process as above) has revealed some interesting facts.

The natives of Ghatgaon near Jashpur of Bastar region practice a process which again uses a shaft type furnace made of local mud. It is heated with charcoal (of sal wood) and then iron ore fines (it is a washed iron ore fines which they collect from the bed of an adjacent river) is spinkled into the furnace from the top along with some more charcoal. The end product is a sintered sponge iron and some liquid slag. The process may be compared with the modern fluidized bed process where gases are used to serve the dual purpose of fluidizing media and the reducing agent.

In description the Bishunpur and the Jashpur processes may appear similar. But the tribals of Jashpur are able to use this as a means of earning due to one important difference. They are able to perform it as a part time job. While they attend to their main occupation during the day time, in the evening they fill the furnace with charcoal and allows it to burn slowly by natural draught during the night. In the early hours, around the 4 or 5 a.m. the furnace is hot and as soon as they start blowing with the feet operated bellows, the fresh charges of charcoal and iron ore start reacting. In a matter of 2 to 3 hours, not only the lumpy sintered sponge is taken out, it is also beaten into shape and then they go about engaging themselves in their regular occupation.

DISCUSSION

When we look at these processes of historical and archaeological importance we find that they have a socio-cultural message to convey. Their evolution and development speak highly about the creative nature of our ancestors and tells about not only the success story of our self sufficiency which we had achieved in this field but also the excellence we had acquired. It is definitely of archaeological interest to document this ancient technique and preserve this technological heritage of the country. But question arises—should we stop here or should we do something more ? And what should be this "something more" ? Should we try to help the tribals in such a way that they are able to make a livelihood out of this practice ? In other words should we not try and improve these processes which they are using ?

As regards the last question, two attempts have been made. RDCIS, SAIL made detailed study and put forth some recommendations (Prasad, 1990) regarding the Bishunpur process. Although the implementation have not been exactly what we would have desired, but certainly some attempts were made and none of the results so far have shown any encouraging trend.

On the other hand I.I.T., Bombay tried to reconstruct the furnace, being used at Bishnupur, using fired fireclay bricks. But only after a very thorough insulation with Cerwool blankets could they achieve results somewhat close to what the tribals achieved (Ballal, 1990).

The above observations indicate that the modern materials, techniques and concepts need not be effective in improving such a process. And seeing the Jashpur experience, it appears that adoption of good points of one of such processes in another may be more effective.

The processes have another message. Although great developments are taking place in the field of materials, iron and/or iron based materials would probably remain important for civilised society even after the fossil fuels have been exhausted. And the means of iron extraction should be either the nuclear steelmaking concept or a shaft process using charcoal (renewable energy source), developed from what we have discussed above.

CONCLUSIONS

1. It is quite clear that all these ancient ironmaking processes are no match with the modern processes, mainly because of their highly energy intensive nature.
2. While considering modifications/upgradations of these processes one distinct feature has to be kept in mind. We cannot afford to change the basic features of these processes. The basic strength is utilisation (though not in a very efficient manner) of the locally available and free of cost raw materials. As soon as we start suggesting upgradation of raw materials or import of better material from other sources, all the processes will be unviable, as the cost of the product will not compare with what is available from modern technology.
3. The solution possibly lies in marginal improvement in the local practices so as to make iron available to small communities to take care of their day to day basic needs at a cheaper price as compared to what they can procure from the commercial places. Horizontal transfer of technology i.e. good points of one tribal practice, if implemented in other places may bring advantages.

ACKNOWLEDGEMENT

The authors wish to record their thanks to all those who assisted in the investigations. Special thanks are due to Sri T.M. Mukundan and Sri S. Suresh of PPST, Madras without whose assistance the work would not have been possible.

REFERENCES

Agrawal, B.B. and Prasad, K.K., 1988. Evolution of Sponge Iron Technology, *Invention - Intellegence,* Vol. 23, No. 8, Aug ' 88, p. 211.

Ballal, N.B. *et al.,* 1990. *Interim Report on Indigenous Iron and Steelmaking Technology,* Deptt. of Metallurgy, I.I.T., Bombay May.

Bronson, Bennet, 1986. The making and selling of Wootz - crucible steel of India, *Archaeomaterials,* Vol. 1, No. 1 (Fall 1986).

Coomaraswamy, A.K., 1908. *Mediaeval Sinhalese Art.*

Crysko, W.W., 1980. The Brith of Metallurgy, *J. of Metals,* July'80, p. 44.

Needham, J., 1980. The coming of the Age of Iron, p. 538 T.A. Wertime and J.D. Muhly, Eds., Yale University Press, New Haven, CT.

Noten, F.V. and Raymackers, J., 1988. Early Iron Smelting in Central Africa, *Scientific American,* June, p. 84.

Prakash, B., 1983. *Bull. Met. Museum* 8(1983), p. 23.

Prakash, B. and Tripathi, V., *Metals and Materials,* Sept. (1986), p. 586.

Prakash, B. and Igaki, K., Ancient Ironmaking in Bastar District, *Indian Journal of History of Science,* 19(2), pp. 172-185.

Prasad, K.K. *et al.,* 1990. An attempt on revival of rural Iron-making, *Metal News,* Vol. 12, No. 1, Feb, pp. 1-6.

Tylecote, R.F., Austin, J.N. and Wraith, A.E., 1971. The mechanism of the Bloomery process in shaft furnace, *Jl. of Iron & Steel,* May, p. 342.

Verhoeven, J.D., 1987. Damascus Steel. Part - I: Indian Wootz Steel, *Metallography* 20, pp. 145-151.

Yater, W.M., 1983. The Legendary Steel of Damascus, Part - III, *The Anvil's Ring* 11 (2): 2-13 (Summer).

Revival of Rural Ironmaking in Orissa

N.R. SRINIVASAN

Preamble

Iron and more particularly steel, has been in the service of mankind for a very long time. The art of making iron in primitive furnaces was practised in many parts of the world including India much before the dawn of the Christian era. It was essentially a small scale craft which was meant to meet the needs of agriculture and also war, when occasion arose. The emergence of the Industrial Revolution in Europe sowed the seeds for the birth of the iron and steel industry and tonnage production. The early decades of this century witnessed the setting up of an organised iron and steel industry in this country. Despite this, primitive iron making continued in the hinterlands, till even the early sixties. Thereafter, the curtain came down on ironmaking in rural India, because of its uneconomic nature coupled with the easy availability of the metal on the market and at an affordable price.

In recent years, the Government have taken up some measures to improve the lot of the villagers, primarily through employment oriented schemes. Moreover, scientific interest has been evinced by some institutions, to gain an understanding of the ancient process behind the art of making iron, which has produced monuments of curiosity. Some social organisations in the country also have launched programmes on reviving this ancient art.

Rourkela Steel Plant, at the instance of Dr. S.K. Gupta, then Managing Director, initiated a project in December 1990, on the revival of the iron making practice, with the view that given the benefits of marginal technological inputs, the ancient art could be made a viable rural occupation. It could provide a means of additional rural employment, meet local needs of the metal and generate some worthwhile income to the rural community, in Orissa, once conversant with the making and shaping of iron.

Ancient Practice

The antiquity of iron in India has been well established through literary evidence and archaeological finds, the latter with the aid of radiocarbon dating. It has been shown that the early iron age culture began around 1,000 B.C. and flourished for several centuries thereafter (Niyogi, 1914; Macdonell *et al.*, 1912 and D.P. Agarwal).

In Orissa

Orissa, as the other parts of India, was in the mainstream of ironmaking for several centuries. The local need of iron for making agricultural implements mainly was met by smelting in primitive furnaces, dispersed in many villages in Orissa. In addition, rural iron-making, shaping and treating it to the cause of weaponry, such as at the times of the Kalinga wars, were executed under royal patronage. It has been catalogued that some 17 types of weapons had been used in ancient Orissa (Mitra, 1961). These had been made out of iron smelted locally and treated and shaped suitably, to lend themselves for use in war.

Demonstration

Tribal iron making was demonstrated in a village near Baipariguda, in Jeypore subdivision of Koraput district, to a team led by Mr. S.K. Bagchi, Director, Birla Industrial and Technological Museum, Calcutta, in 1960. With the help of the village Sarbaseva Sangha, a clay furnace was constructed, the process of iron-making, including the preparation of the raw materials, was demonstrated. Photos of tribal ironmaking were also taken. It was the intention to follow-up this work but could not be done, owing to lack of financial support (Personal Comm.).

Around Rourkela

In a survey conducted around Rourkela in December 1990, it was found at Bangurkela village, which is about 20 kilometers from Rourkela on the rail route to Ranchi, that a small Lohar Community lives there. The people had migrated from Ranchi about three generations back. The Lohars here had practised iron smelting till about the time of the World War II. A significant fact that came to light was that they had used local "Kankar" or limestone as flux. However, the operations came to a close, as the men found that the profession was no longer able to sustain them. (Data was personally collected by the author at Bangurkela village from Sri Begila Lohar in 1990 and reported in Report No. PRS/7.0/14 dated 22.12.1990).

In the context of the project at Rourkela, it would be eminently relevent to highlight the archaeological finds and some unique artifacts observed and details on iron smelting in Orissa, which was practised as a traditional craft in many centres of the State.

Archaeological Finds

Archaeological excavations had been conducted in the post-Independence era, in many places of Orissa and specific references, relating to iron objects unearthed, are mentioned in brief here.

Sisupalgarh

At Sisupalgarh, a historical fort near Bhubaneswar, the Archaeological Survey of India, brought to light in 1948, 197 iron objects, such as nails, spikes, sickles, knife blades, spearheads, daggers, arrowheads and others. The objects have been dated and cover a period of over 5 centuries from 200 B.C. to 350 A.D. (Lal, 1949), revealing the existence of a flourishing craft in days of yore.

Gudavella

23 iron objects were excavated from Gudavella in the Bolangir district in 1970. Besides, 13 iron implements were collected from Kharligarh during 1967-68 and they are now kept in the Tribal and Harijan Research Centre of the Government of Orissa. The materials from both the finds are sickles, axes, spades, daggers, knife, nails, latch, chisel and objects of household utility. These are identified as being roughly contemporaneous with those of Sisupalgarh. These iron relics of the past in Orissa from the two sites, have been assigned a timeframe of 200 B.C. to 400 A.D. (Dash, 1982). The above archaeological finds tend to show the existence of an active iron working phase in Orissa in the remote past.

Balasore

Recently in June 1990, a joint excavation conducted by the Orissa Institute of Maritime and South East Asian Studies and the State Archaeology Department of the Govenrment of Orissa, at a site in Balasore district, has unearthed a Danish ship belonging to the 16th century A.D. The excavation was done under the supervision of Dr. C.B. Patel of the State Archaeology Department (*Indian Express*, 1990). A sample of a rusted iron nail taken from the ship, furnished by Dr. Patel, was analysed at the Research and Control Laboratory of the Rourkela Steel Plant, in January 1991. The chemical composition and the micro-structure have indicated that the nail was made of a type of wrought iron (See Annex. 1). It is learnt that further investigations at the site are in progress.

Artifacts

Konarak

There are some interesting artifacts of iron in Orissa. Iron beams of wrought iron as also iron pieces had been used in the construction of the famous Sun Temple at Konarak in Orissa. The forging of the beams is not very perfect and they are in a highly rusted condition, which has been ascribed to accelerated corrosion by the saline environment. The Konarak beams are dated to the 13th century A.D. Thus, the technology of iron making and especially the forging of large size blooms had been prevalent in Orissa at that time.

Huma

It has been recorded in the District Gazetteer, Sambalpur, that an iron spike is present in the Samalai Temple (Senapati, 1971). Moreover, iron Trisuls, nearly one meter high have been observed, in front of the Bhairava temple at Huma, 32 kilometers from Sambalpur by this author, as also on the top of the temple towers. The Bhairava temple is at the entrance to the Siva Temple complex. An astonishing fact little known to the outside world is that the Huma Temple is akin to the Leaning Tower of Pisa. Its inclination is a significant 47 degrees, but the remarkable fact is that the pinnacle is vertical to the ground. Another unique feature is that the "Lingam" in the Sanctum Sanctorum is also seen to be leaning. The shrine, on the banks of the Mahanadi, is learnt to have been built in 1670 A.D. (*The Telegraph* Calcutta, 1991). Therefore, the iron artifacts at Huma are over three centuries old.

Sambalpur

Moreover, some unique specimens of iron artifacts in the form of bells have been inspected by this author at the Samaleswari Temple at Sambalpur, which name is derived from the Goddess. The iron bells, numbering over a hundred, are long and cylindrical in shape and vary in size from about 80mm in diameter to 120mm in height to around 160mm in diameter to 250mm in height. They were in a partly rusted condition. They had rings on top for the passage of cord and wooden rods in the centre. It is learnt from the temple authorites that the bells were donated to the temple as offerings for keeping vows and tied around the temple pillar. Oral tradition claims that the bells could be around five hundred years old (Senapati, 1971). It is also learnt that iron bells exist at the Gauteswari Temple at Chipilma and at Bonaigarh and that iron objects were worshipped at Bonaigarh and Deogarh. The use of iron objects for offerings and worship seems to be peculiar to Orissa but does not appear to be in vogue today. The precise purpose of inducting iron into such religious worship remains obscure at this distance of time.

Traditional Craft

A survey of the traditional crafts, which were persisting at the time of the Census Operations in Orissa in 1961, was instituted by the Government. As a consequence, a survey on the status of ironmaking in Orissa was undertaken in 1962. A report of the survey (Census, 1961) had noted that the Lohars of Koraput and Sundergarh and the Kamars of Sambalpur districts derived their sole means of livelihood from the hereditary profession of iron smelting and smithy. Their women folk also assisted in the operations. It is learnt that during the passage of time an element of specialisation was introduced into the profession. People primarily engaged in smelting were called Lohars, while those who shaped the metal came to be known as Kamars. However, the terms were mutually interchangeable and denoted those practising the rural craft of iron making and forgoing.

Places

According to the Survey, the craft was carried on in many of the villages in the interiors of the districts of Koraput, Kalahandi, Bolangir, Sambalpur, Sundergarh,

Keonjhar and Dhenkanal. Koraput and Sambalpur were predominant, 50 villages in the former district were engaged in ironmaking with Singelbecha as the most important centre. In Sambalpur, at least 15 villages were on this job and Pentabahal was singled out for making tools and implements of iron. A map of these villages connected with ancient ironmaking in Orissa is appended at *Map - I*.

The Practice

Primitive ironmaking was done in clay furnaces, called 'Bhati' in Sambalpur and 'Gaana' in Koraput. While the furnace in Sambalpur was built above the ground, that at Koraput was underground and claimed to give better yield. Low grade sized iron ore was charged in layers with wood charcoal and foot bellows were operated by women. Each charge was called a "Bhadi" and the entire operation done with 12 charges of ore and 9 of charcoal. The heating was done for about two hours with slag discharge and when the Kamar judged the metal to be ready, the blowing was stopped. The blazing body inside the furnace was pulled out and taken to the forge. Three operations were done in a day.

The metal from Penthabahal was analysed at the R & C Laboratory at Rourkela and found to be wrought iron, with 0.3 per cent carbon and non-uniform. However, it was good enough to be shaped into tools and implements. The metal was carburized to impart some hardness. The yield was very poor and each operation resulted only in about 2 "seers" of sponge iron.

The Economics

It was calculated that, at the time of the survey, a 7-day week with 3 persons at work resulted in 5 "seers" of saleable products. At the then selling rate of Rs. 1.37 per "seer", the total earnings came to Rs. 7.50. This worked out to only 50 paise per worker per day, as against the then prevailing rate for casual labour at Rs. 1.50 per head per day. Though the Kamars were committed to this craft, in the absence of any other vocation, it was totally unremunerative. But the patronage extended by the rulers of the erstwhile small states and Zamindars made them continue with the profession. However, after Independence, the Kamars were subjected to considerable harassment by petty officials and the administration took no notice of their plight, thus denying them the very living existence. Therefore, the ironsmiths had no option but to close shop and rural iron making became a relic of the past from early 1962, in Orissa.

Revival

In taking up the Project on the *Revival of Rural Ironmaking*, Rourkela Steel Plant has been inspired by larger considerations, in an endeavour to harness the large rural human resource potential and its skills, by providing an additional/alternate source of income, by bringing in marginal technological inputs, so as to aid in the resurgence of a vibrant rural economy in Orissa.

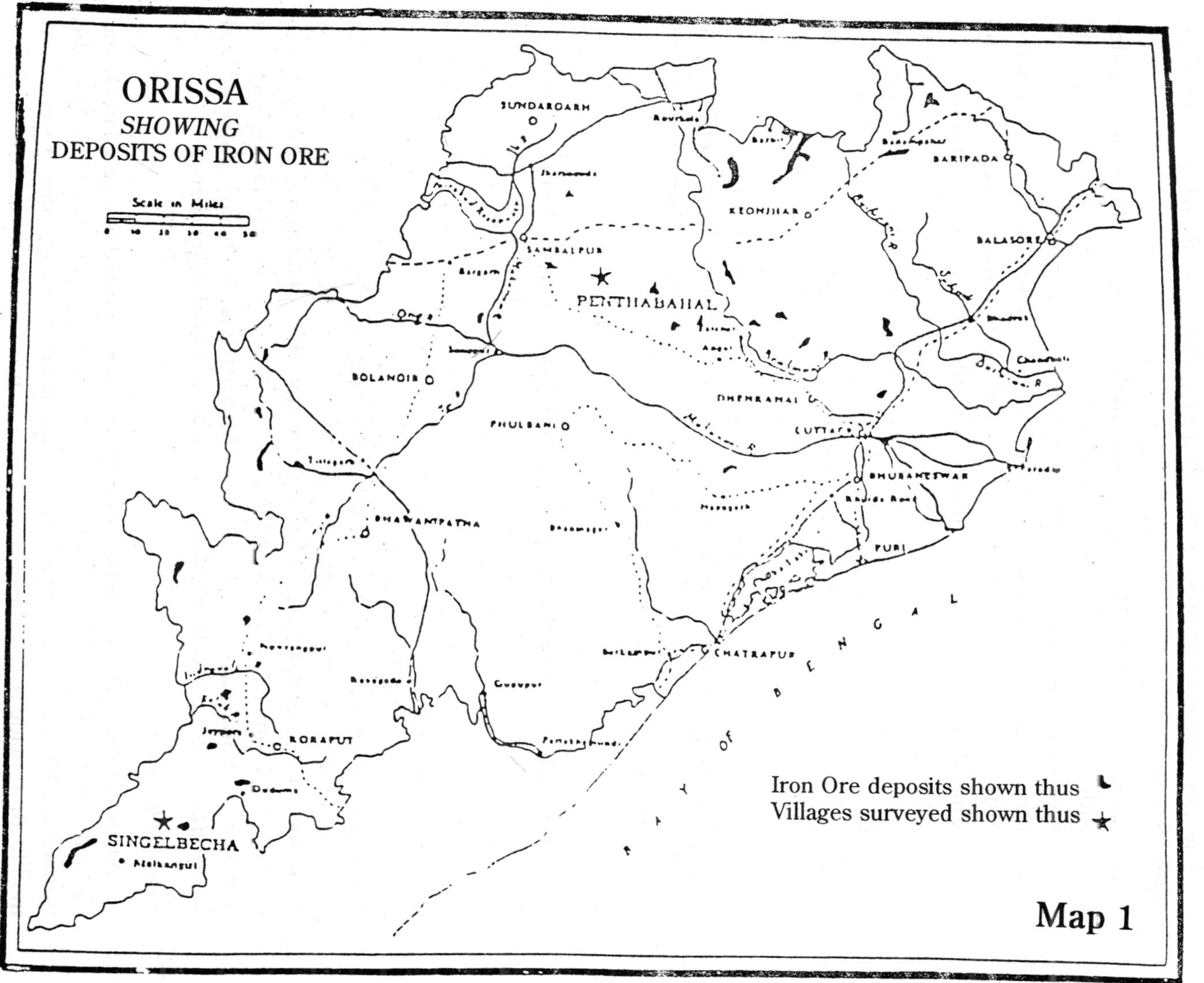
ORISSA
SHOWING
DEPOSITS OF IRON ORE
Scale in Miles
SUNDARGARH
SAMBALPUR
PENTHABAHAL
KEONJHAR
BARIPADA
BALASORE
BOLANGIR
PHULBANI
DHENKANAL
CUTTACK
BHUBANESWAR
PURI
BHAWANIPATNA
CHATRAPUR
KORAPUT
SINGELBECHA
BAY OF BENGAL
Iron Ore deposits shown thus
Villages surveyed shown thus
Map 1

Objectives

The immediate objectives of the Project initiated at Rourkela have been outlined as the:

(i) Setting up of a Pilot/Prototype ironmaking furnace on the ancient model, with the active collaboration of Lohars currently employed at the plant and reproduce the making of iron, as practised in the past

(ii) Establishment of a demonstration Furnace at the Industrial and Safety Museum of the Township and invite participation from interested local and adjacent rural population with a view to the dissemination of information

(iii) Providing of marginal technological input with a view to make the ancient craft energy intensive and cost effective, through experimentation support from the Research and Control Laboratory

(iv) Selling the idea of rural ironmaking to selected villages initially in the Sundergarh district and assist in setting up furnaces with active local participation

(v) Inviting the cooperation and soliciting the goodwill of the District and State Administration and exploring ways and means of aiding the setting up of furnaces in rural Orissa and encouraging the rural populace into adopting the ancient but now more viable process.

Schedule

A core group was constituted with Mr. B.R. Satpathi, Assistant General Managar, Technical Development as the team leader for spearheading the project work. This author was assigned the charge of Documentation, while Mr. U.C. Patra, Sr. Manager, was in charge of process development. Seven selected Lohars working in the plant were inducted to familiarise them with the craft, with the view that they would be deputed to the villages, for helping the construction of furnaces and teaching the men the forgotten art. A tight schedule was drawn up, whereby, work would be undertaken in two phases, and culminate by March 31, 1991. The first phase was the setting up of a modified clay furnace at the plant and establishing the parameters of improved operation and the second phase was the putting up of a demonstration furnace at the Township. This work has now been completed and salient details of the results obtained are furnished hereunder, in brief.

Furnace Operation

A typical furnace, based on the design of the one reported to be used in Orissa, was constructed. The furnace details are shown in Fig. 1. The total height of the furnace is 1,050 mm, of which about 200 mm is below the ground. A significant modification made is that the furnace wall is in two layers with an annular space, filled with sand, for better heat conservation. A fluted tuyere is placed at an angle

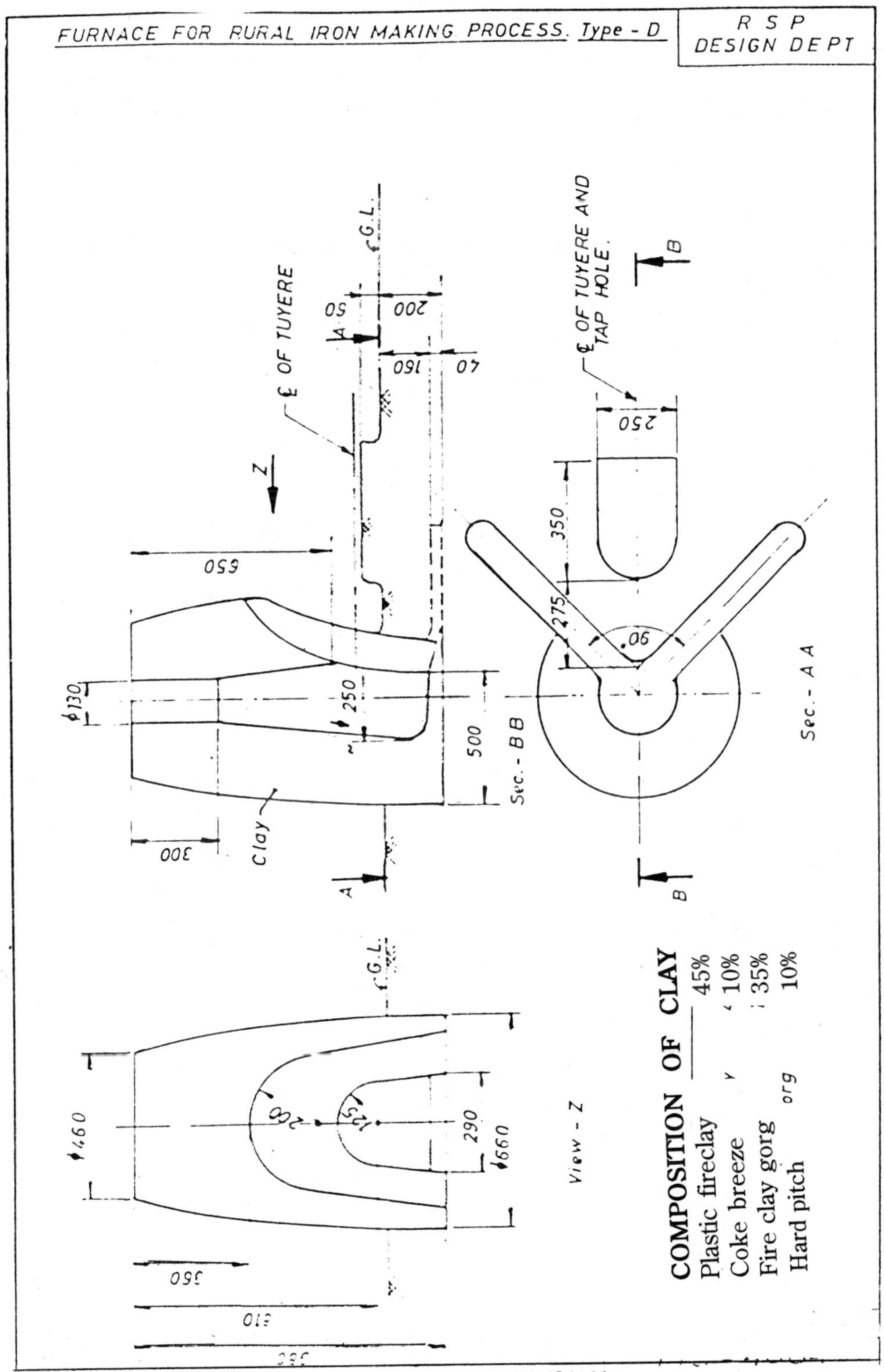
FURNACE FOR RURAL IRON MAKING PROCESS. Type - D
R S P
DESIGN DEPT
℄ OF TUYERE
G.L.
℄ OF TUYERE AND TAP HOLE.
Clay
Sec. - BB
Sec. - AA
View - Z
COMPOSITION OF CLAY
Plastic fireclay 45%
Coke breeze 10%
Fire clay gorg 35%
Hard pitch 10%

Fig. 1.

of 15-20° to the horizontal. Preheating of the charge is done (The picture show the air blowing into the furnace Pl. 47).

The furnace is started by charging charcoal from the top and lighting it. The burden consists of 10 charges of 15 kg. sized ore and 25 kg. charcoal each, with some lime-stone. Air blast is maintained continuously by a hand blower, which has been found to be better than foot operated bellows. Normally, 4-5 deslagging operations are done. The blowing is stopped when the charge has descended to the critical level. The sponge iron lump is pulled out with a pair of tongs. It is subjected to hammer blows to consolidate it and remove slag inclusions. Then, it is transferred to the forge for shaping into implements. The parameters of the operations have been perfected during nearly two months of trials. A photo micrograph of the wrought iron (Pl. 48) shows the uniform texture, with very little of inclusions.

Viability

The crux of the entire scheme is its viability, on which the following details are furnished. The cost of the raw materials and incidentals are only reckoned, while the labour put in by two men is not accounted for separately, as they are deemed to be self-engaged and hence the beneficiaries. The preliminary costing is shown in Table - 1.

TABLE 1

Preliminary Details (SAIL)

Charge weight		
Iron ore	-	20 kg.
Charcoal	-	20 kg.
Limestone	-	03 kg.
Total time taken		
for tapping	-	04 hours
for forging	-	04 hours
Wrought iron product	-	06 kg.
Finished weight	-	04 kg.
Number of implements made	-	20
Number of persons engaged		Two
Cost		*Rs.*
Cost of mineral inputs		4.00
Cost of charcoal		30.00
Incidentals		6.00
Total		**40.00**
Value of output reckoned at current price		100.00
Available surplus		60.00
Earning per head/.day		30.00

PHOTO - 1
As received.

PHOTO - 2
M=100 X
Unetched

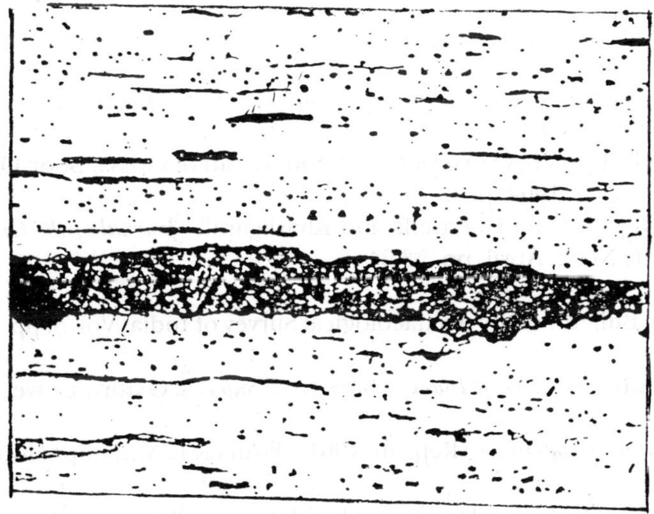

PHOTO - 3
M=100 X
Etched
(Nitlal 2%)

Fig. 2:

It would thus be seen that the return is commensurate with the labour put in and is at par with the current wage, in the region.

Technology Transfar

The project was aimed at improving the viability of the ancient practice, with induction of marginal technological input. It has now been established that this could prove to be a viable rural occupation, especially in areas, where raw materials are available. A preliminary report on the project has since been submitted to the State Government authorities at Rourkela and the district-headquarters at Sundergarh. The Community Development department of the steel plant is anxious to take the technology to the rural villages and aid in the establishment of modified clay furnaces in them. The transfer of technology is sought to be achieved through Lohars, now thoroughly conversant with the process. For the scheme to succeed, it is indispensable that the State Government authorities extend their cooperation and willing involvement. Once the idea catches up at least in one or two selected villages near Rourkela, then it can be expected to be emulated elsewhere in the State. Thus, Rourkela Steel Plant has taken the first step in contributing to a resurgent rural Orissa, through the revival of the ancient ironmaking practice.

ACKNOWLEDGEMENT

The author has pleasure in thanking Mr. B.R. Satpathy and Mr. U.C. Patra for making available data on the furnace operation and the economics of the scheme. He thanks Steel Authority of India Limited, for permission to present this paper.

REFERENCES

Agrawal, D.P. *et. al.*, "Chronology of Indian Pre-history from the Mesolithic Period to the Iron Age", *Journal of Human Evolution*, Vol. 7, pp. 37-44.

Census of India, 1961, Vol. XII, Part VIIA Survey of Traditional Craft of Orissa, Book One, *"Indigenous Smelting of Iron"* 1964, p. 63.

Dash, R., 1982. "The Iron Objects from Gudavella and Kharligara", *Journal of the Orissa Research Society*, Vol. 1, No. 2, April, pp. 53-64.

Indian Express, Vizianagaram, June 8, 1990.

Lal, B.B., 1949. *Ancient India*, Bulletin of the Archaeological Survey of India, Vol. 5, pp. 62-105.

Macdonell, A.A. and Keith, A.B., 1912. *Vedic Index of Names and Subjects*, Oxford University Press, London.

Mitra, Rajendralal, 1875. *Antiquities of Orissa* (Reprint. 1961). Firma K.L. Mukhopadhyaya, Calcutta, 1961.

Neogi, P., 1914. *Iron in Ancient India*, The Indian Association for the Cultivation of Science, Calcutta.

Personal Communication from Director, Birla Industrial and Technological Museum, Calcutta, No. BM-34 (44)/SC/90 dated 24th November 1990 and enclosure.

Report of the R & C Laboratory, Rourkela, dated January 1st, 1991, appended as Annex-I.

SAIL, Data furnished by Messrs B.R. Satpathy and U.C. Patra. Technology Development Department, Rourkela Steel Plant.

Senapati, N. (ed.), 1971. *Gazetteer of India, District Gazetteer, Sambalpur*, Orissa Government Press, Cuttack, p. 548.

The Telegraph, Calcutta, 15 August, 1991, p. 7.

Tour Report on visit to Sambalpur by Research Fellow, SAIL, vide P.R. S/5.0/17 dated 13.12.1990.

ANNEX - I

Sub : Report on Rusted Iron Nail received from Dr. N.R. Srinivasan from Rourkela - Retrospect and Prospect Study.

Ref : Letter No. PRS/3.0/11 dated 18th December'90.

1.0 One rusted iron-nail sample was received by us on 19.12.'90 from Dr. N.R. Srinivasan, Research Fellow, SAIL, RSP, Rourkela. As informed by him, this nail was found in Balasore excavation by Dr. C.B. Patel and belonged to 16th century A.D.

2.0 A detailed metallurgical study was conducted in our laboratory to ascertain the material of the nail. The results are as follows:

2.1 *Visual Examination :*

The sample nail was found to be heavily corroded (Photo 1).

2.2 *Chemical Compositions* :

Sample drillings for chemical analysis was obtained after removing surface scales and the analysis was done by conventional wet method.

C	:	0.06 %	S	:	0.100 %
Mn	:	0.04 %	Si	:	0.23 %
P	:	0.167 %			

2.3 *Hardness* : VPN 148-152 (HV5)

2.4 *Microscopic Examina ion* :

Two samples were cut from the nail - one along the length of the nail and the other across. The results are given below :

Longitudinal Sample (Photo 2 and 3 at 100 X):

Cleanliness	:	Plenty of slag, glassy silicates and iron sulphide inclusions.
Microstructure	:	Thick stringers of slag and inclusions in the direction of metal working. Ferritic matrix.

Transverse Sample :

Cleanliness	:	Patches of slag in the matrix.
Microstructure	:	Equiaxed grains of ferrite, with globular silicates.

3.0 *Remarks :*

The chemical composition and microstructure indicated that the nail was made of a type of wrought iron.

Chief Superintendent (Met.)

Sd.
(K. Bishoyi)
Superintendent (Met.)

High Tin Bronze Working in India
The Bowl Makers of Kerala

SHARADA SRINIVASAN

Archaeo-Metallurgical Back-Ground

The earliest report on bronzes with a high percentage of tin in South India comes from the study by Breeks (1873: 94) on the bowls uncovered from the cairn burials of the Nilgiris in Tamil Nadu. A few bronze bowls were found to contain about 23% and others 30% tin. Tin bronze vessels of 22-30% tin have been found in the iron age megalithic burials of Tamil Nadu of the first millennium B.C. from Adichanallur, Maula Ali, Coimbatore, and also the Nilgiris (Leshnik, 1974). As early as 1940's Paramasivan (1953: 418) conducted a metallurgical investigation on a bowl from Adichanallur, Tirunelveli district. This was found to contain 23% tin, with the micro-structure substantially different from ordinary lower tin bronze. Marshall (1954) also reported the occurrence of bronze vessels from the Gandharan capital of Taxila in North West Pakistan datable to mid 1st millenium B.C. with upto 25% tin.

A common feature of these bowls, usually shallow and perfectly concave in shape, is their thinness, often only a couple of millimeters. Some are even much less than a millimeter, as in the case of the strainers from Adichanallur (Rea, 1915). Such thinness would have been very difficult to achieve using the traditional cast lost wax process or a cast piece moulding process. If the bowls were to be wrought, ordinary alpha tin bronze (i.e. a solid solution of tin upto 10-15%) would have been inadequate as it can be worked to only a limited extent after annealing. Above 15% the hard and brittle delta-component of the eutectoid forms in cast bronze at ambient temperature (Fig. 1), which is very difficult to work. However, at 22.5% tin in the copper-tin equilibrium system a metastable beta phase is formed between the temperature range of 755-586° C, which is a superplastic alloy known as high tin bronze or beta bronze. This plastic beta phase can be worked to a considerable degree and brought within the temperature range of beta formation i.e. 586° - 755° C. Hence the presence of about 22-25% tin in the bronzes would suggest that the ancient metallurgist may have had the empirical knowledge to work within the beta phase range of temperature.

Rajpatik and Seeley (1982: 26-31) comprehensively investigated similar bronze bowls from the burials of Ban Don Ta Phet, Thailand, datable to about 3rd-

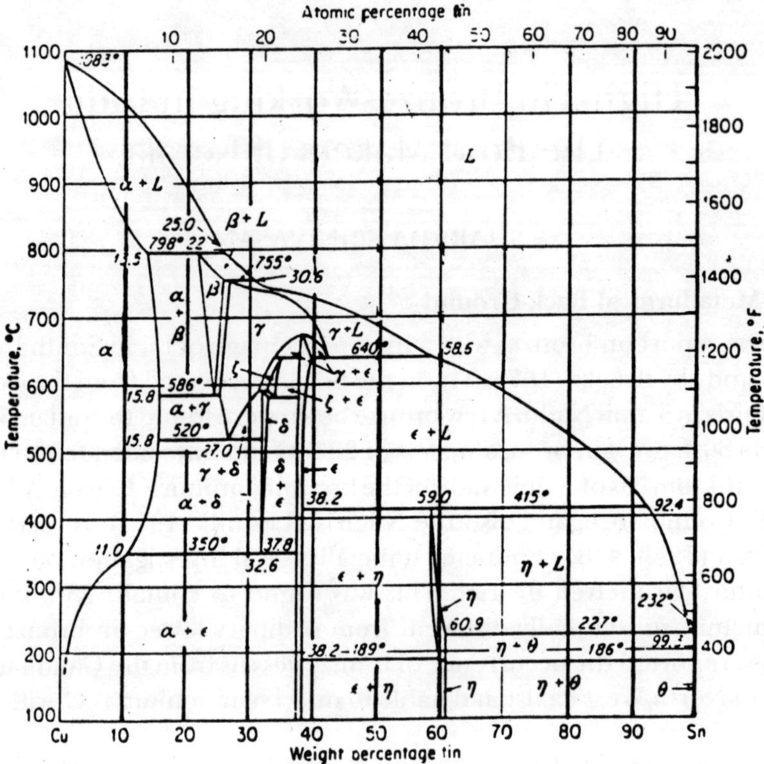

Fig.1 Copper-Tin Equilibrium System. Note the presence of the ß -phase at 22.5% tin between 586-755° c (American Society for Metals Handbook)

2nd Cent. B.C. on the basis of associated finds of characteristic Indian beads. These bowl fragments showed 23% tin and were much less than a millimetre thick. The most interesting feature was the prevalence of the metastable ß-phase in the microstructure as elongated acicular martensitic needles. This indicated that the martensitic beta phase had been 'frozen' as it were by rapidly quenching in water after the alloy had been heated to the super-plastic beta temperature and worked to a certain degree. Quenching would have served to retain the metastable ß-phase and thereby prevent the formation of the brittle delta-eutectoid which would have formed if the alloy was allowed to slowly cool down to ambient temperatures. The extreme fragility and brittleness of such bronzes in an archaeological context may be due to the formation of the brittle delta phase component over time as a corrosive process.

Yet another interesting feature noted by Seeley and Rajpatik of quenched beta bronze is that when polished, it acquired a bright golden colour which would have been prized in antiquity. There is evidence that this alloy was used quite abundantly in coinage by the Satavahanas of the Krishna Valley in Andhra, in the early centuries A.D. with alloying proportions around the beta range i.e. 19-23% tin (Shastri, 1980; Allan Horsler). This preference may have been due to the golden colour and super-plasticity well suited to die striking, around the ß temperature range.

Moreover wrought bronze metal has a high ringing tone when struck, exploiting the inherent plasticity of the wrought alloy. This tone is quite distinct and much higher than the more resonant sound obtained from the popular alloy used for bells i.e. cast bell metal consisting of 25%-30% tin to 70% copper. Hence wrought and quenched beta bronze, also known as gong metal, seems to have been used to make cymbals and gongs as observed by Martha Goodway (1988) in a study on the makers of gongs in the Philippines.

In the Indian context, Rajpatik and Seeley found that the micro-structure of the Adichanallur bronze studied by Paramasivan compared well with those studied from Thailand. They concluded that similar bowls as found from Adichanallur, Nilgiris and Taxila were probably imported from the tin-rich regions of South East Asia as there appeared to be no metallurgical evidence for working high tin bronze in India. Compared to South East Asia, the occurrence of bronze in South India, and India in general, is quite sporadic which may be attributed to the paucity of tin reserves and dependence on external trade to import tin. There is no evidence in antiquity for the use of ß-bronze outside the Indian sub-continent, South East Asia and the Indo-Tibetan regions (Goodway, 1988). However, cymbals and gongs have been part of the rich musical tradition of the sub-continent. Further more, scattered literary and anthropological data may be found related to bell metal working in general, which do suggest that the metallurgical traditions of working high tin bronze may be more deeply entrenched in the sub-continent than hitherto explored.

Literary and Ethnographic Evidence

The occassional literary reference does throw light on the use of high tin bronze and bowls in India. There is an important description in the Greek account of Nearchus in the 4th cent. B.C. (Strabo, 1115.1.67) (Rajpatik, 1982) that the Indians used vessels which were beaten rather than cast and which broke like pottery when dropped. This clearly indicates the use of beta-bronze in this period, whether imported or of local manufacture; as corroborated by the finds from Taxila and the South Indian megaliths of the mid 1st millennium B.C.

The *Arthashastra* (2.12.14), the noted Mauryan Sanskrit treatise on economics (3rd cent. B.C.) gives insights into a well organised mining and metalworking industry. The exploitation of what has been interpreted as tin ores ie. *trapu dhatu* is referred to therein (Kangle, 1972). Tin ores proximal to the Indo-Gangetic plain could have been those of the Bactrian region in Afghanistan or even the uneconomic but significant cassiterite deposits in Bastar region of Madhya Pradesh and the Chhota Nagpur plateau in Bihar—which is closer to the Mauryan capital of Pataliputra in Bihar. There is also the possibility that the more abundant South East Asian tin was accessible.

The *Arthashastra* (2.12.23) also mentions that the Director of Metals should establish factories for copper (*tamra*), lead (*sisa*), tin (*trapu*), brass (*arakuta*), Steel (*vaikṛntaka*), iron (*loha*) and bronze/bell metal (*kamsa tala*). *Kamsa tala* is interpreted here as bronze of various proportions (*tala*: measure). The Tamil classic of the Sangam era, the *Silapadikaram,* loosely dated coeval to the megalithic period (3rd cent. B.C. - 3rd cent A.D.), mentions the use of *kanca-tala* or cymbals for musical accompaniment (Srinivasan, 1989: 9-16). *Kanca* is interpreted as a corruption of the Sanskrit *kamsa* (*Tamil Lexicon,* p. 43, Col. 2). Hence *kamsatala* may also perhaps be taking to signify cymbal metal, as *tala* also denotes rythm, in the context of *Arthashastra.* The *Rasaratnasamuchchaya,* the 12th cent. alchemical text (Ray, 1956), refers to *kamsa* as an alloy of 80% copper and 20% tin; which is close to the composition of beta bronze.

The words *kanjam, kancam* and *kanjaram* in Tamil seem to be used to describe a range of bell metal alloys. (Bell metal is used in this section to broadly denote bronze alloys with a high proportion of tin). (*Tamil Lexicon,* p. 43, Col. 2). *Kanjam* is used to describe cymbals, also a bright polished speculum mirror (which is a copper tin alloy; as still made in Aranmula, Kerala) in the *Arichantira Purana* (*Vivaka* of 123), and bell metal censor (*Civakacintamani,* V 2140). *Kancam* denotes bell metal, musical cymbal, drinking vessel and also that which is golden; while *kanjara* denotes tambourine. Hence *kanjam/kancam* suffices etymologically to describe most properties of high tin bronze.

Although the significance of ß-bronze working has not hitherto been specifically identified in the ethnographic record, mention has been made in general by Meera Mukherjee (1978), of artisans making 'wrought bell metal bowls' in various parts of India, which may very likely indicate ß-bronze bowl making.

Many such centres described by her are in West Bengal and Orissa and some in South India. The Kansaris or bell metal artisans of Orissa made rice bowls called *kansa* and gongs called *kansaris* by hammering out ingots (Mukherjee, 1978: 103; 101). In West Bengal, the *Kangsabaniks* made dishes of *Mihi Kansa* or fine bell metal and vessels by the wrought metal process from ingots alloyed to specifications (Mukherjee, 1978: 91; 92; 89). *'Kansa'* in use appears to be a corruption of the Sanskrit *Kamsya.* Goodway (1988) also mentions that the Philippines word for high tin bronze was a corruption of the Sanskrit *Kamsya.* Further South Mukherjee (1978: 127) mentions the wrought bell metal bowls in parts of Andhra Pradesh and

Tamilnadu made by the Kammalars or artisans.

Coastal Orissa, Andhra, Tamil Nadu and Bengal were part of an organised maritime trade network with South East Asia from the Christian era as pointed out by Himanshu Ray (1990) which could account for import of tin. But it is also significant that the bell metal centres of Orissa and West Bengal are located near the small cassiterite deposits of the Hazaribagh and Bastar regions of Chhotanagpur where some pre-industrial mining by local tribals is reported. (Mallet, 1874: 23-24). Bronzes of a higher tin content were also noted from 1st cent A.D. Assur graves of Chhotanagpur (Chakravarti, 1978), throwing up the possibility of limited local exploitation in antiquity.

About the Nilgiri material, Breeks (1873: 94) report observes that the bowls from the megalithic burials compared with those from local bazars in Calicut in Kerala with about 23% tin, while some had 30% tin. In this context, the customs of the Toda tribes of the Nilgiris—who can claim the closest descendance from the megalithic peoples of the Nilgiris—could throw light. Indeed Toda anthropologist Evam Piljian (Pers. Comm.) observed that in Toda custom, bowls of 'vengalai' (colloquial Tamil for bell metal) were received as sacred offerings. This was because the Todas were revered as the descendants of the Pandavas who, according to folklore, were believed to have set up the megaliths. These bowls she said came from Kerala in the plains. Vengalai is a term broadly used to describe bronze alloys and more particularly cast bell metal with about 30% tin.

Commenting on the subject my grandmother from Tamil Nadu mentioned that in her heydays, Trichur, the temple heartland of Kerala, was renowned for its quality vengalai, although it was also procured in Tamil Nadu. Significantly she mentioned the vessels called *Olavettu* and *Talavettu* which 'shattered when dropped' like Nearchus' description. A speciman of this, that I examined, bore out a resemblance to wrought ß-bronze bowls. While *tala* stood for musical, *vettu* she said was old Tamil for plate or vessel. It is noteworthy that the articles in public use mentioned in Chola inscription (*South Indian Temple Ins.*, Vol.13, No. 35) included *vettil* i.e. cup or vessel.

Olavettu and *Talavettu* preserved curds and sour things better than *Vengalai* or bell metal, she said, but would burst when heated. These would be in keeping with the properties of the unstable quenched ß-phase of high tin bronze. She also thought that in her knowledge, these preceded the use of *Pittalai* or brass, which was not as useful for cooking purposes due to its corrosive tendencies.

High Tin Bronze bowl making in Kerala

Based on the clues that emerged, the starting point of this investigation was the metalware shops of Trichur in Kerala. Naduvarambu, about one kilometre from Irinjalakuda in Trichur district, has a thriving traditional bell metal casting society.

Here traditional vessels and cauldrons, called *uralis* and lamps are made by shaping the mould on a lathe, casting the item using the lost wax method and finishing the cast item on a lathe. In the bell metalware shops of Irinjalakuda were several gongs and bowls, old and new. These the shop keeper confirmed were made—not by casting—but by the wrought process and then quenched in water, i.e. beta-bronze bowls. He knew of only two places in Kerala where this formerly thriving industry survived; in Payangadi in Palghat District and Mallapuram in South Kerala.

Payangadi, about fifteen kilometers from Pattambi, Palghat district, is a small hamlet where eight people were identified from the Kammalar caste, jointly working on making wrought beta-bronze bowls by hammering ingots. The Kammalars were an important traditional caste of artisans said to have migrated from Tamil Nadu to Kerala (Mukherjee, 1978). They have been variously identified as braziers (Thurston, 1909) and smiths. They are mentioned in medieval Tamil (*South Indian Temples Ins.,* Vol. 3 pt. 2 No. CLVII) and Kannada inscriptions (*Epigraphia Carnatika* V(R), Ka. 116).

The Kammalars of Payangadi were engaged in fashioning the ingots into bowls when I visited them. The alloy used was mentioned by Kammalar Bhaskaran to be of 10 kg. of pure copper and 2.750 kg. of 'velliam' or white lead i.e. the colloquial word for tin. This works out to a rather precise alloy proportion of tin to copper of 22.4%; very nearly the beta bronze proportion of 22.5% tin. He also noted that about 500 gms. of alloy was lost due to heating.

The ingot to be wrought was made by melting copper and tin in the above proportion or scrap of the same alloy in a crucible heated uniformly with cakes of cowdung fuel, and then poured into a circular depression in the earthern floor. This gave a flat circular ingot about 8" in diameter and about a centimetre thick. To hammer this ingot into a concave bowl, the superplasticity of the alloy between 586°-755° C was exploited. The ingot was put on a low charcoal fire and heated to a rosy pink temperature which the Kammalar insisted had to be the right temperature—not red hot nor too cold—which is we may presume is the beta range of temperature.

To hammer the ingot, four men were positioned equispaced around the ingot. A large wedge-shaped iron hammer fixed in a square wooden frame called *cherungalam* was slung over their shoulders and used to hammer the ingot in quick successive blows (Pl. 50). After hearing all four blows the man supervising the process expertly turned the ingot a bit with iron tongs. The blows were then repeated moving outwards from the centre of the ingot each time.

Thereafter the ingot was once again heated to raise it to the beta temperature from which it would have cooled down. The ingot was placed convex on the flame and heated mainly on the concave side which was to be hammered. The plasticity of the alloy was indicated in the ability to beat out an ingot 8" in diameter and 1 cm thick; into a perfectly shaped concave bowl all of 14" in diameter with rim thickness of upto 3 mm. The whole process of hammering was remarkably skilled and controlled and took about three hours of sustained effort.

Thereafter the bowl was once more heated to the rosy pink metastable beta temperature range and plunged into water quickly to quench it. Kammalar Bhaskaran demonstrated his empirical understanding by stating that it was necessary to immerse the bowl in water after working because otherwise it would be very breakable. This would be due to the formation of the brittle delta component if allowed to cool slowly. (Pl. 51)

In the finishing stages, the rough flans of the rim of the bowl were chiselled off and filed down to a smoothness. The inner surface of bowl, blackened due to hot working, was scraped, polished and filed down using steel scrapers to reveal a brilliant golden lustre. (Pl. 52) A heavy wooden mallet was used for smoothening. Then the artisan proceeded to use two scribers like a divider affixing one at the centre and the other describing a pattern of a series of arcs. The outer convex surface was left blackened in this case which contrasted rather beautifully with the inner golden surface. (Pl. 53) The steel/iron implements were sharpened by tempering and quenching.

An interesting feature was the use of a circular wooden lathe board with four long iron nails rivetted along it. These nails had curved clamps affixed to them so that the bowl could be held vertically on the board. When the lathe was rotated by hand the scribers were used to make concentric circles at the centre or along the rim and rear of the bowl.

Another type of lathe in use in Nadavarambu consisted of a heated resin mount for the lamp base, rotated by pulling ropes affixed around the wheel. Either of these mounting methods could perhaps have been used by Nilgiri artisans to make the concentric rings that number seven to nine on the bottom of the bowls. The fluted and knob based features of some of the Nilgiri bowls [datable to perhaps as late as mid 1st millenium A.D. (Leshnik, 1974)] could also perhaps have been achieved by skillfully manipulating the ingot when hammering to leave a central rise, and hammering out the fluted regions around.

Kammalar Bhaskaran said that the bright golden colour of the bowls could preserve for long with regular washing with ash. This was unlike pittalai or brass which required regular tamarind rinse to remove the tarnish he said. He said that among the articles made by this method were *Kinnam* (vessels) *Idatala* or different musical cymbals including *Panchavadvam* Cymbals, Chenda, Jalana, Chetulam and Yaganda—a flat circular gong used in temples. They also made a spatula called shatuvam which could be reheated, reworked, quenched and polished again if it tarnished he said.

The Kammalar claimed that this tradition was very old and could not recall any interaction with South East Asia on this score. He lamented the fact the craft which had been a thriving industry in the past was dying out. There was no longer demand for traditional vessels with the invasion of stainless steel and aluminium, a fact noted by Edger Thurston (1909). Only eight people from eight families were today engaged in the crafts with others seeking alternative employment. As it was impossible to purchase raw materials at the market price today the artisans are increasingly dependent on the barter trade in old vessels of the same kind; to recycle them into new ones.

Similar bowls were spotted in various scrap merchants shops in Kumbakonam and Madurai, sometimes with a circular stand at the base like the Nilgiri bowls. These were said to have come from Palghat in Kerala, Rajahmundry in Andhra Pradesh and Nacharkoil in Tanjore District, Tamil Nadu. The shop keeper in Kumbakonam also claimed that the alloy was used to make certain Ayurvedic preparations in Palghat.

Sr. No.	Object	Find spot	Me-thod	Cu (wt %)	Sn (wt %)	Zn (wt %)	Pb (wt %)	Fe (wt %)	Ni (wt %)	As (wt %)	Sb (wt %)	Ag (wt %)	Co (wt %)	S (wt %)
1.	Bronze bowl Wrought	Payangadi Palghat Dist. Kerala	ICP	80.5	22.864	0.373	0.304	0.04	0.06	0.04	0.181	0.081	0.001	0.05
2.	Bronze bowl Wrought	Irinjilakuda Trichur Dist. Kerala	ASS	81.25	20.33	0.064	n.d	0.132	n.a	n.a	0.123	0.021	n.a	n.a
3.	Bronze Wrought	Madurai Tamil Nadu	ASS	82.01	21.76	0.14	n.d	0.174	n.a	n.a	0.16	n.a	n.a	n.a

n.d: not detected (trace)
n.a: not analysed (trace)

DISCUSSION

The discovery of a metallurgical tradition of high tin beta-bronze working in Kerala raises possibilities of its more widespread occurrence in other parts of India. Given the continuity over countries of traditional skills that is often seen in India, as with the lost wax casting process, a new dimension i.e. of indigenous provenance is added to the question of the source of the high tin bronze bowls of the South Indian burials such as the Adichanallur and Nilgiri material. The Palghat area in Kerala where such bowls are made today is proximal to the Nilgiris via Wynad, and also to parts of Tamil Nadu via the Palghat pass.

Of course it is much easier and perhaps less far fetched to attribute the origin of such bowls to South East Asia: given the abundance of tin, the commercial links through maritime trade and the much larger volume of finds of South East Asian bronze age materials is the first millennium B.C. (Rajpatik and Seeley : 26-31). Nevertheless, while trade in tin ingots is a strong possibility there is perhaps not enough convincing evidence to emphatically conclude that the technique originated in South East Asia. Besides, the bowls from Ban Don Ta Phet, Thailand, cannot be positively dated much earlier than those from Taxila or the South Indian megaliths.

The specialized metallurgical practice of water quenching was not unknown to South India in antiquity, which according to literary sources was the producer of the finest and perhaps earliest known steel in antiquity. Accounts from the court of Darius of Persia in the first millennium B.C. and Roman accounts of the early centuries A.D. speaks of the fame of Chera steel (Warmington, 1928).

Wootz steel was produced as late as the 19th century in South India by carburizing ingots of iron with prolonged heating under highly reducing conditions in highly refractory crucibles (Thomas, 1856: 1-17). These ingots were said to have been used to make the famed Damascene swords of antiquity by a process of tempering, working and finally water quenching (Rao and Lahiri, 1975).

The excavated site of Kodumanal, close to the ancient Chera capital and gemstone centre of Karur throws some interesting evidence in this regard. Period I at Kodumanal is a burial cum-habitation site of the Iron age megalithic period datable from 250 B.C.- 100 A.D. (Rajan, 1991). Here iron slags were found in an oval furnace with in situ vitrified crucibles; in the shape of wootz ingots as described by 18th cent. accounts. These were found in same horizon as the thin bowls and bowl fragments from the megaliths, comparable to the Adichanallur material which on cursory examination appear to be wrought. Hence the technology of quenching, required to make the martensitic ß-bronze bowls would appear to have been known by this period.

The *Arthashastra* also mentions workshop and ores to make *vaikṛntaka,* interpreted as steel (Kangle, *Arthashastra,* 2.12.16; 2.12.23). Hence the metallurgical background to make such bowls by empirical experiment was perhaps known in India rather than South East Asia. The Satavahanas of the Krishna Valley also appear to have uniquely used the technology of ß-bronze working in the early centuries A.D. to make die struck potin coins.

More investigations into vessels, musical instruments and coins of various periods is required to explore the questions regarding a continuing tradition of ß-bronze working and the possibilities of its origin in the Indian subcontinent. Unfortunately the absence of a continuing burial tradition in India coupled with the organised bartering of old vessels and scrap to make new metallic items over time renders such a study incomplete. A fuller documentation into such ethnographic practices in India would also serve to record and perhaps preserve a remarkable and endangered metallurgical tradition of the Orient.

REFERENCES

Allan, Horslar, Personal Communication, Institute of Archaeology, London.

Breeks, James Wilkinson, 1873. *An account of the Primitive Tribes and Monuments of the Nilgiris*, London, India Museum, p. 94.

Chakravarti, D.K., 1978. 'The problem of Tin in India', *Man and Environment*, Vol. III, 1979.

Goodway, Martha, March 1988. 'High tin bronze Gong making', *Journal of Metals.*

Leshnik, L., 1974. *The Pancukal Complex-South Indian megalithic burials*, Weisbaden.

Kangle, R.P., 1972. *Kautilya Arthasastra*, Part II, Motilal Banarsidass, Delhi.

Kuppuram, G., 1986. *Imperial Cholas of Karnataka*, Sandeep Prakashan.

Kuppuram, G., 1989. *Ancient Indian Mining, Metallurgy and Metal Industry*, Vol. II, Sandeep Prakashan.

Mallet, F.R., 1874. 'Geological notes on Northern Hazaribagh', *Records Geological Survey of India*, 7:23-24.

Marshall, 1951. *Taxila*, 3rd Vol. Cambridge.

Mukherjee, Meera, 1978. *Metalcraftsmen of India*, Anthropological Survey of India.Calcutta.

Rajan K., 1991. 'New Light on the Megalithic culture of the Kongu region Tamil Nadu', *Man and Environment*, Vol. XVI, No. 1, pp. 93-100.

Rajpatik, W., Unpublished thesis, Institute of Archaeology, University College, London.

Rajpatik, W. & Seeley, N.J., 1982. 'The bronze bowls from Ban Don Ta Phet, Thailand', *World Archaeology*, 11, pp. 26-31.

Rao, K.N.P. and Lahiri, A.K., 1975. 'Iron and Steel Technology in India in Ancient and Medieval period', *Metallurgists and Metal Technologists.*

Ray, Himanshu, August-September 1990. 'In search of Suvarnabhumi—Early sailing networks in the Bay of Bengal', *IPPA Yogya*, pp. 1-7.

Ray, P. (ed.) 1956. *History of Chemistry in Ancient and Medieval India*, Calcutta.

Sastry, V.V.K., Purushottam, D., Rao, G.R., Subbaiah, K.V. *et al.*, 1990. 'Chemical composition of some Satavahana coins', *Journal of Archaeological Chemistry*, Birla Institute of Archaeological Chemistry.

Srinivasan, K.R., 1958-59. 'The megalithic burials and urn-fields of South India in the light of Tamil Literature and tradition' - *Ancient India* 2, pp. 9-16.

Thomas, E.C.G., 1856. 'On the iron ores and manufacture of Iron and Steel in Coimbatore *Report of Government Museum Madras*, Appendix, pp. 1-17.

Thurston, Edgar, 1909. *Castes and Tribes of South India*, Vol. K.

Warmington, E.H., 1928. *Commerce between Roman Empire and India*, Vol. II, Cambridge.

Metal Working in Medieval Assam

R. DASGUPTA

The metal ware industry of Assam was flourishing under Ahom rule. But some new elements were introduced by them in the shapes and designs of the untensils and jewellery. These were different from the standard Indian shapes and had parallels in the neighbouring countries of Tibet, Burma, Siam and even in distant Indonesia. The main difference was that each plate or bowl had a little stand at the bottom, ranging from about a couple of inches to about a foot in height. Thus, these new vessels had the look of chalices and goblets rather than bowls and dishes. These shapes were introduced by the Tais and such vessels had a general prefix "Ban" before the actual name of the vessel, e.g. Bān-bāṭi, (Pl. 54) (Bowl, with stand), Ban-thal (Dish with stand), etc. The earliest example of the dish-on-stand in India is the wheel turned terracotta piece from Cemetery R-37 at Harappa (Wheeler, 1960). Here, as in Assam the plate has raised brims and the stand has a parallel in the Ahom tumbler though the latter has a more slender stand. The general Assamese term Ban has an equivalent in Thai as Pan or a bowl. A large sized Pan is known 'Toke' (pronounced Tou) in Thai which too has an Assamese equivalent in Tou, - a large brass basin for washing rice and other edibles before cooking. Some of the dishes-on-stand had hemispherical lids with a long tapering handle. These were known as Sarai or a receptacle and the name is derived from the Sanskrit Saravam or a bowl. The shape of a Sarai (Pl. 56) is exactly like a miniature Stupa, complete with its 'Hti' finial. Some of the elaborate Sarais had circular discs attached to the finial and they represent the Chattravali of a stupa finial. The number of such discs range from 3 or 5 or 7 according to the size of the vessel. The shapes of the Sarai clearly shows that it is derived from the stupa-shaped metal reliqueries of the Buddhist period of Indian culture. The earliest example is a bronze relic casket from Manikial (Fergusson). The wall paintings from the Central Asian shrine of Bezeklik (shrine III) show a devotee holding before the Buddha on his out-stretched hand "an imposing censer". (Andrews, 1948). It is very much like the later day Baṭās and Sarais of Assam and even the etched designs showing full-blown lotus petals are also common to both. This censer also has little round baubles hanging around the main member of the vessel. The only difference being that the Bezeklik vessel has a large platter for its base as in Tibetan ritual vessels, while the Assamese ones have a plain lotus petal bases. Such vessels have been found from Ceylon and Eastern India. The Sarai is known in Ceylon as the Pātra while the lidless Bata as the Bulathwaṭṭiya. Both the Bata and the Bulathwaṭṭiya have been found from Ruvanaveli and Tiriyay belonging to a period ranging from the 6th to the 8th cent. A.D. (Devendra, 1958). But these examples carved out of single blocks of stones served as flower alters in ancient Ceylon. Here too the lotus dominated the design and shape of the vessels since the lotus form is popular in the Buddhist art of Ceylon according to D.T. Devendra. 'Lotus stalks, intertwining with leaves most realistically sculptured, end in an open flower which is the "table-top" of the flower (Devender, 1958). Colombo Museum has some very interesting examples of these from Anuradhapura which belong to a period between the 8th and the 9th centuries A.D. Bata like forms have also been found in Eastern Indian sculptures of the Pala-Sena period. Four actual example in copper were found by R.D. Banerji from Naihati, in Burdwan district, Western Bengal. He identifies them as being stands for 'Pani-sankha' or water-filled conches used in pujas. But it does not seem probable since the 'Pani-sankha' has a round bottom and would not stay on a platter-like object. To this day the Pani-sankha is in use but it has a ring-like frame on stand wherein the Pani-sankha fits properly and does not topple. Thus it is clear that the above vessels were used for offering eatables to the deities and not for holding the 'Pani-sankha'. "The four stands are very simple in design. They consist of a round base and lathe turned or plain round shafts". (Banerjee, 1933). These belong to the 11th year of the reign of Vallalasena as the accompanying copper plate grant would suggest. The Bata (Pl. 55) and the Sarai (Pl. 56) are similar in shape, except that the shaft is more elaborate and consists of an "amalaka" in the centre flanked by two discs on either sides. The top and the bottom of the shaft and in full-blown lotuses with flared mouths. The members of the shaft are however all detachable and a rod passes through the holes in the centres of the members. They are held together by a screw system.

Other important shapes in metal ware are the Bāṭi or Bān-Bāṭi, i.e. smaller bowls for curries, the Jara and the Bhog-jara, i.e., ewers and jugs, and the Gacha and the Chaki, i.e. lamp stands and lamps respectively.

The Bān-bāṭi (Pl. 54) is of Ahom derivation as the name suggests. These resemble in shape the 'Tibetan butterlamps' lit in the monasteries (Jisl). Here too the lotus flower and the trumpet-like base is present. But the Bān-Bāṭi has more the look of a chalice with its hemispherical bowl standing upon a simple Cylindrical stand with a disc-like base. The only decorations are the concentric circles incised on the disc of the base. The brim outside is also decorated with a couple of incised lines running around. What Tucci (1967) says about the symbolism of the shapes of the Tibetan butter lamps, tea-cup and condensers also apply to these vessels from

Assam. He says, "The shape is that of a chalice, either hemispherical or slightly flared out as though to symbolise a blossoming flower resting upon a high stand which incorporates the symbolism the 'bumpa'. The latter is the vase that holds the water of immortality, a form of ritual and decoration that has travelled from Mesopotamia to India and to many other parts of Asia". In a like manner the shape had entered Assam. Vessels very similar in shape to the Bān-Bāṭi have been found at Amlash in Iran belonging to the period 9th to 8th centuries B.C. Both are in silver but they have a handle each like in our modern tea-cups. These are in private collections in Tehran and in New York (Ghirshman, 1964).

The Bhogjarās and Jarās are of east metal, mainly of bronze, gold and silver with stout bodies which are deeply incised. In this respect the Assamese wares are more like the Tibetan tea-pots and water-vessels of Arunachal region (N.E.F.A.) known as the Khoding. According to Tucci (1967) such vessels do not have the lightness of Chinese and Persian wares. Like the Tibetan wares, the Assamese wares also have bulging bodies and stout necks. The Bhogjarās have intricate spouts ending in 'makara' and animal heads. They have the same beauty as the Tibetan tea pots have in their handles. The length of the spout is sometimes like the scaly bodies of serpents and other aquatic animals in Tibet the Makara shape is married to the Chinese dragon in Assam. Lamp stands or Gachās are all cast in bronze and usually have a single deeply chiselled figure carrying the actual lamp. These figures are placed on elaborate and bold lotus pedestals. Rampant Ahom lions (Pl. 57), elephants and Garuḍas (Pl. 58) are the usual figures carrying the lamp. The central figure, the lamp-base and the base of the figure were separately carved in the cire perdue technique and then soldered together. The lamp was either built in the stand with one or multiple mouths or a shallow depression served to hold the separate Chaki or lamp. Two exquisite examples of the 17th century from Prof. S.K. Saraswati's collection, Calcutta, serve to illustrate the popular types of lamp stands. The one with Garuḍa has a high lotus base on which sits a kneeling Garuḍa with outstretched wings and hands folded in Añjali pose. A cylindrical member stands from the head of Garuḍa on which a perforated lotus base for holding the Chāki. The Garuḍa figure has a flowing uttariya and the front pleat of the dhoti dangling in front. The crown and kundalas are typical of the Ahom metal images. Feathers of the wings are stylistically and conventionally treated. Garuḍa is completely human except for the beak-like nose suggesting his bird character. The other lamp-stand shows a rampant Ahom-lion among boughs emerging from a pot carried by deeply cut lotus base. This design in reminiscent of a carving on the Um-nyakneh bridge between Jariain and Syndai in the Khasi-Jaintiya Hills. The lamp base is soldered between the curled up tail of the lion, its ears and manes. The lamps mentioned above were found in north Bengal where they had found their way through export. Assamese brass and bell metal goods were in great demand in the adjoining areas of Bengal till the 19th century as we find from the accounts of Sir B. Hamilton. Similar Garuḍa and Lion lamps are seen in the Assam State Museum, Gauhati, Bharat Kala Bhawan, Banaras and in the personal collection of the Maharaja of Banaras. The Assam State Museum also had some bird shaped Sarotas (betel nut slicer) and votive bells of exquisite design. King Krishna Chandra of Kachar was converted to Hinduism in 1790 and the Hiraṇyagarbha ceremony was performed to proclaim him born of a cow, a deity. But the effigy of the cow was made of copper and not of gold as is usual for such ceremonies (Gait).

Technique

In Assam metal ware are made both from brass (Pital) and Bell-metal (Kāñh). While the former is beaten from thin sheets of brass the latter is cast in moulds. The sheet metal is worked by the Mariās (the beaters) who are a Muslim community, but who have eschewed their orthodox customs like circumcission and the Namaz and freely take wine. The brass sheets are cut and beaten to shape over the anvil known as Belmuri Chutuli. Other implements of work are the hammers, pincers, chisels of various sizes. Goat skin bellows attached to a hole in the floor of the workers' hut complete the simple furnace. "When it is desired to join two sheets of brass together, nicks are cut in one edge, into which the other edge is filled, and the two are then beaten flat. A rough paste is made of 'pān', a substance which consists of three parts of sheet brass and one part of solder, and borax is then smeared over the joint. The metal is heated, the pān melts, and the union is complete.

Objects

A good account of the traditional metal ware of Assam is found in Gait's (1896) account in the *Journal of Indian Art.* The list of articles given below is after Gait.

Brass

1. *Berhā* : It is a tripod consisting of a ring about a foot in diameter, supported on three hollow legs, the interior of which is filled in with pieces of broken brass which produces a jingling sound. The tripod stands about 14" high and is used for supporting flat dishes out of which the upper class Assamese eat their meal.

2. *Chāki* : It is a lamp and sometimes a lamp and a stand combined. In a combined one, the top is a cup-shaped hollow for serving as a lamp which holds the oil and the wick.

3. *Chariyā* : It is a wash-hand basin and is also used as a common bowl for keeping sundry things. It is also made of bell metal. In Surma valley it is known as Tagar.

4. *Don* : It is a measure to hold five seers of rice or other grains.

5. *Ghāgarī* : It is a decanter-shaped vessel with an ornamental mouth, shaped like an open eight petalled lotus. It is meant for carrying and storing water.

6. *Ghaṭi* : It is a common pot for holding liquids as is similar to the Lota of north India.

7. *Gachā* : It is a lamp stand sometimes with a built-in lamp bowl on the top.

8. *Kalas or Kalsi* : These, like the Ghagari are water containers but are provided with rims. The top is either plain or flower shaped. These are also made of bell metal and copper, the latter being used for ritualistic purposes for containing water. In Surma valley also they are known as Kalsi and in Mainpur as Tukna.

9. *Kariā* : It is a milking pot.

10. *Kharāhi* : It is a sieve for rinsing rice and is also used in Surma valley as the Jhānjarī.

11. *Khundonā* : It is a handy pestle and morter, shaped like a thimble and is used by old folks for pounding pān and tāmol or betel nut.

12. *Loṭā* : It is a small vessel for holding water and also for drinking. It is also known as Lota in Surma valley and as Khujai in Manipur. This is also made with bell metal.

13. *Śarāī* : It is a round tray with a stand used by Hindus for offering edibles as religious offering and is also made of copper.

14. *Śetā* : "It is a large spoon with a handle, thin towards the bowls and thick towards the opposite extremity and is used for distributing rice and curry". Gait seems to have confused the spelling. In Assamese it is actually pronounced as Heta and is written with the aspirant "h" and not the sibilant "s" although in Assamese normally the sibilant is changed into the aspirant in pronunciation. In the Surma valley it is known as the Hata.

15. *Ṭemā* : It is a small round box for keeping arecanut and is the same as the Dibi of the Surma valley. It is also made of bell-metal. Those for royal use were of gold set with jewels. Indian Museum, Calcutta, has some exquisite specimens of Temā and Ṭemī.

16. *Ṭemī* : (Pl. 59) Made both in brass and bell-metal, it is smaller in size than the Ṭemā and is used for keeping lime to be eaten with pan. It is known as the Chunaṭī in Surma valley.

17. *Thālī* or *Jhakā* : It is a large boiling vessel. Thālī is derived from Sanskrit 'Sthālī' or a cooking vessel and the actually pronounced as Jhakā in Assamese.

18. *Tou* : It is a basin-shaped vessel for boiling paddy, cooking large quantity of food and also used for keeping water and other things. Its affinity with a vessel of the same name from Thailand had already been mentioned. In the Surma valley the name is Taslā.

Bell metal

19. *Bānbāṭī* : It is cup joined to a circular rim at the bottom for holding curry. (In Thai it is known as Pan).

20. *Bānthāl* : It is a concave shaped round dish joined to a circular rim at the bottom for holding rice.

21. *Barkāh* : It is a gong of bell-metal or Kah and hence the name.

22. *Bāṭī* : Made both in bell-metal and copper, it is a round bottomed cup for holding curry. In Surma valley it is known as Bati and as Tengkol in Manipur.

23. *Baṭā* : It is a flat circular basin with raised edge and is generally ornamented. The upper basin is joined to a circular rim at the bottom and is used for betel-nut.

24. *Dog-dogī* : It is similar to a Lota.

25. *Kāhī* : It is circular Thal with a raised rim and is used for keeping rice. The name is derived from Kanh or bell-metal of which the objects are made. In Manipur it is known as the Pukam and as Thal in the Surma valley.

26. *Maipong* : This vessel is similar to the Banthal but has a cover in addition.

27. *Pāndhoā* : It is a flat basin with an ornamental raised rim and is also used for keeping rice.

28. *Pikdān* : It is spittoon but as the name suggests it might have been introduced from mainland during medieval period. In Manipur it is known as the Kwasen.

29. *Tāl* : These are brass cymbals used in religious orchestra of Assam and are known as Kartal in the Surma valley and as Kardan in Manipur. The tāl is also known as the Bhor-tāl or Bhot-tāl, i.e., Tibetan or Bhutanese cymbals. These vary in sizes from about six inches to eighteen inches in diameter and give a deep sonorous note when struck together. Exactly similar cymbals are used in Tibetan Lamaistic worship and hence the name, both the Tibetans and the Bhutanese being designated by the generic name Bhota in Assam. These people still have contacts with Assam due to their annual visits to the temple of Hayagriva Madhava in Kamrup district. Over and above, the annual winter fairs held on the Bhutan frontiers or Duars of Assam are also places of economic as well as of cultural exchanges between the two countries.

30. *Modharā* : These are narrow necked water vessel used by the Khasis and the Jaintias. The Modhara is placed in conical baskets of bamboo which are slung from the forehead of women with straps. The Khasis and Jaintias, however, imported these vessels from Sylhet.

Centres of production

During the Ahom period this industry was mainly concentrated near the capital and also in Kamrup district. But other centres also flourished. Since the Khasis did not produce any wares themselves they imported articles from Habiganj and Panchakhanda in Sylhet. Nijpat Jaintiapur became a trade centre for such imported articles (Gait, 1896). The centres in Assam continued to flourish as before and were recorded district-wise during the early British rule.

Centres in Sibsagar

Namdang, Dopdar, Salmara, Titabar, Kakajan mauza and Chariali in Jorhat sub-division. Kacharihat in Golaghat sub-division. The Moria craftsmen of Sibsagar and Jorhat also migrated to Mangaldai and Silghat respectively (Allen, 1906).

Centres in Lakhimpur

The centres are inhabited by Morias only (Allen, 1906). There are Japaragaon, Tekelasising, Japihajia, Ghorbund and Jarwa. The usual articles are Hata (vessels for boiling rice) and Daos or native choppers (Allen 1906). The workers are engaged in the industry only when there are no agricultural activities (Allen, 1906).

Centres in Darrang

"The number of braziers is, however, very small, and of workers in bell-metal smaller still". The chief centres are Chutia tehsil, Becheria, Modopi, Bihuguri, Mangaldoi and Patharughat (Allen, 1906).

Centres in Nowgong

Here centres are many though the products are not as remarkable as those of Kamrup and Sibsagar. The centres are Kuwaritol, Raha, Jagial, Samaguri, Morigaon in Sarbhagia mauza, Sutargaon in Khatoalgaon mauza and Kahargaon in the Raha tehsil (Allen, 1906).

Centres in Kamrup

Kamrup excel in its metal-wares. The brass industry is concentrated in Jaho and Gauhati while that of bell-metal at Sartheibari. Iron wares and implements like plough-shares, 'daos', knives and sickles are also produced at different places. Shantipur, a suburb of Gauhati and situated just at the foot of the Kamakhya hill, has a settlement of bell metal workers (Allen, 1906).

Weapons of War

Being "war-like people" the Assamese had developed many engines of warfare and had also excelled in the use and manufacture of guns and canons of a varied nature. Guns and muskets were known as Hiloi and canons as Top or bar-top, — the prefix 'Bar' meaning great. The sword was the Khāndā and the shield the Bāru. Arrowheads were varied, i.e., plain, barbed etc. Manucci (1907) describes a peculiar weapon of torture used by the Ahoms against the straglers of the Moghul army who had accompanied Mir Jumla during his invasion of Assam. "To as many as they caught they inserted in the anus an iron like a Turkish padlock, which once in, could never be withdrawn; its shape was thus. It shut up as it went in, and ... had
inside it opened out, leaving the handle projecting. The poor thin...m. When
the army lamenting, and died in a desperate state". ...nons as booty
...76). The records

The French traveller Travernier refers to a ... 'On tient que c'est ce
learnt the use of guns and gun powder at an ...
Mir Jumla attacked Assam he received a ...
which have been preserved in the M...
of Travernier clearly mention ...

meme peuple qui a trouve anciennement l'invention de la poudre et du canon, laquelle, a passe d'Asem au Pegu du Pegu a la Chine, ce qui est cause qude d'ordinaire on l'attribue aux Chinols". In China, says Maclagan (1876) that, petroleum from the Western Provinces was used for a Greek Fire as a combustible material. "Assam also, it may be observed, is a petroleum country". Hence in Assam natural petroleum as well as Khar or gun-powder might have been used as combustible materials. But Ahom records do not mention the use of petroleum for the purpose.

The guns and canons were of a great variety and were often artistically cast having mouths shaped like the Makara (Pl. 60), the tiger and the elephant. The more important ones were highly polished and damascened or inlaid with gold wire. Gold inlay work was known in Assamese as 'son-khatowa'. Some good specimens of Ahom Bar-tops can be seen on the bank of the Sibsagar tank in front of the Sibsagar court.

Types of canons and their names

Guns (Hiloi) and canons (Top and Bar-top) of many descriptions were made in mediaeval Assam....... Even the inimical Mughals were full of praise for the Assamese guns and canons in their chronicles like the Fatiyah-i-Ibraiah and the Baharistan-i-Ghaibi. Sahabuddin Talish mentions the famous Basādārī Hilois (guns) which were finely enamelled and decorated. Under king Gadadhar Singha a muslim gunsmith of North Gauhati, named Gerela cast the famous Ripuñjaya canon in 1703 A.D. (Handiqui). During the temporary success of Mir Jumla, the Mughals captured in Assam about 8,000 guns and canons of various shapes and sizes. The Bar-Hiloi was a large cannon which used canon balls 3 maunds in weight. Other kinds were Ramsangi, Kāmāyana, Jhumur, Miṭhāholeng and Ṭubukī. Some of the canons were decorated with damascened work and brass relief work, designs of gods and goddesses (Dasgupta, 1966).

The Muslim inhabitants of Garhgaon, the Ahom capital had informed Mir Jumla of the huge dump of guns and canons in the Padum Pukhuri tank which the Raja and the Phukans had thrown in it. The finest Rāmchāñgis and Zamburaks were collected from there by Farhad Khan and Mir Sayyid Muhammad of the Mughal army (Sarkar, 1951). According to Irvine the Rāmchāñgi is a light field piece. Hodivala opines that Rāmchāñgi (Ramsangi in Assamese pronunciation) is [illegible]ed from Persian "Ramjānī" or a 'pleasure-girl' who dances at command. Its Man[illegible] the ball fired, resembling a dance in the mind of a love-lorn gunner was and croco[illegible] unusual nomenclature. Describing the war-boats of the Assamese, [illegible]s) and petrechos of bronze, of which the muzzles whence [illegible] into shapes of animals—tigers, lions, dogs, elephants [illegible] Such was the admiration of the Mughals for

Assamese gunsmiths that Mir Jumla ordered that the Assamese experts in gun making were to be sent to Dacca, the provincial Mughal capital to be employed by the Mughal army (Sarkar, 1951). The warboats filled with artillery were given the general term Hiloi-chara-Nao. Famous Assamese canons were often captured by the Mughals while the reverse was also happening. Two historic canons with inscriptions were found by Simson in the Dikhou river (Bandopadhyay, 1909). The larger of the two was made of brass. The older of the two has a Sanskrit inscription in Assamese script. It runs as follows:

> "Saumārésvara Sri Sri Sa ....
> tra .... svarganarayanasya Sake 1560".

It is dated 1560 saka or 1638 A.D. when the Ahom king Susenpha (Pratap Singha) or Buddhisvarganarayana was ruling. The name of the king Satra-savarganarayana, might have been another name of Susenpha. It cannot be Satrajit, the Hindu Jamindar of Dacca who accompanied the Mughal invading army under the command of Sayyad Hakim and Abu Bakr. The two titles Saumaresvara and Svarganarayana were exclusively used by the Ahom monarchs.

The other gun which is of brass is bigger and has two inscriptions in Assamese script and Sanskrit language and in Persian. The Sanskrit inscription is in incorrect spelling and is dated 1604 Saka (1682 A.D.). It is deeply incised and runs as follows:

> Sri Sri Suragganarayana deva saumaresvara Gadadhar Simhena Ya —
> Vanam jitva Gurakahatyam idam astram praptam sake 1604.

The Persian inscription on this gun is in a bad state of preservation. The English rendering of it by R.D. Banerji is as follows.

> "The just king Shah Alam Bahadur ... Year 1124".

The era used is the official Mughal era Hejira and it comes to 1716 A.D. Thus the king mentioned is none but the eldest son of Aurangzeb.

The guns and canons were of great variety regarding size and potency. Apart from Hiloi and Bartop other fire arms and artillery weapons were Pātharkalai, Kāmāyana, Jamur, Kesai, Towā, Gathiā-jumar, Pāhalangi, Khokā, Hāth-hiloi Rāmśañgi, Ṭubuki, Mithā-holong. Hāthi-muriā, and Bāgh-murā (Gogoi, 1961). Of these the Rāmśañgi has already been discussed above. Kāmāyana seems to be a Sanskritised form of the Perisan Kaman, i.e., a missile. The Hath-hiloi as the name suggested was either a gun or a small field piece which could be held in the hands, or may be it was one cubit long. The last two in the list are named after the heads of the elephant and the tiger, shapes in which their muzzles were designed.

Swords and sabres also had varied names and shapes like the Tarol, Héngdān, Mit-dā, Kapi-dā, Mesi-dā, Nakoi-dā, Narā-dā etc. The Taroal was the local form of the Talwar or sword of Hindusthan while the Héngdān had a special place among the Ahom royal and upper classes. The Héngdān was a ceremonial sword which was held only by the king and his courtiers (Gogoi, 1961). In about 1864 A.D. when the European tea-planters occupied the palace of Garhgaon found among many objects a fine sword with a silver hilt and its scabbard was made of gold and ivory (Foster, 1872). The "da" was the native chopper which was heavy and curved with a wooden hilt. It was an all-purpose weapon and was also used for cutting fire wood, slicing vegetables and meat or even for cutting bamboo and chopping straw for housebuilding. The "dā" of the Nagas and of the Narā tribe were famous and were often used in war. In the Buranjis the sabre and shield were commonly termed as Khāndā and Bāru respectively. Spears and javelines were known as Jāthī or staff, evidently derived from the Sanskrit Yasti. Those used by the Nagas were often decorated at two places of the pole with tufts of human hair dyed crimson. These not only added to the beauty of the weapons but also aided them in their flight when hurled.

Gold and silver were not used much for making utensils and every day wares. "Although it would not be correct to say that the gold and silver 'wares' of Assam comprised exclusively articles of dress, that is jewellery, yet the number of gold and silver products which are not jewellery is so small as to be quite insignificant .... Whatever may be the case Assam has very little to offer but jewellery .... The Assamese who is very luxurious, keeps his betel nut in a silver box called 'tema' and has a plate (baṭā) or bowl (bāṭī) of silver, but few or none of these things show any artistic design .... " (Henniker 1905). The above observations of Henniker only state the situation when the artistic and economic life of Assam had completely broken down due to the Burmese wars and plunders. Even with the establishment of the British rule, the Assamese craftsmen did not react favourably. They receded from the forefront and their places were taken by immigrant workers from Bengal and Hindustan as far as the towns were concerned. But true and native Assamese forms and designs were continued by the Assamese craftsmen in the rural areas.

Gold washing

Gold was obtained in Assam mainly by gold washing technique. During the rule of Rajeswar Singha (1751-1769 A.D.) Upper Assam thus produce 2,500 ozs. of gold dust for the royal treasury (Allen, 1906). The Ahom king Sukrangpha (1259-1335 A.D.) organised on a grand scale the guilds of Sonwal and Rupwal who washed gold and silver dust from the sands of the rivers like Subarasiri, Dikrong, Lohit, Disoi, Jglo etc. Gold washing was practised more in the north bank of the Brahmaputra while in lower Assam Dhing, Chatia, Ghatguri and other places sent the annual requisite of gold dust to the court at Garhgaon. The Paiks of the individual gold washers worked under a Kheldar Barua or a superintendent of the guilds. The Ahom kings engaged a large number of Chutiyas and Kacharis since they had better knowledge of the technique of production. Every year the Paiks worked from the months of Kartika to Vaisakha and returned home for agriculture in the month of Jaistha. Each man had to produce $2\frac{1}{2}$ tolas of gold and if he produced anything more he kept it (Dasgupta, 1966).

A detailed account of the process of gold washing is given in the Assam District Gazetteer for Lakhimpur district. "The gold washers selected a gravely shore opposite to a falling bank, where the current flowed fast round a corner. The stream was diverted over the selected spot till the surface sand had been washed away and the graveley strata exposed, and was then turned back into its former channel. The auriferous sand was thrown on to a bamboo strainer placed over a wooden trough, and water was poured over it, the trough being tilted to a slight angle, so that the water and lighter substances ran off, while the gold and the heavier matter remained at the bottom. When a sufficient quantity of this matter had been collected, it was placed in another vessel and quick-silver added to attract the grains of gold. The mixture of quick-silver and gold was then put in a shell and placed on a fire. The shell was converted into lime, the quick silver evaporated and the gold was left lying in the lime. The lime and gold were then thrown into a vessel full of water, when the precious metal at once sank to the bottom" (Allen, 1906).

Silver was collected in a different manner. "Silver or Rup was extracted out of Rupmāṭī from the slopes of the Garo-Khasi hills by the Rupwāl Khēl". The Chalāpthār of Sibsagar derives its name from the straining of Rupmaṭī to get silver. (Dasgupta, 1966). Rupmaṭī was the name given to the silver in one form mixed with earth and other minerals.

REFERENCES

Allen, B.C., *Assam District Gazetteer,* Kamrup, Vol. IV, p. 161.

Allen, B.C., *Assam District Gazetteer,* Nowgong, Vol. VI, p. 148.

Allen, B.C., 1906. *Assam District Gazetteer,* Darrang, Ch. V, p. 158.

Allen, B.C., 1906. *Assam District Gazetteer,* Vol. VII, p. 162.

Allen, B.C., *Assam District Gazetteeer,* Lakhimpur, Vol. VIII, Ch. VI, p. 207.

Andrews, F.H., 1948. *Wall Paintings from Ancient Shrines in Central Asia,* p. 70, Text volume, Pl. XVIII, No. BEZ III, S.U. Oxford University Press, London, Plate volume.

Bandopadhyay, R.D., 1909. Two inscribed guns from Assam, *Journal of the Asiatic Society of Bengal,* p. 465.

Banerjee, R.D., 1933. Eastern Indian School of Medieval Sculpture, Metal Objects. Chapter VII, p. 143 Pl. LXXIV. Archaeological Survey of India, New Imperial Series, Vol. XLVII, Delhi.

Bhuyan, S.K., 1962. *Tripura,* Buranji (ed.) p. 15, 2nd edition.

Dasgupta, R., 1966. Metal ware of Assam, *Art in Industry,* Vol. VIII, No. 2, pp. 12-13, Calcutta.

Devendra, D.T., 1958. *Classical Sinhalese Sculpture* (300 B.C. to A.D. 1000), p. 40, pl. 119, Alec Tiranti (ed.), London.

Fergusson, J. *History of Indian and Eastern Architecture.* p. 94, A.S.I.R., Vol. II, p. 167, Pl. 65, Relic Casket with Five Chhatras from Burj-i-Yak-Derch Tope East of Kabul.

Foster, J.M., 1872. Note on Garhgaon, *Journal of the Asiatic Society of Bengal,* No.1, Pt. 1, p. 40.

Gait, E.A., *History of Assam,* p. 309.

Gait, E.A., 1896. The Manufacture of Brass and Copper Wares in Assam, *Journal of Indian Art and Industry,* vol. VII, No. 54, pp. 1 & 3.

Ghrishman, R., 1964. Persia, From the Orissa to Alexander the Great. *The Art of Mankind* Series.

Gogoi, L., 1961. *Ahom Jati Aru Asamiya Samskriti,* Sibsagar, p. 179.

Hamilton, B., *An Account of Assam,* p. 60.

Handiqui, B.C. *Purani Asamar Shilpa,* p. 25.

Henniker, F.C., 1905. *The Gold and Silver Wares of Assam,* Shillong, p. 1.

Jisl, L., *Tibetan Art,* p. 166. Spring Books, London.

Maclagan, R., 1876. On Early Asiatic Fire Weapons, *Journal of the Asiatic Society of Bengal,* Pt. 1, p. 62.

Manucci, N. 1907. Storia Do Mogor, *Mogul India* (1653-1708 A.D.) Indian Texts Series. Translated with Introduction and Notes by William Irvine, London, p. 99. Vol. II, p. 100.

Sarkar, J.N., 1951. *Life of Mir Jumla.* Calcutta, p. 247.

Tucci, G., 1967. *Tibet : Land of Snow,* Art p. 121, Oxford and IBH Publishing Company, Calcutta.

Voyages, de J.B. Travernier, II, p. 427.

Wheeler, R.E.M., 1960. *The Indus Civilization,* p. 77, Second Edition, Figs. 1 and 2.

Is Iron Appearing in Archaeological Context in Eastern India A Product of Tribal Expertise ? An Ethno-Archaeological Enquiry

BHUPENDRA PAL SINGH

Iron has a long history in India that extends over 3000 years. The origin of iron, its production and the nature of its spread have been active topics of debate among the archaeologists. The archaeological evidence unearthed during last three decades has established beyond doubt that iron working developed indigenously in the country. An enquiry based on the study of diversity in the cultural millieu of various iron bearing areas concludes that iron technology developed independently at several centres in the country (Prakash and Tripathi, 1986: 568-79). However, results of this and similar other investigations lacked success in suggesting how, where and when iron was first used or what were those factors and forces which helped in the spread of iron technology that ultimately ushered Iron Age around 1000-800 B.C.

Although, recent archaeological findings are very specific about a late 2nd millennium B.C. date for the presence of iron in India, but a hitherto unknown evidence suggests that date of breakthrough might go even earlier. The eastern India with its natural wealth of iron ore and fuel is one of the most potential areas where an early breakthrough can be looked for. This is not merely a coincidence that an indigenous iron technology exists with certain tribes and we have failed to take note of this evidence only because it does not fall into archaeological context. The unique situation in eastern India where iron figures both in folk and urban traditions, provides opportunity to understand the nature of an ancient metallurgy. This ethnographic source can yield significant insights for archaeologists to reconstruct different aspects of a metallurgy of early times, particularly the cultural context in which it actually operated.

A question is often raised that inspite of early presence why there is a time lag between innovation of iron and spread of its technology. Probably the reasons for the said delay, particularly the production of significant quantity of iron in archaeological context can be attributed to certain long existing factors. For instance, the sources of iron ore in the country were as plentiful as in any other iron producing culture of ancient east but the question of availability and accessibility made the all important differences. The sources of high yielding iron ores are confined mainly in the densely forested and hilly tracts of eastern India and those present outside are mostly of low grade. The difficult access to high grade ores must have created multitude of problems for an early breakthrough of iron. This situation was likely to deprive ancient pyrotechnologists of the facility to make liberal use of ores for their experiments to achieve iron. In fact, except for a few areas, rich in high grade ores, use of low grade ores was an inevitability for the early smelters. The inadequate quantity of iron extracted from the low grade ores not only restricted the production but adversely affected the spread of technology in a wide area. The conditions were such that even if the iron was discovered, it probably remained confined within the area of its emergence for considerable period of time. I feel that a fast spread of iron technology was not on cards until a stage was reached when the smelters started employing high grade ores to extract good amount of iron in a single operation and became affluent enough to afford distribution of the metal to other areas or when smelters became aware of the art of adjusting the size of furnace according to the quality of iron ore used and the metal was made hard enough to compete with the existing copper/bronze.

Almost half a century ago Forde (1949: 386) wrote about occasional recovery of iron nodules embedded in slag from copper furnace in which an ore containing both copper and iron had been smelted. Later, Wertime (1980: 3) and Charls (1980: 151-81) expressed possibility of accidental discovery of iron technology through copper smelting. Lured by this novel suggestion some archaeologists came out with a theory that Indian iron was a by-product of copper technology (Sahi, 1979: 365-366; Chakrabarti, 1984: 81-85; Tripathi, 1985: 75-79). The presence of iron in Ahar Copper (Hegde: 1969) was made the basis to advance this theory. However, the scant consideration shown for technical and chemical implications and gross negligence for the stratigraphical position of the evidence lead us to believe that only the urgency to enlist an example of indigenous iron of little early date rather than proposing any worthwhile evidence of discovery of iron led these archaeologists to emphasize this irrelevant evidence.[1]

In this paper, I intend to propose that the art of iron smelting was known to certain tribes of eastern India much before the pottery was introduced in the region. The use and later actual smelting of iron evidenced on settlements of eastern India and in some other areas owes its beginning to the iron technology existing with the tribes. We will also direct attention to examine whether the specific choices and decisions featuring prominently in the smelting process convey any message regarding its antiquity. The relevant archaeological data is also brought into focus to check the validity of the proposed hypothesis.

The Smelters

Though, tribal population of the hilly and forested tract of eastern India is considerable, it is the iron smelters that claim our attention. The people like Asur,

Birjia, Karmali, Agaria and Barae are the one intimately associated with the craft and have not shown much inclination towards cultivation of soil. These human groups silently developed the craft and formed a cultural level which had probably established at unknown times in antiquity.

Although, the craft of iron smelting emerged among tribals in remote past, the difficult nature of their country delayed the knowledge of their existence till late 19th century. The most remarkable feature is miraculous presence of iron with the people who had no previous experience of any other metal. It is likely that at one period of time, the technologies of copper and iron existed simultaneously with people of two different traditions who for obvious reasons never crossed each other.

A natural wealth of ore and abundance of fuel made the tribal metallurgy easy, but without a permanent need of metal for its numerous users such intricate craft would probably not have survived through centuries after initial breakthrough. Normally, the perpetuation of local iron technology depends on the demand for iron tools generated from local economic success and productivity (Childs, 1991: 338). Iron and iron producers are important for the existence and vitality of farmers, hunters and foragers who needed supply of iron for their tools to promote productivity and population growth. It is highly paradoxical that the possession of iron could not induce these tribals to claim land from the forest to indulge into cultivation, and thus, raises doubts about the oft-repeated claim of archaeologists that iron is responsible for creating a condition for large scale cultivation which in turn was responsible for enhanced productivity and urbanization.

Antiquity of Iron Smelting in Eastern Indian Tribes

It is quite presumable that in a jungle fire some hunting-forging people of eastern India might have witnessed the flow of black molten material coming out from the accumulated decomposed ore at the base of some rain gully. Before they took the process to their own experimentation, they should have seen it happening time and again in similar fires and observed that burning of certain kind of earth produces a black material which is harder than stone when cool and can be fashioned into varied shapes when heated. The same jungle fires also made them aware of the importance of fuel and specially the wind blast for activation of good fire.

However, the question that at which point of time the tribals might have discovered the art of iron smelting remains to be settled. There is strong body of evidence to prove that the iron smelting was discovered and established in eastern India at an early stage possibly much before the appearance of pottery in the region. The archaeological opinion generally concedes that pottery is older than metal. Since, pottery preceeds metal, it is likely that pottery firing technique prevalent in the region might have inspired the development of metallurgical gear employed in local smelting. A micro-level ethnographic survey of firing techniques of peasant potters of India has revealed the existence of three principal firing zones in the country (Saraswati and Behura, 1966: 103-129). These zones were stratified in ancient times and have survived through centuries to present day. The open-firing is confined mainly in north and north-western parts of the country (Rajasthan, Punjab, Haryana, western Uttar Pradesh and north-western parts of Madhya Pradesh). Clearly, this tradition is stratified mainly in the areas which have figured prominently in the process of first urbanization in the country. This method requires less wood and more straw, dry leaves and brush wood for firing. This firing method is consistent with the archaeological evidence as the forest cover in these areas was constantly under pressure due to activities associated with urbanization. The open firing is done in apical or horizontal fire channels. With highly oxidizing conditions it is expected to produce more red pots than black ones. Incidentally, the areas coming under this firing tradition have been associated with copper metallurgy and dominant production of red pots, both employing an open-firing technique. The firing in oven is prevalent in south India spreading eastward upto Orissa, where it meets yet another firing tradition i.e., Kiln firing which occupies practically whole of eastern India. In oven firing wood fuel (including green wood) and cattle dung is employed which emits lot of smoke in the process. The fire is kindled from a stoke hole at the bottom. The kiln firing requires abundance of wood fuel to burn the pots. Among the latter two firing traditions, the oven firing suggests comparatively less wooded country. The formation of these two firing zones depended largely on the ecological and cultural factors. These firing methods, particularly the oven burning affords opportunity to make a combination of black-and-red pots at will. Incidentally, the areas coming under oven firing have shown that ancient potters were producing black or black-and-red pots.

Unlike other parts of the country where metallurgy followed pottery, the process seems to have moved in reverse direction in eastern India. This is testified by scores of tribal societies of the region, among whom the practice of potter's craft is a taboo. Pottery making is still an exclusive domain of caste potters who were late arrivals in the region. There is strong possibility that earliest pottery of eastern India was a product of oven firing tradition reaching there through incoming potter groups from south India. Later, however, an adoption of kiln firing equipped the emigrant ancient potters with a better firing device for generation of high temperature and ability to control it. Generally, the kiln firing is taken as improvement over oven firing.

A technical fraternity between pottery and the iron smelting kilns can be traced on the basis of certain striking similarities in their structure and working. Both show presence of stoke holes and an identical method of regulation of draught, particularly the practice of fuel feeding at regular intervals. In the vertical updraught kiln or in a ground level stoke hole, much help is taken from external air similarly as in iron smelting. The word *ava* (a kiln) is derived from the word *va* which means to blow. The evidence suggests that tribal smelters motivated the early emigrant potters to develop their kilns on the lines of iron smelting kilns and the same has survived to present day to become dominant firing tradition of the region. It is certain that this transformation could only take place in eastern India because the less wooded country in the south not only rendered the adoption of kiln firing of pottery impracticable but created problems for iron smelting too. The evidence discussed above favours a pre-pottery date for the existence of iron in tribal communities, implying that the iron metallurgy was already well established before this metal was first recorded in the archaeological context of eastern India.

Manifestation of Choice, Decisions and Selections in the Iron Smelting

Far from primitiveness of their economy and limited material culture, the tribal smelters make most elaborate arrangements for their craft. The physical manifestation of choices and selections have shown amazing quality of ingenuity. These complex choices and selections are local growths, the fact that they have taken roots after long period of experimentation is a point of great importance. The fundamental situation which has made the whole process possible is abundance of ore and wood in their physical habitat affording them to make experiments with selected and chosen material at every stage of developing suitable method and kit for their craft.

The specific choices, decisions and connected rites in every department of iron smelting reflects craving for good iron but hazards and uncertainty associated with the process feature prominently in them. The magical sanctions and set of prohibitory taboos to protect iron makers from any deviation from the set norms points to the very inherent fear of uncertainty that prevailed in their minds since they learnt the craft. Throughout the hill-forest belt of eastern India the magical value of iron overrides its functional value. The smelters are honoured and are accredited with the knowledge of special magic for their work and their fire has sacred qualities. The traditional stringencies are all compelling on every smelting tribe. The complete rejection of all traits outside the traditional structure acted as great cohesive force in protecting this system from disintegration and made its survival possible in unadulterated form upto recent times.

The degree of technological conservatism and behavioural constraints exhibited in the metallurgical process offers an excellent medium to add a cultural perspective to the understanding of an ancient metallurgy.

Iron Smelting in the Tribes of Eastern India

The various aspects of tribal iron smelting would reveal that physical manifestation of decisions, entrenched in tradition, always produced desired results. The same has helped smelters to achieve high degree of proficiency in obtaining purest iron and competency to produce 'kindred' forms too.

The first step in smelting is selection of site where burning of ore in open fire or in furnace can take place smoothly. The selection of site is often affected by availability and accessibility to at least one raw material e.g., ore, charcoal, clay or water and the place should also provide perfect conditions for wind blast.

Selection of ore

It is not without significance that tribal smelters avoid using massive ore particularly that of magnetite as it involves labour of mining and breaking by hand beside causing difficulty in smelting. They mostly select weathered ore and even go for unconventional ores such as ferruginous shales or ferruginous sands and even laterite if availability is easy (Ball, 1881: 335). Tribals often cover long distances in search of decomposed ores which are found in the form of small pieces on the surface of deposit. In Khasi hills the ore occurs in the form of a fine sand consisting of minute crystals of titaniferous magnetic oxide which is regularly distributed in the mass of decomposed granite. The soft and easy yielding rock is not mined but merely raked by the Khasi women during the rainy season into a small stream at the base of the rock from which ore is obtained (Ball, 1881: 413). In Bihar, three varieties of ore are recognized by the Asurs. One is magnetite called *Pola*, the hematite extracted from Banded-Hematite-Quartzite veins known as *Bichi* and a laterite ore *Gata* (Leuva, 1963: 147). They, however, prefer powdery iron ore (blue dust) available at a low altitude in the hematite formations. Some times, decomposed ore is located at ground level which is extracted and the resulting pit is used as hearth for open firing.

Although the tribal smelters are completely ignorant about the chemical implications but they are aware of the fact that the use of decomposed ore raises fusibility of the charge on a low temperature with less fuel in a shorter period and gives high percentage of pure iron (+ 70%). It also ensures reduction of ore into metal with little loss to slag. The functional importance of the weathered ore is best exemplified in the case of ore around Jabalpur which contains high amount of phosphorus and is unsuitable for industrial smelting but tribals of Bastar smelt it suitably and are able to remove considerable amount of phosphorus content.

Fuel wood and charcoal

The smelters are aware of the fact that iron suitable to produce tool steel could only be obtained after burning of magnetic ore with charcoal. The knowledge of wood charcoal as an accidental by-product of combustion can not be ruled out which was later used for its own sake (Horne, 1982: 6). In fact, the plentiful availability of wood in their physical habitat helped the tribal iron smelters to learn about the qualities of various woods and charcoals. Generally *Sal* wood is preferred and considered best fuel for smelting (Leuva, 1963: 148). In the absence of better fuel, the bamboo charcoal, though considered an inferior one, is the most popular choice. The chemical implication involved with bamboo fuel is the fact that its ash content (8.3%) acts as a good flux and the high carbon (83.3%) helps in achieving fair amount of carburization in metal (Turner, 1920: 326).

Shape and size of furnace

The smelters use small furnaces which is indicative of employment of high quality iron ores. It has been proved by certain studies (Van der Merwe and Avery, 1985) that taller furnaces are required to smelt low grade ores. The height of furnace is increased in the case where natural draft is to be utilized for slow reduction of large volume of low grade ore to get an adequate yield of bloom. Some other studies in Africa (Childs, 1991: 345) have shown that larger furnaces yielding higher blooms require a stable work force and were used only in densely populated regions.

Clay for furnace

It is significant that most iron smelting tribes are very selective about the clay they use in building a furnace. They prefer clay from ant hills for making furnaces after mixing it with good amount of rice husk (Turner, 1920: 326). The smelters may not be aware of the chemical implication involved here but it is a fact that the clay produced out of cellulosic intake of the ants provides them a medium free from contaminating minerals and impurities present in common clay. Naturally, the iron produced in such furnaces is many grade purer and free from impurities. Moreover, burning of rice husk acts as an additional source for carburization. This clay affords resistance to thermal shocks and is refractory to collapse of high temperature.

Method of smelting

Tribals employ easier methods suitable for obtaining iron from the ore in single operation. In a simpler process, the smelting takes place in the forest near the ore, charcoal, and water in an open firing. The washed decomposed ore is thrown in fire at a fixed interval. After three to four hours a few small rounded balls of iron ore formed at the base of slaggy pile. These balls of iron are recovered and compounded into a bigger mass. In a second method, the smelting is accomplished in an open headed short furnace with successive layers of ore and charcoal. The twin bellows, covered with cowhide are attached to the furnace through bamboo or clay tuyeres to supply draught.

Use of flux

Since ore and fuel is available in plenty in the area, most smelters do not have direct knowledge about the use of flux in the way it is known to non-tribal smelters.

Nevertheless, they know that some ores produce good iron or certain fuel helps in the easy recovery of iron from slag. It is logical to assume that they must have learnt it by long experience that burning of ferruginous shales or ferruginous sands produces good iron. The bamboo charcoal is generally regarded as an inferior fuel but smelters use it with some advantage. In case of the former, the ores are rich in calcium and silica respectively which makes the charge a self fluxing one whereas the bamboo charcoal acts as a good flux (Turner, 1920: 326).

Hardening of metal

There are evidences to prove that smelters were fully aware of devices used to harden the metal. Their capability to make carburized iron suitable for quenching is confirmed by various taboos observed by tribal smelters. The Agaria forbids menstruating women to bring water for quenching the metal (Elwin, 1942). The extreme sense of caution suggests the importance of whole process. The practice of compounding the bits of iron recovered from the firing in the forest through repeated heating with charcoal gives fair amount of carburization to the iron and makes it low carbon steel (0.35 to 1.5% carbon). Tribal smelters are also skilled enough to make hard iron either by direct burning of manganiferous ore or by mixing certain amount of locally available manganese in the charge. The resultant white iron is much harder than the bronze and the tools fabricated out of it have smoother surface. The 1.0 to 1.5% manganese content in iron provides appreciable hardness to tools (Turner, 1920: 267). In fact, the presence of 1 to 2% manganese in iron is more beneficial than silica as it produces desulphuring effect and the iron consequently is white. This iron had a good market at ancient urban centres where it was used as an alternative to bronze. The ethnographic evidence tells us that the Korwas of south Mirzapur (U.P.) were making white iron from local ores and it was in great demand for making best grade axes and arrowheads (Dalton, 1978: 228-29).

The tribal smelters insist on using stone anvil as the iron implements shaped on it did not rust easily because the practice siliconizes the outer skin of the metal (Leuva, 1963: 148).

Archaeological Evidence of Iron in Eastern India

Lately, the discovery of iron objects within Chalcolithic levels at Pandu Rajar Dhibi (De and Chattopadhyay, 1989: 40-41) and Mangalkot have pushed back the dates of introduction of iron in archaeological context of eastern India. We feel that the presence of iron objects within the copper dominated levels of these settlements fulfills one important precondition that these objects must have been hard enough to compete with the metal already present on the site. Otherwise it is difficult to believe that people were eager to adopt a new metal in the form of wrought iron which is less hard than copper and can not be used to fabricate ornaments. Nevertheless, the adoption of iron for their tools shows that it must have been equally hard or harder than their previous metal. The evidence from Barudih (Ghosh and Chattopadhyay, 1982: 63) and Dhuliapar (De and Chattopadhyay, unpublished paper) shows sudden appearance of iron in an assemblage of Neolithic nature. The element of suddenness and cultural priority hints that iron recovered here was of tribal origin. There are many settlements in eastern India where local smelting is suggested on the basis of recovery of slags but do not register a significant use of iron. These archaeological situations are indicative of the fact that a stage was reached when tribal iron in smelted form started getting access on many settlements of eastern India who became iron users without being smelters. The presence of slags on a settlement is not always a sure indicator of local smelting. The tribal smelters in practice supply metal in the form of bloom or *giri* (consisting iron, slag and charcoal) to its users. The metal is recovered by local smiths after removing the slag through repeated beating and hammering. The heaps formed by the accumulation of separated slag pieces on the settlement gives a false indication of local smelting.

CONCLUSION

The ethnographic evidence discussed above demonstrates that craft of iron technology held by certain societies in eastern India existed from remote times perhaps much earlier than the introduction of pottery in the region. The total performance of iron smelting is very complex and involves many decisions and choices at every stage of the process.

In all likelihood the well established tribal iron technology was prime motivational factor behind the spread of the craft in whole of eastern India. For instance, first it helped in establishing a stage of use of iron by affording supply of metal in smelted form. In fact, the practice of tribal smelters confining only to the production of bloom or *giri* which is passed on to the users, always harboured chances of metal being used beyond the borders of tribal belt. Later, however, a stage was reached when smelting became possible on many settlements either by importing expertise or by migration of tribal smelters in person to different areas to help in utilizing the locally available ore. For instance, the Agari or Gole Agarias of Kumaon Hills trace their origin from Agarias of eastern India. Similarly, the Mahle Asurs of central India are an outgoing section of Asurs of Chhota Nagpur. In Kumaon Hills, the Gole Agarias are associated with the production of metal while Lohars do the smithy. It is archaeologically significant that in Kumaon Hills exist three different classes of metal workers viz., Agaria for iron, Tamta for copper and Tirwa for tin.[2] It can be gathered that eastern India with its tribal iron smelters stands crucial for the spread of iron technology in a wide area.

Since, the Asurs, the leading iron smelters of eastern India, prefer to live in hills, people like Karmali and Barae emerged in the plains and performed the additional duty of smiths.

NOTES

1. It is to be noted that chemical implications associated with copper objects and slags from Ahar do not necessarily confirm the conclusions that have been derived. For instance, the presence of 6.48% iron as impurity is an inevitability where an ore containing both copper and iron is smelted. One can only regret the technological inability of the Aharian coppersmith to check the flow of iron from passing down to slag and for not being skilled enough to develop a kit to separate it from a combined ore of copper and iron (if at all they knew iron in elemental form and wanted to retain it). Nevertheless, one can always ask as to why pyro-technologists were using Chalcopyrite to extract copper when other high yielding copper ores were available in the country. Was it a deliberate effort in order to achieve iron ? But the fact is that the reason for the use of this ore was entirely different. In fact, the copper ore which is found in workable deposits in India was only Chalcopyrite and thus remains the biggest source of copper from ancient times. Obviously there would have been many centres in the country where burning of Chalcopyrite was a routine practice and always harboured the chances of emergence of Ahar like iron as by-product of copper metallurgy. Unfortunately, inspite of increased chances neither iron technology appeared anywhere through the furnace of coppersmith nor people have shown any craving for the use of iron. Thus any insistence for the emergence of iron as a by-product of copper technology does not prove itself on chemical and technological grounds.
2. I am thankful to Dr. P.K. Agrawala, Department of Ancient Indian History, Culture and Archaeology, B.H.U., for drawing my attention towards possible equation of Sanskrit words *Tamra* and *Trapu* with the Tamta and Tirwa tribes of Kumaon Hills who are associated with copper and tin respectively.

REFERENCES

Ball, V.A., 1881. *Manual of the Geology of India: Economic Geology*. III. Calcutta.

Chakrabarti. D.K., 1984. Study of Iron Age in India. *Puratattva*, 13-14: pp. 81-85.

Charls, J.A., 1980. The coming of copper and copper base alloys and Iron: A metallurgical sequence. In *Coming of the Age of Iron* (eds. T.A. Wertime and J.D. Muhly), pp. 151-180.

Childs, S.T., 1991. Style, technology and iron smelting in Bantu speaking Africa. *Journal of Anthropological Archaeology*, 10.4: 332-359.

Dalton, E.T., 1978. *Tribal History of Eastern India.* (Reprinted). Cosmo Publications, New Delhi.

De, S. and Chattopadhyaya, P.K., Early evidence of tuyeres from Dhuliapar. West Bengal: Archaeometallurgical Studies. Neolithic to Iron Age Transition. Typed copy. pp. 1-4.

De, S. and Chattopadhyaya, P.K., 1989. Iron objects from Pandu Rajar Dhibi: Archaeometallurgical studies. *Steel India.*, 12.1: pp. 40-41.

Elwin, V., 1942. *The Agaria.* Calcutta.

Forde, C.D., 1949. *Habitat, Economy and Society,* London.

Ghosh, A.K. and Chattopadhyaya, P.K., 1982. An early steel implement from Baruadih, Bihar Province, India: Metallurgical Studies. *Masca Journal,* 2:2: pp. 63-64.

Hegde, K.T.M., 1969. Technical studies in copper artifacts from Ahar. In *Excavation at Ahar* (Tambavati). Pune, pp. 225-228.

Horne, Lee, 1982. Fuel for metal worker (The role of charcoal and charcoal production in ancient metallurgy). *Expedition,* 25.1: p. 6.

Leuva, K.K., 1963. *The Asur,* New Delhi.

Van der Merwe, Nicholas and Avery, Donald, 1985. Science and magic in African technology: Traditional iron smelting in Malawi. *Africa* 57.2: pp. 143-172.

Prakash, B. and Tripathi, V., 1986. Iron technology in ancient India. *Metals and Materials,* 2.9: pp. 568-579.

Sahi, M.D.N., 1979. Iron in Ahar. In *Essays in Indian Protohistory,* (eds. D.P. Agrawal and D.K. Chakrabarti), pp. 365-366.

Saraswati, B. and Behura, N.K., 1966. *Pottery Techniques in Peasant India.* Anthropological Survey of India Memoirs No. 13, Calcutta.

Tripathi, V., 1985. From copper to Iron - A Transition. *Puratattva.* 15: pp. 75-79.

Turner, Thomas, 1920. *The Metallurgy of Iron.* London.

Wertime, T.A., 1980. The Pyrotechnologic Background. In *Coming of the Age of Iron* (eds. T.A. Wertime and J.D. Muhly), pp. 1-24.

Iron Metallurgy in Ancient Madhya Pradesh

K.D. BAJPAI

Madhya Pradesh is extremely rich in natural resources. Its major part is covered with the Vindhyan and other hills. Its thick vegetation has retained, to some extent, the old glories of *Vindhyāṭavī* and *Daṇḍakāraṇya*. The Vindhya mountain was one of the eight *kula-parvatas* of the country. Its several ranges cover an extensive region north of the Narmada. Over the western part of Madhya Pradesh is spread the Aravalī range. The region south of the Narmada has two ranges, Mahadeo and Satpura.

The State is considerably rich in minerals. The chief minerals are: iron, coal, manganese, limestone, bauxite, clays, other corundums, sillimanate, silica sand, dolomite, steatite, red oxide of iron ore, mica, graphite, diamond and lapidolite.

Iron-ore occurs in Durg, Jabalpur, Bastar and Gwalior districts. In Durg the deposit is confined to the Dallī-Rajharā range of hills with an estimated resource of 114 million tons and is linked to Bhilai. In the Bastar district 14 ore-bodies occur in isolated hills. Bailadila (Tahsil Dantewara) has an estimated conservative reserve of 610 million tons of high grade, which has subsequently been revised to 3,600 million tons of all grades.

Madhya Pradesh possesses a very high grade iron ore, with blue dust being as much as 75% pure. The Hematite iron ore of Bailadila averages a purity of 64%. Occurring as it does in close proximity to coal, limestone and dolomite, Madhya Pradesh iron ore can feed a steel industry and the ferro-alloy industry, which can raise the State to a level of development prevailing in advanced western countries. The combination of limestone and coal also provides a base for a major cement industry, which is fast expanding, and also a coal-based fertilizer industry.

The Balaghat district of Madhya Pradesh possess copper-ore. From a site called Gungeria in the district a large number of copper implements was obtained several years ago. A few copper celts are known from Eraṇ in the Sagar district. During the period ranging between *c.* 300 B.C. and 12th century A.D. thousands of copper coins were minted in the mints at Ujjain, Vidisha, Kutwar, Eraṇ, Tripuri, Nandner and several other sites.

The Proto-historic Phase

The proto-historic phase of Madhya Pradesh is undoubtedly of unusual importance. Apart from the painted rock-shelters which have furnished a rich source for our knowledge, the excavations and explorations conducted at several sites brought to light valuable material pertaining to this phase. The recent studies have shown that it was on the banks of the rivers—Chambal, Betwa and Narmada—that various facets of culture could cherish an unhampered growth from about the beginning of the 2nd millennium B.C. to 600 B.C. The important Chalcolithic sites of Madhya Pradesh so far excavated or explored are: Maheshwar, Navadātolī, Kāyathā, Eraṇ, Nāgdā and Pipalyā Lorkā.

Out of these, Maheshwar-Navadātoli, on the river Narmada were the first Chalcolithic sites of Madhya Pradesh which were excavated under the direction of Sankalia. Period I at these sites goes back to about 1700 B.C. The typical pottery of this period has been called the Mālwā Ware, which is akin to the one found at Kāyathā, Eraṇ, Pipalyā Lorkā and a few other sites.

Kāyathā, near Ujjain is an important Chalcolithic site. The excavator of Kāyathā, Wakankar, has designated the earliest phase at Kāyathā as 'Pre-Harappan Culture-Complex'. It may, however, be pointed out that the relevant material, as known to us from Harappā, Kālībangā and some other sites, has practically nothing in common with the material of period I at Kāyathā. The earliest phase at Kāyathā, in my opinion, represents the basic ingredients of the Malwa Chalcolithic culture. It can better be termed as 'Mālwā Chalcolithic Complex'. The regional variations in fabric, types and painted designs, known from Kāyathā and other Chalcolithic sites, should be studied on a comparative basis before reaching any conclusions pertaining to nomenclature,technique and cultural diffusion. It is clearly discernible that the first phase at Kāyathā represents the earliest Ware which was succeeded by the later Mālwā Ware.

'De Luxe' Culture of Eraṇ

Eraṇ (District Sagar) was excavated under the direction of the present author. Some important Carbon-14 dates of its Chalcolithic phases are now known to us vis-a-vis the dates of other sites in M.P. and Rajasthan. I have made a detailed study of the cultural sequence at Eraṇ, with reference to the Chalcolithic phase. The two earlier Carbon-14 dates of iron known from Eraṇ, are quite significant in this respect. It may be mentioned that in the Eraṇ excavations, iron has been found in almost all the layers upto layer no. 16. Its early phase can be bracketed between 1300 (or 1250) B.C. and 700 B.C. The associated objects are unpainted black and red ware, beads of terracotta and shell, paste and semi-precious stones, terracotta human figures, shell bangles and microlithic tools of various types.

The introduction of iron at Eraṇ should be taken as a distinctive feature to mark the beginning of a new cultural phase. It has, therefore, become necessary to put the early culture-phases at Eraṇ under the following categories:

Period:				
Period:	**I** (A) C. 2100	-	1700 B.C.	Chalcolithic
	I (B) C. 1700	-	1300 B.C.	Chalcolithic
	II (A) C. 1300	-	700 B.C.	Introduction of Iron
	II (B) C. 700	-	200 B.C.	
	II (C) C. 200 B.C.	-	100 A.D.	

Period **I** (both A and B) at Eraṇ has furnished painted black-on-red ware, whicl in fabric and shape is much superior to the wares at Maheshwar-Navadāṭolī, Kāyathā and other Chalcolithic sites in Madhya Pradesh. This superior ware at Eraṇ can be termed as 'de luxe' Mālwā Ware. Besides this, Period **I** at Eraṇ has yielded Grey Ware with paintings, thick ware similar to that of Kāyathā (but not actual Kāyathā) and black-and-red Ware similar to that of Ahar near Udaipur in Rajasthan. The Painted Grey Ware at Eraṇ is different from that of the Gaṅgā valley and has its distinct fabric and shapes.

A remarkable feature of the Chalcolithic culture at Eraṇ is the construction of a thick mud defence wall and a moat in period **I** (B). The defence wall measures 46.79 metres X 6.41 metres. The important objects found in the wall, apart from the pot-sherds, were a copper celt, various types of beads and microliths. All these belong to the Chalcolithic phase only. The defence wall protected the town on the south side. The width and the maximum depth of the moat around the defence wall were 36.60 metres and 5.40 metres respectively.

Pottery

It is interesting to note that the potters of Eraṇ during the Chalcolithic phase were past masters in preparing excellent wheel-made pottery. Both from the point of view of fabric and designs, the 'de luxe' pots of Eraṇ rank high and can well be compared with those found at Harappa, Mohenjodaro and other Chalcolithic sites of Punjab, Sind and Baluchistan. Most of the wares of the early phase at Eraṇ are beautifully painted in black-on-red exterior. Apart from the geometrical designs, other motifs on the outer parts represent the flora and fauna of eastern Malwa portrayed tastefully by the potter-artist. Some of the red ware sherds have incised decorations.

The developed devices of Chalcolithic pottery were continued by the people of Pipalyā-Lorkā, another important site near Bhopal, close to the source of the river Betwa. The site has been excavated by the State Department of Archaeology recently and has yielded interesting evidence of the material culture of the early Chalcolithic phase.

Inter-Relations

The pottery-types indicate that there were contacts between the people of Kāyathā and those of Ahar (near Udaipur in Rajasthan). The Ahar people, after establishing themselves at Kāyathā, may have migrated to Eraṇ. The inter-relations among these people can be ascertained after a thorough comparative study of the material culture known from the various Chalcolithic sites in Madhya Pradesh and the contiguous regions of other State.

In Central India the eastern and south-eastern parts of Malwa made a more rapid cultural development during the proto-historic times than other parts of the region. Apart from enjoying a bracing climate, these parts of Malwa were located on the direct north-western links. From Harappa, the chief centre of the proto-historic culture, one main route seems to have passed through Kālībangā, Ahār and other sites in the Rajasthan-Malwa region.

Madhya Pradesh provides a rich field for the study of pre and proto-historic cultures in respect of their nature, expansion and chronology. Several problems pertaining to the early Indian environments, human habitations and cultural migrations can be solved with the help of the material obtained in different parts of this region.

In regard to the metallurgic technique, it may be pointed out that, due to scarcity of proto-historic copper and iron objects from the Betwa, Narmada and Chambal valleys, the artistic nature of the implements can not be precisely determined. But for the early historical period (*c.* 300 B.C. to the 4th cent. A.D.) we have enormous quantity of artifacts and coins for study. From excavations conducted at Eraṇ, Kāyathā, Navadātolī, Malhar, Tripuri, Tumain and several other sites in M.P., a large number of copper and iron tools and agricultural implements have been obtained. Their workmanship can well be compared with the contemporary metal objects discovered at other areas of northern India.

In so far as the copper coins of the historical period are concerned, they undoubtedly show a more developed technique and artistic excellence.

REFERENCES

Ansari, Z.D. and Dhavalikar, M.K., 1975. *Excavations at Kayatha,* Pune.

Bajpai, K.D., 1978. *Indian Numismatic Studies,* Delhi, pp. 1-17, 108-13.

Bajpai, K.D., 1978. *Malhār,* Sagar.

Bajpai, K.D., 1984. *History & Culture of Madhya Pradesh,* Ahmedabad, pp. 1-8; 24-29.

Bajpai, K.D., 1985. *Cultural History of India,* Vol. I, Kanpur. Ch. II and Ch. VIII (pages 166-69).

Sankalia, H.D., 1974. *Prehistory and Protohistory in India and Pakistan,* Bombay.

Wakankar, V.S. *Kayatha Report,* Ujjain.

Metal Implements of Northern India (C. 1000 B.C. to 500 B.C.)

J.P. UPADHYAYA

Introduction of iron and its early history in the Indian sub-continent has attracted much attention of archaeologists for quite some time now. As a result, much field data has come to light, which gives details regarding the chronology and spatial distribution of early Iron Age cultures on the one hand, and has helped in the accumulation of rich cultural material, on the other. On the basis of available data and also metallurgical studies of the metal objects, much has been written on various issues of the early stages of iron technology in India.[1] Inspite of these important studies, there is dearth of synthetic accounts of individual category of archaeological material in general and that of antiquities other than pottery in particular. There cannot be two opinions that such accounts are essential for proper understanding of the living pattern of the ancient societies.

In the present paper, an attempt is made to put together all the available information on implements of the early iron using-communities of north India. It is now well established that the initial stages of the Iron Age in this region persisted during first half of the first millennium B.C. Thus, I have selected three cultural horizons, viz., Painted Grey Ware, Black-and-Red Ware and overlap phase of Painted Grey Ware and Northern Black Polished Ware, all of which fall within the time bracket of C. 1000 B.C. to 500 B.C. All the metal implements used for hunting, fishing, war, agriculture and domestic purposes from these phases have been studied and catalogued in the form of three tables.

Table 1 deals with the implements of Painted Grey Ware culture,

Table 2 deals with the implements of overlapped phase of PGW and NBPW culture, and

Table 3 deals with that of Black-and-Red Ware culture

Objects which are frequent in these phases include arrowheads, spearheads, knives and daggers, supplemented by other implements like nails, rods, fish-hooks, pins, needles and antimony rods and a few celts, axes, hoes, sickles, chisels, borers and a single ploughshare. The typological and numerical details mentioned herein are based primarily on the published accounts of important excavated sites in various volumes of *Indian Archaeology : A Review* and detailed reports, wherever available. Comparative account of the objects made on copper and iron in my study brings forth some noteworthy patterns regarding the preference for one over the other metal in particular regions and also for certain categories of implements. Some of these facts seem to open new avenues for cultural interpretations, which are attempted below:

TABLE 1

Sl. No.	*Site*	*State*	*Tool-type*	*No. of Tools*	*Metal*
1.	Atranjikhera	U.P.	Arrow-head[2]	1	Iron
2.	Alamgirpur	U.P.	Arrow-head[3]	1	Iron
3.	Hastinapura	U.P.	Arrow-head[4]	1	Copper
4.	Allahapur	U.P.	Arrow-head[5]	1	Iron
5.	Jakhera	U.P.	Arrow-head[6]	More than one	Iron
6.	Atranjikhera	U.P.	Spear-heard[+]	8	Iron
7.	Alamgirpur	U.P.	Spear-head[8]	1	Iron
8.	Allahapur	U.P.	Spear-head[9]	1	Iron
9.	Jakhera	U.P.	Spear-head[10]	More than one	Iron
10.	Atranjikhera	U.P.	Knife[11]	3	Iron
11.	Jakhera	U.P.	Dagger[12]	1	Iron
12.	Atranjikhera	U.P.	Axe[13]	1	Iron
13.	Jakhera	U.P.	Axe[14]	1	Iron
14.	Atranjikhera	U.P.	Chisel[15]	6	Iron
15.	Jakhera	U.P.	Sickle[16]	1	Iron
16.	Atranjikhera	U.P.	Celt[17]	1	Copper
17.	Allahapur	U.P.	Celt[18]	1	Iron
18.	Jakhera	U.P.	Hoe[19]	1	Iron
19.	Jakhera	U.P.	Plough-share[20].	1	Iron
20.	Bhagwanpura	Haryana	Rod[21]	More than one	Copper
21.	Atranjikhera	U.P.	Rod[22]	7	Iron
22.	Allahapur	U.P.	Rod[23]	1	Iron
23.	Jakhera	U.P.	Rod[24]	1	Iron
24.	Daulatpura	Haryana	Antimony Rod[25]	1	Copper
25.	Atranjikhera	U.P.	Antimony Rod[26]	2	Copper
26.	Jakhera	U.P.	Antimony Rod[27]	1	Copper
27.	Alamgirpur	U.P.	Pin (?)[28]	1	Copper

(Contd.)

Sl. No.	*Site*	*State*	*Tool-type*	*No. of Tools*	*Metal*
28.	Ahichchhatra	U.P.	Pin[29]	More than one	Copper
29.	Atranjikhera	U.P.	Pin[30]	2	Copper
30.	Hastinapura	U.P.	Borer[31]	1	Copper
31.	Atranjikhera	U.P.	Borer[32]	6	Iron
32.	Ahichchhatra	U.P.	Nail[33]	More than one	Copper
33.	Jakhera	U.P.	Nail[34]	More than one	Iron
34.	Atranjikhera	U.P.	Nail[35]	20	Iron
35.	Atranjikhera	U.P.	Nail parer[36]	2	Copper
36.	Allahapur	U.P.	Nail parer[37]	1	Copper
37.	Atranjikhera	U.P.	Fish-hook[38]	2	Copper
38.	Atranjikhera	U.P.	Hook[39]	7	Iron
39.	Jakhera	U.P.	Needle[40]	1	Copper
40.	Atranjikhera	U.P.	Needle[41]	1	Iron
41.	Atranjikhera	U.P.	Antimony[42] rod-cum-nail parer	1	Copper
42.	Atranjikhera	U.P.	Clamp[43]	1+21 (22)	Copper Iron
43.	Atranjikhera	U.P.	Shaft[44]	10	Iron
44.	Atranjikhera	U.P.	Tong[45]	1	Iron
45.	Atranjikhera	U.P.	Slag*[46]	7	Iron
46.	Atranjikhera	U.P.	Bangle*[47]	5+2	Cooper Iron
47.	Atranjikhera	U.P.	Ring*[48]	2	Copper
48.	Atranjikhera	U.P. object	Indeterminate[49] (17)	3+14 Iron	Copper
49.	Singh, Bhagwanpur	Punjab	Slags*,[50] Clamp, Pieces of spearhead and javalins	-	Iron
50.	Hulas	U.P. Copper, antimony rods and Styluses	Slags*,[51]	-	Copper, Iron

* Though these metals are not implements, these have been shown in the Table to show the richness of metal.

TABLE 2

S. No.	*Site*	*State*	*Tool-type*	*No. of Tools*	*Metal*
1.	Noh	Rajasthan	Arrow-head[52]	More than one	Iron
2.	Noh	Rajasthan	Spread-head[53]	More than one	Iron
3.	Jajmau	U.P.	Arrow-head[54]	1	Iron
4.	Piprahwa and Ganwaria	U.P.	Arrow-head[55]	More than one	Iron
5.	Piprahwa and Ganwaria	U.P.	Spread-head[56]	More than one	Iron
6.	Jajmau	U.P.	Nail[57]	More than one	Iron
7.	Piprahwa and Ganwaria	U.P.	Nail[58]	More than one	Copper
8.	Jajmau	U.P.	Fish-hook[59]	More than one	Iron
9.	Piprahwa and Ganwaria	U.P.	Knife[60]	1	Iron
10.	Piprahwa and Ganwaria	U.P.	Antimony[61]	More than one	Copper
11.	Piprahwa and Ganwaria	U.P.	Sickle[62]	1	Iron
12.	Piprahwa and Ganwaria	U.P.	Ring*[63]	1	Copper

* Though these metals are not implements, these have been shown in the Table to show the richness of metal.

TABLE 3

S. No.	*Site*	*State*	*Tool-type*	*No. of Tools*	*Metal*
1.	Mahishdal	Bengal	Arrow-head[64]	More than one	Iron
2.	Mahishdal	Bengal	Spear-head[65]	More than one	Iron
3.	Mahishdal	Bengal	Chisel[66]	More than one	Iron
4.	Chandraketu-garh	Bengal	Axe[67]	1	Iron
5.	Chandraketu-garh	Bengal	Awl[68]	1	Copper
6.	Mahishdal	Bengal	Slag*[69]	More than one	Iron
7.	Chirand	Bihar	Piece of Blade[70]	1	Iron
8.	Sonpur	Bihar	Wire[71]	1	Copper
9.	Oriup	Bihar	Bangle*[72]	More than one	Copper

* Though these metals are not implements, these have been shown in the Table to show the richness of metal.

The area occupied by aforesaid three culture-phases is considerably wide, and it falls within the present administrative boundaries of six states, viz., Punjab, Haryana, Rajasthan, Uttar Pradesh, Bihar, and Bengal. It is noted that with regard to the yield of metal objects, Gangetic doab seems to be the most promising area, having a dense concentration of sites in this region yielding good amount of metal tools. The potentiality of this area becomes much more pronounced when it is compared with the western and eastern regions, which have comparatively poor yield. For instance, though a number of important sites, e.g. Ropar, Dadheri and Sanghol etc., of Painted Grey Ware phase have been excavated, yet no metal tool is reported from these sites. Singh Bhagwanpur, Distt. Rupanagar, however, yielded a few iron objects from Punjab.

The reported metal objects are not many. The primary reason for such a situation seems to be that due to high corrosion, particularly in case of iron, a good proportion of these have got damaged, and also as a result of this the small fragments are difficult to identify. Another reason for the limited number of objects may be the fact that the metal was used and reused and was not frequently discarded.

DISCUSSION

The above tables clearly indicate that in comparison to iron, copper was strikingly less in use in all the three culture horizons under discussion. Preference for iron, is further suggested by the fact that iron is more liable to damage from weathering processes than copper. Thus, it is possible to postulate that the introduction of iron technology considerably minimized the use of copper, as iron was found suitable for implements in general and heavy-duty tools, in particular. Our synthetic account of the metal tools clearly reveals that during first half of the first millennium B.C., practically all the heavy-duty implements for agriculture and hunting purposes were made of iron. Besides, copper in this stage is utilized mostly for small objects of daily use, like antimony rods, etc. Another noteworthy observation is that not a single piece of bronze has been reported from the three phases.

The use of iron and copper for two different categories of artefacts during the Painted Grey Ware, Black-and-Red Ware and the overlapped phase of Painted Grey Ware and Northern Black Polished Ware Culture phases, appears to depend on the nature of the metals. The use of copper for small objects of personal use is not without significance. Since copper does not rust like iron, even today, pots for the storage of drinking water, tooth picks, nail-parers, etc., are made of copper. Iron is avoided for such use as rusted iron is likely to infuse such diseases as Titanus, etc., to the user. It seems that the advantage of copper as safe metal for the objects to be used directly on the body was realized during this period.

On the other hand, iron was found suitable for heavyduty tools of agriculture and hunting, etc. due to its being harder than copper. Another factor which seems to be responsible for it is that iron was cheaper and easily available in the area. Rich iron-ore deposits are found in Rajasthan, Singhbhum (Bihar) and Bengal, which even now are exploited for iron. Exploitation of iron-ores, smelting and the preparation of objects by forging in habitational areas by the PGW population is well attested from the evidence of two furnaces excavated at Suneri, in Jhunjhunu district of Rajasthan.[73] Moreover, iron is repeatedly used a number of times. My observation of blacksmithy in todays villages of Uttar Pradesh may be useful to note in this regard. The heavy-duty tools like ploughshare, etc., when they cease to serve their normal purpose, are not discarded, but are shaped into other new forms like sickle, khurpi, knife, nails, awls, needles, etc. Whether this practice was in existence in the past is difficult to ascertain, but its possibility cannot be ruled out completely.

ACKNOWLEDGEMENT

I acknowledge with thanks the help given by Dr. (Miss) Vidula Jaiswal, Reader, in the department of A.I.H.C. and Archaeology, Banaras Hindu University in the preparation of this paper and in correcting the manuscript.

NOTES

1. Banerjee, N.R., 1965. *The Iron Age in India,* New Delhi.
   ———, 1984. *The Impact of the Introduction of Iron on Material culture in India;* paper presented at the Indo-Soviet Symposium on Iron Age and State Formations upto 300 B.C., Ashkabad, U.S.S.R.
   Lal, B.B., 1980. Did the Painted Grey Ware Continued upto Mauryan Times ? *Puratattva,* No. 9, New Delhi.
   Mehta, R.N. 1980. The Economic Pattern of India During Early Iron Age (1000 B.C. to 100 A.D.), *Puratattva* No. 9, New Delhi
   Ray, Niharranjan, 1976-77. Technology and Social Change in Early Indian History. A Note Posing a Theoretical Question, *Puratattva,* No. 8, pp. 132-138.
   Shahi, M.D.N., 1978. New light on the life of the Painted Grey Ware people as revealed from excavations at Jakhera (Distt. Etah), *Man and Environment,* Vol. II, pp, 101-103.
   Sharma, R.S., 1983. *Material Culture and Social Formations in Ancient India,* New Delhi.
   Tripathi, V., 1976. *The Painted Grey Ware — An Iron Age Culture of Northern India,* New Delhi.
   ———, 1985. From Copper to Iron — A Transition, *Puratattva,* No. 15, New Delhi.
2. Gaur, R.C., 1983. *Excavations at Atranjikhera,* New Delhi, p. 219.
3. *IAR,* 1958-59, p. 54.

4. Lal, B.B., 1954-55. Excavations at Hastinapura and other explorations 1950-52, *Ancient India*, Nos. 10-11, New Delhi, pp. 95-99.
5. *IAR*, 1970-71, p. 41.
6. *IAR*, 1974-75, p. 44; 1975-76, p. 51.
7. Gaur, *Op. cit.*, p. 219.
8. *IAR*, 1958-59, p. 54.
9. *IAR*, 1970-71, p. 41.
10. *IAR*, 1974-75, p. 44; 1975-76, p. 51.
11. Gaur, *Op. cit.*, p. 219.
12. *IAR*, 1975-76, p. 51.
13. Gaur, *Op. cit.*, p. 219.
14. *IAR*, 1975-76. p. 51.
15. Gaur, *Op. cit.*, p. 219.
16. *IAR*, 1975-76, p. 51.
17. Gaur, *Op. cit.*, p. 232.
18. *IAR*, 1970-71, p. 41.
19. *IAR*, 1974-75, p. 44.
20. *Ibid.*
21. *IAR*, 1975-76, p. 17.
22. Gaur, *Op. cit.*, p. 219.
23. *IAR*, 197071, p. 41.
24. *IAR*, 1975-76, p. 51.
25. *IAR*, 1968-69, p. 8.
26. Gaur, *Op. cit.*, p. 232.
27. *IAR*, 1975-76, p. 51.
28. *IAR*, 1958-59, p. 54.
29. *IAR*, 1963-64, p. 44.
30. Gaur, *Op. cit.*, p. 232.
31. Lal, *Op. cit.*, pp. 95-99.
32. Gaur, *Op. cit.*, p. 219.
33. *IAR*, 1963-64, p. 44.
34. *IAR*, 1975-76, p. 51.
35. Gaur, *Op. cit.*, p. 219.
36. *Ibid.*, p. 232.
37. *IAR*, 1970-71, p. 41.
38. Gaur, *Op. cit.*, p. 232.
39. *Ibid.*, p. 219.
40. *IAR*, 1975-76, p. 51.
41. Gaur, *Op. cit.*, p. 219.
42. *Ibid.*, p. 232.
43. *Ibid.*, pp. 219, 232.
44. *Ibid.*, p. 219.
45. *Ibid.*
46. *Ibid.*

47. *Ibid.*, pp. 219, 232.
48. *Ibid.*, p. 232.
49. Gaur, *Op. cit.*, pp. 219, 232.
50. *IAR*, 1980-81, pp. 49-51.
51. *IAR*, 1978-79, pp. 60-61; 1980-81, p. 76.
52. *IAR*, 1963-64, p. 29; 1971-72, p. 32.
53. *IAR*, 1971-72, p. 42.
54. *IAR*, 1963-64, p. 31
55. *IAR*, 1976-77, pp. 50-52.
56. *Ibid.*
57. *IAR*, 1963-64, p. 31.
58. *IAR*, 1976-77, pp. 50-52.
59. *IAR*, 1963-64, p. 31.
60. *IAR*, 1976-77, pp. 50-52.
61. *Ibid.*
62. *Ibid.*
63. *IAR*, 1974-75, p. 40.
64. *IAR*, 1963-64, p. 60.
65. *Ibid.*
66. *Ibid.*
67. *Ibid.*, p. 65.
68. *IAR*, 1960-61, p. 39.
69. *IAR*, 1963-64, p. 60.
70. *IAR*, 1968-69, p. 5.
71. Sinha, B.P. and Verma, B.S., 1977. *Sonepur Excavations,* Patna, pp. 126-27.
72. *IAR*, 1966-67, p. 6.
73. *IAR*, 1980-81, pp. 55-56.

Newly Discovered Anthropomorphic Figures from Nurpur, U.P.

MADHURI SHARMA AND D.P. SHARMA

The anthropomorphic figure of copper has been considered by some scholars as religious objects (Gupta, 1974). Some scholars have described them as fertility or Earth Goddess (Jayakar, 1982). The antiquity of this object in human form may be assigned to the upper palaeolithic period as the identical bone object datable to 1900 B.C. has been reported from Londana Nala in Belan Velley, Allahabad (Sharma, 1980). Some scholars consider this as Harpoon (Gupta, 1982); whereas Sankalia and Sharma on the basis of the typical shape considered this figure as diety of fertility or mother/earth goddess (Sharma, 1980: 60). Similar objects made of stone have been discovered from Mathura and Karkabhat both being Indian megalithic sites of south and central India, dated to C. 1000 B.C. (Pl. 61; Narsimhiah, 1980). The anthropomorphic form of taurine or Bull head of clay, stone, and metal pertaining to Indus valley period to early historic period are reported from many regions of India (Postel, 1989). Copper amulet of anthropomorphic shape datable to 1000 B.C. reported from Bihar is now in the private collection (Jayakar, 1982). Solid cast copper mixed with gold anthropomorphic figure which resemble with the Yaksha figure dated 2nd cent. B.C. is now in G.K. Kanoria collection in Patna. This was informed by Vinod Kanoria that this object was acquired from Kaushambi, district Allahabad. As a matter of fact anthropomorphic figure of metal is still found in north and west India (Dikshit, 1969) and also a local god of Birkhaur tribes of eastern South Asia.

All the anthropomorphic figures of the copper are geometrical in shape; however, one such object found at Manbhum in Bihar is the only exception. The upper most thick circular part of this object is like a human head. Its forearm makes almost cresent shape, the outer edge of which are sharp and legs are generally shaped in the form of an angle of 90°. In some cases it forms a curve. The torso resembles a human body having prominent shoulder and narrow waist. These objects were chance discoveries. Beside some finds from well stratified deposit of Lothal (dated 1900 B.C., Rao, 1979) so far 20 anthropomorphic figures of copper have been reported of which 17 are from the middle and upper Ganga Yamuna

Doab in U.P. and the remaining ones are from Bihar, Gujarat and Indus region. Middle and Upper Ganga Yamuna Doab in U.P. may be considered as the sources of these finds. All these figures may be grouped on the basis of their sizes and shapes, mainly into three varieties :

Type I

A sort of stocky anthropomorphic figure is fashioned from thick metal having long arms and short legs as well as the trunk (Pl. 62 and 65). The front and the back of this figure bears chisel marks prominently. The upper part of the head is mostly thick (diamond) but there are few examples with the thin head also. The height of type I ranges from 23.2 to 24.1 cm and the average mean weight is 1,260 grams. Out of nine figures seven are from the north west localities of U.P., one from Lothal and another one from Indus region, now in Brooklyn Museum U.S.A. (Chakraborti, 1977). These anthropomorphic figures were found from the following localities.

Name of Localities	*Nos. of finds*
Nurpur (Sharma, 1991)	2
Mathura (Sharma, 1992)	1
Kiratpur (Gaur, 1981)	1
Bisauli (Lal, 1951)	1
Etawah (Yule, 1982)	1
Targawa (Etawah) (Yule, 1982)	1
Unknown locality in Etawah (Yule, 1982)	1
Lothal (Rao, 1970)	1
Indus region (Pakistan) (Chakraborti, 1977)	1

Type II

Anthropomorphic figure of this type has short arms and long legs and its trunk portion is also comparatively longer than type I. All the figures have thick head (diamond) showing distinctive thickening of metal of the upper margin of head. The figures bears vertical chisel marking. The height ranges from 30.2 to 47.9 cm and its mean is 39.19 cm. The weight varies from 3000 to 6000 gm. and its average mean is 4500 gms. In all, we have nine specimens and most of these were discovered near the river Ganga in Kanpur and its adjoining region of lower-middle Doab of U.P. The figures were reported from the following sites: 1 from Mathura, 2 from Bisauli (Muller, 1980), 1 from Chandausi (Gupta, 1986), 1 from Fatehgarh (Yule, 1982), 1 fragment from Ambala (Yule, 1982), 3 from Shiv-Raspur (Kanpur) (Pl. 63) (Lal, 1951), and one from unknown provenance of U.P., now in the private collection of Bombay (Jayakar, 1982). The anthropomorphic figure from Fatehgarh now in Indian Museum, Calcutta, differs in shape with other figures. National Museum has only one anthropomorphic figure from Chandausi in its collection

(Pl. 64) and this belongs to type II. The upper part of this figure is intact but the two legs are broken.

Type III

The Anthropomorphic figure of this type has proportionately broader and circular head and its body is thicker than the above mentioned two types, and it appears to be more archaic (Pl. 66). Its short legs are rather straight and this type seems to be very crude and rough cast. There are indentations on its upper extremity. In this figure male organ is also depicted. This is the only figure found from Manbhum, district Dhanbad (Bihar) which is exhibited in Patna Museum (Yule, 1982). The measurement of this figure is 16.9 x 14.5 x 1.3 cm and its weight is 1230 gms. Agarwal described this figure as *Shrivats* (Agrawal, 1978). The authors however feel it to be an archaic form of taurine or *Nandipada* (Nandya-varna) and are of the opinion that this object of Patna Museum can be excluded from the other groups of copper hoard implements as this is not a true Anthropomorphic figure of copper hoard variety and may belong to an earlier period than the copper hoard period. This figure resembles with the Anthropomorphic figures found in the rock paintings of Isko, in Hazaribagh district in Bihar. The period of Isko rockpainting can be fixed around 2500 B.C. Thus period of this figure of copper can also be dated around 2500 B.C.

The Two Anthropomorphic Figures from Nurpur

The two anthropomorphic figures from Nurpur are of special significance in comparison with the earlier finds. The objects of Nurpur to be discussed here belong to type I variety and are from the private collection of one Mr. Tarique Ahmad Quadari Chisti of Amroha, District Moradabad, U.P.

The first anthropomorphic figure of Mr. Chisti's collection is comparatively bigger in size than the second one. It shows long arms and short legs (Pl. 65). It is fashioned out from thick metal sheet and has a coating of green and grey patina. The length and breadth of this object is 31 and 24 cm respectively. Circular in shape, the head is thick which tapers towards legs. The head is 0.8 cm thick on top but in the middle its thickness ranges from 0.7 to 6 cm. The head is not diamond and does not have the ridge on top. The topmost portion of head bears saw and chisel marks. The two fore arms of this figure are crescent shaped and each has sharp outer working edge. Its inner edge is comparatively thicker. The distances from the joint of shoulder to their outer and inner edges are 12 and 5.5. cms respectively. The thickness of arms tapers towards the ends. The cross-section of two arms are triangular and are at acute angle on the outside. They bear slanting chisel marks on both the sides. The middle portion is of short trunk shape and both the side edges are concave. The thickness of flat middle portion is almost equal. It also has vertical chisel marks on both sides. Its triangular legs are flat and form an angle of 120° with each other. The outer edge of leg is slightly curved upwards. It is very interesting to note that the chisel marks are vertical on its head and arms but they are slanting on the legs. This figure was first cast and flattened by hammer beating. The figure has some resemblance with the three anthropomorphic figures, another one from Nurpur in Chisti collection, one from Bisauli, now in Allahabad Museum collection and third one from unknown provenance now in Brooklyn Museum in U.S.A. (Chakraborti, 1977).

The second anthropomorphic figure from Nurpur, also a chance discovery is now in Chisti collection. This figure is comparatively smaller in size than the one described above. Its length and breadth is 19 and 28 cms respectively. The head is rather smaller in size and the arms are only slightly incurved. The remaining features are comparable with the former one. The distance between its two arms and shoulders are unequal. The end portion of legs are straight. The technique of manufacturing appears to comparatively cruder than the first type.

INTERPRETATIONS

The anthropomorphic figures of Nurpur are typical religious objects. These figures have some resemblance with the shape of *Taurine* and *Nandi-pada* and ancient Brahmi script letter 'ma ' (Postel, 1989). *Nandipada* symbol of Anthropomorphic form and Bull head shape has been found in Indian art covering a period from the Indus Valley (Circa 2700 B.C.) to early historic period and it continues even in modern jewellery such as *Mangal-Sutra* (Postel, 1989). It was noticed in the form of hukan faced bead with its curved legs in the large necklace of *Yaksha* from Pitalkhora (1st century B.C. old) now exhibited in the National Museum, New Delhi (Pl. 67) (Deshpande, 1989) and in the form of an ear-ring of *Yaksha* from Kausambi now housed in Lucknow Museum and also as bead in the necklace of *Yakshi* from Bharhut, now in Indian Museum, Calcutta (Srivastava, 1991). The figure of *Gomukha* or *Nandi-Mukha* is made of thin silver from Gungeria, M.P. have resemblance with the anthropomorphic figure of Copper Hoard and these objects are dated around 2000 B.C. (Bloomfield, 1870). In the anthropomorphic form of *Nandipada*, arms are absent and legs are incurved like the legs of anthropomorphic figures. This further strengthens our view that anthropomorphic figures of Copper Hoard and *Nandipada* symbol can be linked with some cult or auspicious objects or good omen. In northern India tiny anthropomorphic metal objects are still worshipped in the form of *shani* god and the devotee offers oil and coins on this object as they believe that by doing so *shani* god will protect them from ill omen. According to Dikshit these anthropomorphic figures of Copper Hoard culture are cult objects and the present worship of *shani* god shows the continuance of a tradition (Srivastava, 1991). Therefore, there are sufficient grounds to associate this object with some cult or auspicious or *mangal* object rather than branding it as missile weapon. Some scholars opine that the arms and side edges of this figure are sharp probably for cutting sacrificial animal or human flesh. One group of scholars and present authors are however of the opinion that the use of such heavy object as a missile was not only difficult but inconvenient too, secondly innovation of such missile weapons by the ancient Copper Hoard people of that age was beyond the imagination of their mind and such type of missile have not been reported from any contemporary bronze age site (Agrawal, 1971). Kumar is of the view that anthropomorphic figure had been worshipped in the form of Vedic god *Indra* (Kumar, 1987). In South East Asia and Eastern India the anthropomorphic figure is still being used by local tribes to symbolize their ancestor. The anthropomorphic figures were also noticed in the Mesolithic-Neolithic period rock painting of Bhimbetka and Iska in Hazaribagh district of Bihar (Wakankar, 1978). The anthropomorphic figure were also found in the Neolithic period rock painting of Italy and Mangolia (Tossi, 1990) and also in the Megalithic burial of south and central India (Chauhan, 1991: 4). An anthropomorphic shaped *Nandipadas* or Bull heads of Terracotta were reported from Mohenjodaro and also from other Indus valley sites (Postel, 1989). Bone figure of upper palaeolithic period (1900 B.C. old) found at Londanala in Belan valley (Allahabad) is archaic form of anthropomorphic figure (Tiwari, 1986). Anthropomorphic clay hearths datable to 1500 B.C. which partly resemble in shape with the taurine (Nandipada) or anthropomorphic figure of Copper Hoard culture of India, were reported from the early bronze age sites of Trans Causia, Anatolia, Syriya, Palestine, Amiranis Gorain Georgia, Armenia, Kultepe in Azerbaidzhan which show some similarities in two regions (Beregovaia, 1966). It seems taurine and anthropomorphic figure are auspicious symbols though they lost their significance in due course of time.

CONCLUSION

All these evidences confirm that the objects of anthropomorphic figures were existing in India during, pre and post Copper Hoard period. Thus it is clear that anthropomorphic figure found during Copper Hoard culture in both eastern and western group has very early antiquity. Anthropomorphic figure in Bone were found in upper Palaeolithic period and continued during mesolithic, Indus valley, Copper Hoard and post Copper Hoard periods and is still in practise among Birpor tribe of Bihar. Therefore, it can be said that anthropomorphic figure of copper does not belong to any group (East or West) of Copper Hoard culture. Probably it is an old phenomenon of pre Copper Hoard period. It seems to be indigenous feature of pre Copper Hoard period which may be dated even upto upper palaeolithic times.

REFERENCES

Agrawal, D.P., 1971. *Copper Bronze Age in India,* Delhi, p. 205.

Agrawal, P.K., 1978. A Prehistoric Anthropomorphic figure in Patna Museum. *Puratattava,* No. 1, New Delhi, pp. 96 & 205.

Beregovia, N.A., 1966. *Contribution of Archaeology of Soviet Union,* Cambridge, U.S.A. P. 37.

Bloomfield, 1970. *Extraordinary finds of Indian copper implements* (Sax mandham Suffolic), p. 81.

Chakraborti, D.K., 1977. Anthropomorphic in the Brooklyn Museum, New York, *Man and Environment,* Poona, p. 101.

Chauhan, R.S., 1991. *National Museum News Letter,* Vol. II, Nov., p. 4.

Dikshit, K.N., 1968. The Copper Hoard in the light of recent discoveries, in *Bulletin of Ancient Indian History and Archaeology* No. 11, Sagar, p. 66.

Deshpande, M.N., 1959. The rock cut cave of Pitalkhora in Deccan, *Ancient India,* No. 15, New Delhi, pp. 66-93.

Gupta, P.L., 1974. *Asian and Pacific Archaeology Series* No. 9, Manoa, pp. 297-333.

Gupta, S.P., 1986. Copper Hoard in National Museum, New Delhi, *Puratattava,* No. 16, New Delhi, p. 80.

Gupta, S.P., 1982. The Alleged Upper Palaeolithic Bone Mother Goddess from Belan, *Puratattava* No. 11, New Delhi, p. 116.

Gaur, R.C., 1981. One more Copper Hoard from the Ganga Basin and a Reassesment of the problem, *Festschrift M.N. Deshpande Vol.* (Madhu) ed. M.S. Nagaraju Rao, New Delhi, pp. 73-74.

Jayakar, Pupul, 1982. *Auditi,* exhibition catalogue, London, p. 89.

Krishna Kumar, The copper Anthropomorphic: A New Identification, paper presented at the 19th Annual Conference of the Indian Archaeological Society held in 1987, Sagar, India.

Lal, B.B., 1951. Further Copper Hoard from the Gangetic basin and a review of the problem, *Ancient India,* No. 7, New Delhi, p. 22.

Srivastava, K.M., 1984. *Settlement, History and Rise of Civilization in Ganga Yamuna doab,* New Delhi.

Mullar, Korpe H., 1980. Handbuch der Vorgesh Shichte IV A. 3 Munich Pl 564.

Narasimhiah, B., 1980. *Neolithic and Megalithic Cultures in Tamil Nadu,* New Delhi, p. 1.

Postel, M., 1989. *The ear ornament of India,* Bombay, p. 196.

Rao, S.R., 1979. *Lothal a Harappan port town,* New Delhi, p. 1.

Sharma, R.C., 1992. *Art and Culture of Brija,* New Delhi, see list.

Sharma, D.P., 1991. Two newly discovered Anthropomorphic figures from Nurpur. *Statesman* 5.10.91, New Delhi, p. 8.

Sharma, G.R., 1980. *History to prehistory:* Archaeology of Ganga Valley and Vindhyas, Allahabad, p. 60.

Srivastava, A.L., 1991. *Nandya Varta,* Allahabad, p. 12.

Tiwari, Rakesh, 1986. Harpoon from Mirzapur, *Dhyanam,* No. 5 & 8, Lucknow (Hindi), p. 12.

Tossi, M., 1990. *Neolithic Culture of Mongalia,* talk in Tibet House. New Delhi, Unpublished.

Yule, Y., 1982. *Metal work of the Bronze age in India,* Munich.

Wakankar, V.S., 1978. *Stone age painting of India,* Bombay.

Two Unique Lugged Shouldered Copper Axes from the National Museum of India Collection

D.P. SHARMA

The National Museum of India, New Delhi has in its possession 191 Copper Hoard implements from several localities in lower middle Ganga Yamuna doab of U.P., and from Mehsana in Gujarat (Sharma, 1989). Of these, 137 implements which were acquired between 1969 and 1987, have been briefly described by S.P. Gupta and the author (Sharma, 1987; Gupta, 1986). The remaining 54 implements acquired in 1988, have been described in detail elsewhere by the author (Sharma, 1989). Detailed notes about antennae swords and also about two anthropomorphic figure in our collection have been published by Yule (Yule, 1982) and the author (Sharma, 1987). Dikshit and Pant have given brief accounts of harpoons and other artifacts in the National Museum collection (Dikshit, 1968 and Pant, 1978)

The National Museum of India has a collection of 37 shouldered axes and 2 lugged shouldered axes (Pls. 68 & 69) from middle doab. The lugged shouldered axes described and illustrated here represent a different variety among the axes generally found in the Copper-Hoards. The lugged shouldered axe is a local innovation as described to the copper hoard people of the middle doab. This technique of manufacturing was probably adopted by population belonging to the eastern group of this culture (Bengal, Orissa, Bihar and Eastern Uttar Pradesh) when they came in contact with the inhabitants of middle doab region of Uttar Pradesh, (West of Kanpur) where double mould casting method of copper was practised. Harpoons, hooked-swords as well as lugged shouldered axes of Western group of Copper Hoard culture were cast by double mould method. The copper hoard people of western part of middle and upper daob region probably learned this method of casting when they came in contact with the late Harappan population of upper doab region of Western Uttar Pradesh.

The lugged axe is a type of shouldered axe with concave or straight edges. It has a prominent elliptical or convex sharp working edge with a straight sided butt-end. This implement was sharpened by abrasion. The weapon is characterised by one pair of lugs projecting from the centre of the two sides. The lugs were meant to facilitate hafting. The first discovery of shouldered lugged axes was reported from Sidhauli in the Sitapur district of U.P. (Singh, 1972). On the basis of variation of the shape of the working edges as well as the side edges, the lugged axes are classified under the following two groups :

Group I

The lugged shouldered axe of this variety has a set of flat rectangular head in the middle on each side. It also has two concave shoulders, above and below the lugs. Its working edge is sharp, elliptical or convex. The butt-end is straight. Axes of this type were reported from Kathmandu Valley in Nepal (Fig. 1), the Kanpur-Unnao border region, Lothal (Fig. 2)and one axe is supposed to be in the private collection of Jagdish Gupta of Allahabad (Yule, 1982, Gupta, 1986 and Singh, 1972).

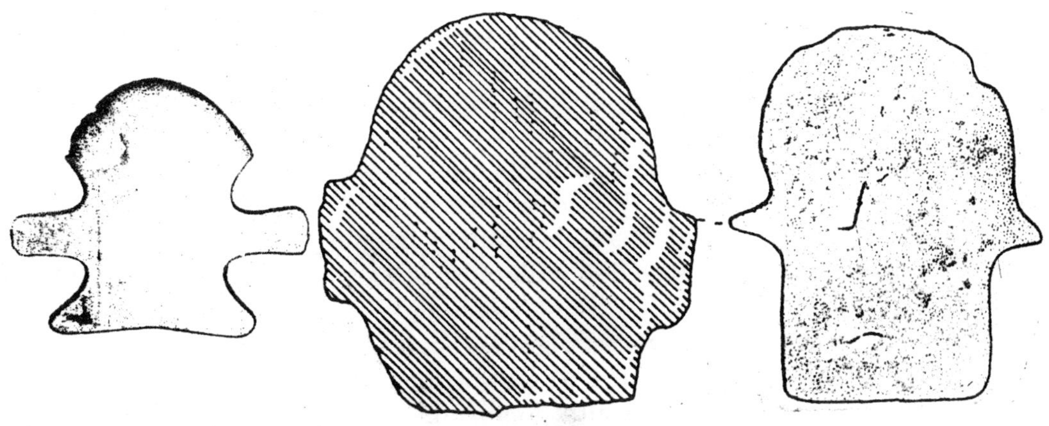

Fig. 1 Fragmentary Lugged Shouldered Axe, Group I, Kathmandu Valley, Nepal (After Yule)

Fig. 2 Copper Lugged Shouldered Axe, Group I (1900 B.C.), Lothal Alleged Anthropomorphic figure.

Fig. 3 Copper Lugged Shouldered Axe, Group II (2000 B.C.), Rewari (After Yule)

Group II

The lugged axe of this variety has a set of flat conical lugs. Its two side edges and butt end are straight. This does not have a concave shoulder near lugs. So far only one lugged axe of this type was discovered from Rewari, Haryana (Fig. 3) (Yule, 1982).

The lugged shouldered axe (Acc. No. 86.59/11) in the National Museum is unbroken and belongs to group I. This object is a chance find from Mohamadabad near Sitapur, U.P. Its length and width are 13.6 cm and 11.3 cm, respectively. The maximum thickness near the middle portion is 1.2 cm. The concave depression on its right corner was meant for hafting purposes. The total length of the elliptical working edge is 16 cm and the width from one corner to the other is 10.4 cm. The elliptical working edge of this axe resembles that of the convex head portion of the anthropomorphic figure. This rather minute difference means that the axe does not possess as thick a head as that of the anthropomorphic figure. The two concave shoulders of the axe have depressions as in the case of the anthropomorphic figure. However, instead of two incurved hands the axe has a set of lugs on each side as well as vertical indentation marks similar to the anthropomorphic figure as these two pieces were probably cast by the same method and were subsequently flattended with the help of a chisel and hammer. The length and width of each lug on the axe is 2.3 cm. and 1.9 cm., respectively.

The thickness of the lugs near the joints is 0.8 cm, while at the ends they are 0.5 cm. The butt-end is straight and its corners are angular with a covering of green patina on its sides.

Another fragmentary lugged shouldered axe (Acc. No. 87.22/14) of National Museum, New Delhi, measures 8.2 x 5.7 x 7 cm and its find spot is Kanpur-Unnao Border area. It also belongs to Group I. Its elliptical working edge and the left part of the shoulder as well as the central portion are intact, but its two lugs and the butt-end are missing. In also has sharp vertical chisel marks on both sides. This piece resembles the lugged shouldered axe described above (Acc. No. 86.59/1) and also the one from Lothal. The author is of the opinion that the (alleged) anthropomorphic figure from Lothal is a lugged shouldered axe of group I type. The axes under discussion have sharp elliptical working edges and concave shoulders. The one from Lothal also has a set of short straight lugs on each side which should not be described as arms. On this basis the Lothal implement can be described as a lugged shouldered axe and not as an anthropomorphic figure. Agrawal's view on the anthropomorphic figure from Lothal (Agrawal, 1982) is significant indeed.

A large number of shouldered axes have been reported from eastern India as well as from Kanpur and the adjoining region. Shouldered axes of eastern India (Bihar and Bengal) are very crude and inferior in technology, whereas similar objects discovered in the lower and middle doab are more refined and exhibit innovative manufacturing techniques. The lugged shouldered axe described and illustrated in this note represent a new variety in the shouldered axe group and this new innovation is attributed to Copper Hoard people in the middle doab (West of Kanpur to east of Bulandshahar, U.P.). This new innovation came about as a result of contacts with the people of second (Western) group of Copper Hoard whose knowledge of copper technology was more advanced. Copper hoard people of middle doab (Western group) were already in contact with the Harappan people who were well versed in double mould casting technology. This explains the absence of this type of copper tools in West of middle doab.

The lugged shouldered axe has its origin in the shouldered axe of eastern group which itself has its origin from South East Asia. Hollow cast copper shouldered axes dated to 2nd millennium B.C. were excavated from Gomuns Dom

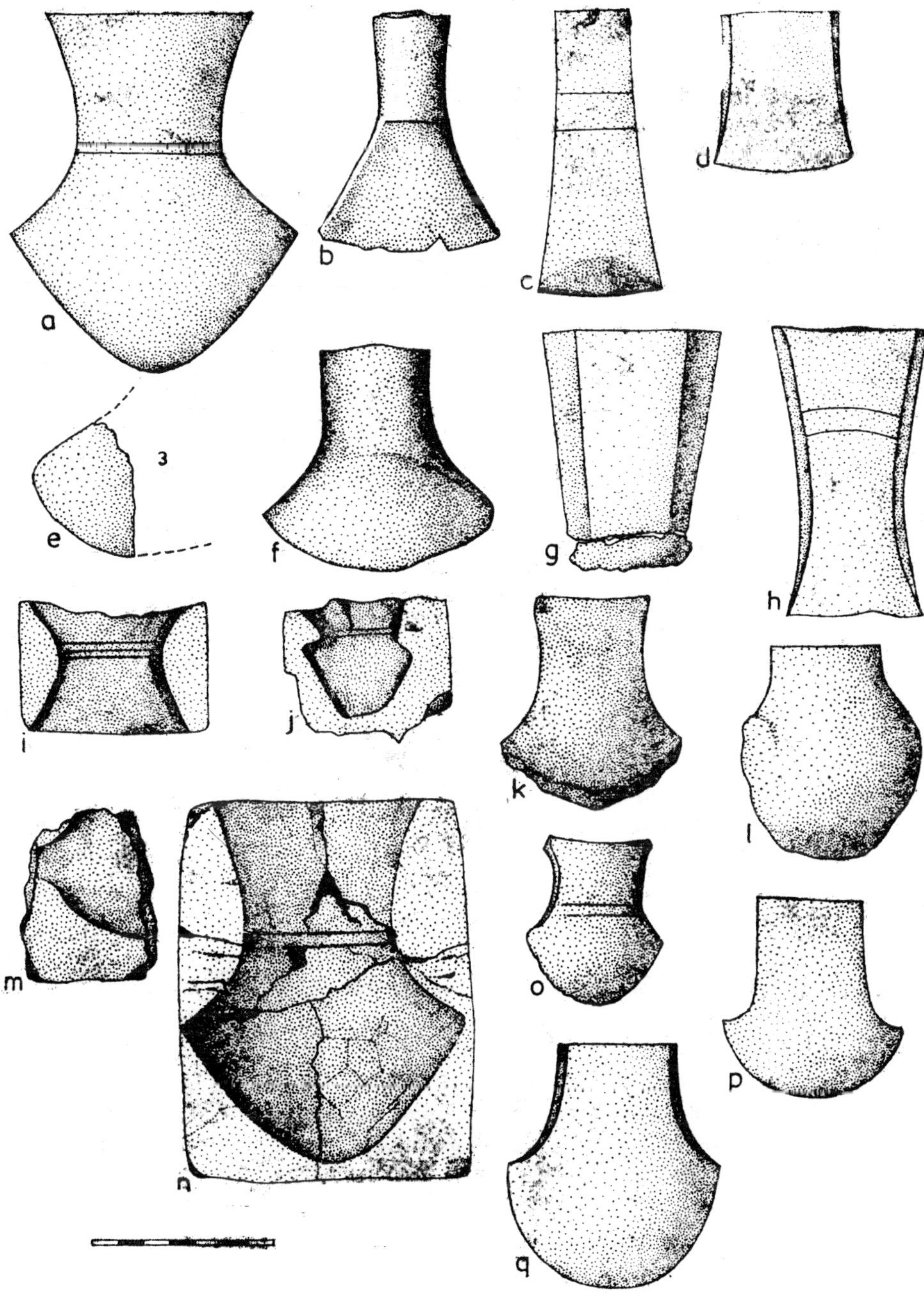

Fig. 4 Copper Shoulder Axe, 2000 B.C., Thailand

in Thailand and also from other localities in South East Asia. (Fig.4) (Sorensen, 1988). In Thailand shouldered axe has an earlier date than the one from eastern India. The earliest copper weapon in Thailand comes from Non Nok which has been dated to 2700 B.C. (Sorensen, 1988). It is possible that copper shouldered axes in eastern India were introduced by (copper age) people from the South-East-Asia. The linguistic similarity of the Mon-khmer tribes of Burma and Thailand with the tribal people of Eastern India also supports the South East Asiatic origin of the shouldered adze. Stone shouldered celts, prototype of copper axes, dated 8000 B.C. were reported from the Hoabinian culture sites of South East Asia. Shouldered axes of copper in this region had evolved out from this prototype of stone shouldered celt of North East India and of South East Asia. Copper shouldered axe of India is solid cast whereas of Thailand is Hollow cast (Sharma, 1991).

REFERENCES

Agarwal, D.P., 1982. *Archaeology of India.* Curzon Press. London,

Dikshit, K.N., 1968. The Copper Hoards in the Light of Recent Discoveries. *Bulletin of Ancient Indian History and Archaeology,* 11: 43-50.

Gupta, S.P., 1986. Copper Hoard Implements in National Museum Collection, *Puratattava,* 16.

Kumar, Krishna, 1972. Some Unpublished Copper Hoard Implements from U.P., *Vishvesvaranand Indological Journal,* XXIV: 11.

Pant, G.N., 1978. *Indian Arms and Armoury,* Vol. I, New Delhi.

Sharma, D.P., 1987. Copper Hoard Implements in the National Museum Collection, *Puratattava* 17.

Sharma, D.P., 1989. Copper Hoard Implements in the National Museum, New Delhi 1988-89, Paper read in Indian Society for Prehistoric and Quarternary Studies, VIIth Annual Conference, 22-24 Dec. 1989. Bhopal.

Sharma, D.P. & Madhuri Sharma, 1991. Comparative Study of Neolithic Culture of Eastern India and South East Asia, Paper read in the confernece of Historian of Asia at Hong Kong University (Press).

Sharma, D.P., 1982. Two Anthropomorphic Figures in the National Museum Collection, *National Museum Bulletin,* New Delhi.

Sorensen, 1988. *Archaeological excavation in Thailand,* London.

Singh, S.B., 1972. *Archaeology of Lucknow Region,* Lucknow.

Yule, R., 1982. *Metal Work of the Bronze Age in India,* Munich.

Section III
Literary Evidence on Archaeometallurgy

A Literary Perspective on Archaeomaterials and Scientific Concepts in Ancient India

SULEKHA BISWAS AND ARUN KUMAR BISWAS

The scope of metallurgy has recently been widened to material science, and therefore it is proper that we discuss archaeomaterials in a comprehensive manner including metals as well as the non-metallic phases of archaeological importance. Elsewhere (Biswas et al., 1990, 1991) we have critically surveyed the Sanskrit literatures referring to gems, non-gem minerals and metals. This paper deals with our observations on (a) correlations of the literary evidences with the archaeological data and other historical facts, and (b) scientific concepts and attitudes in ancient India related to archaeomaterials and material science.

Archaeology and History

Archaeology digs into not only the earth but also the past in a telling manner. In this modern age of sophisticated instrumentation, the ancient artifacts can be accurately analysed and C-14 dated through the accompanying organic residues; as a contrast, there are controversies on the dates of several works of literature.

On the other hand, the archaeological samples of minerals and metals are mute, and as such do not tell whether these were imported or locally made, and in the later case the status of *scientific* knowledge of the indigenous manufacturers. The literary evidences tell us about the state-of-art and the status of concepts and philosophy. Even though the literary evidences on minerals and metals in the Indian scene are scanty, scattered and occasionally of uncertain date, these must be viewed as complementary to the archaeological evidences.

The Ṛgveda describes the transition from the neolithic to the chalcolithic period in the Indian history (earlier than second millennium B.C.), the use of conch-shell and conch-pearl, different ill-defined metals, and widespread trade and navigation. Homer referred to the import of tin and ivory from India.

The post-Ṛgvedic literatures describe purification steps, sieving, winnowing, smelting in furnaces using bellows and different kinds of metals - the ornamental and barter metals of gold and silver, low melting point metals lead and tin, copper and the new metal *śyāma* or iron obtained through solid state carbothermic reduction. The art of forging or joining of metals was mentioned.

The problem of correlating archaeological and literary evidences is distinctly severe with the Ṛgveda and the epic Mahābhārata. The epic was definitely expanded from time to time, and the references to ant-gold, *marakata* or emerald and gem-incrustations must have been later-day insertions. We cannot accept that *marakata* was known to the residents of 1200 B.C. Hastinapura.

On the other hand, the art of making iron, well-known in 1100 B.C. Atranjikhera, was never fully recorded in the literature. High carbon steel was definitely made in the pre-Christian era Taxila, but the scattered literary evidences came much later. Similarly, zinc and brass were known in Taxila and during the Kauṭilya's time, and yet the first clear reference to the making of these items came much later from the celebrated Nāgārjuna's pen.

The problem of correlating archaeological and literary evidences is not so great with Pāṇini and Kauṭilya. The writings of Pāṇini and Kauṭilya clearly reflect the contemporary scenes (5th to 3rd century B.C.) of material civilization in India. The coins, the multi-bellowed furnaces and beryl mentioned by Pāṇini, and *vṛtta*, the circular discs of carburised iron, *tīkṣṇa*, the steel and *ārakūṭa*, the high-zinc brass of Kauṭilya have all been found in the different pre-Christian era strata of Taxila.

The Gupta era splendour was fully reflected in the contemporary literatures. The then uses of gems, coins, copper, zinc, brass, iron and steel, the international trade with Sri Lanka and Rome, and the Graeco-Roman influence in India, etc. were revealed in many writings, *Ratnaśāstras, Aṅgavijjā, Amarakośa, Bṛhatsamhitā* and so on.

The Rasaśāstra texts appeared after the decline of the Gupta era, and portrayed the decadent scenes of the Tāntric period, the continued practices of zinc, brass and ferrous metallurgy, the use of mineral-based medicines and the technological contact with China through Tibet. Ray (1956: 97) believed that the knowledge of zinc metallurgy came to India from China, but this was an erroneous view. India was the first to make zinc and brass; what we learnt from the Chinese were precipitation of copper from solution by scrap iron, the arts of making paper and gunpowder etc.

Scientific Concepts related to Material Science

Notwithstanding the technical excellence related to minerals and metals in ancient India, the question remains as to the level to *scientific* appreciation in the ancient world.

The Harappans were meticulous about measurements and standardisation, and it is a pity that their scripts have not yet been deciphered. The Vedic literatures illustrate basic advancements in linguistics, mathematics, geometry and astronomy which are outside the purview of our studies.

Atharvaveda speculated (4.10.1-7) about the cause of origin of conch-shell and conch-pearl. Along with a mythological theory of these gems originating from 'gods' bones', the other theory proclaimed that the materials could have grown in the sea under atmospheric influences. This mixture of belief in mythology and rational thinking continued for centuries. We find Varāhamihira declaring (BS. 80.3) in the fifth century A.D. that the wonderful variety of gems is caused by the characteristic qualities of the earth, and yet the fourteenth century A.D. text Rasa-Ratna-Samuccaya continued to repeat the mythological theory.

Kauṭilya referred to the scientific approach of categorisation of gems according to colour. The Ratnaśāstra texts brought in the useful criteria of relative hardness, specific gravity and optical properties towards the identification of gems. The reference to *rasapāka* in Kauṭilya shows that *rasaśāstra* studies had been initiated during his time to transform matter to useful varieties. Alchemy gradually evolved from the sustained efforts of Caraka, Suśruta, Patañjali and Nāgārjuna.

In the meantime, the postulate about the atomic constitution of matter, made by Kanāda, was being refined by the followers of the Vaiśeṣika, Buddhist and Jaina philosophies. Yet there was little correlation between the atomic theories (speculative in nature) and the *rasaśāstra* experiments related to distillation, digestion, reduction, oxidation, sulfidisation etc. During the Tāntric period we obtain the concept of *sattva*, denoting metallic essence, the metal extracted from a mineral or ore. There are references to 'mercury from mercuric sulphide', 'copper from copper sulphate' etc., but no categorical statement to the effect that such an extracted metal is indeed the same as the variety available in native state or obtainable from another source. Iatrochemistry in ancient India fell short of the European break-through in Daltonian Chemistry, and yet by the thirteenth century it had outgrown the earlier alchemical extravaganza.

Scientific attitude in Ancient India: The Irrational Trends

Chattopadhyaya has articulated in great detail (1976, 1977, 1982, 1986) what he considers to be 'living (scientific) and what is dead (unscientific) in Indian philosophy'. According to him the Indian materialistic philosophy of the *Lokāyata* was systematically denigrated in India by the anti-scientific and Brahmanic philosophy of the Vedānta. Prafulla Chandra Ray had earlier (1902, 1909) developed the same theme, and criticised the Brahmanic tradition for introducing the pernicious caste system and the anti-reason theological stranglehold in our sub-continent.

Much of their criticisms are valid. The early casteless Ṛgvedic society gradually introduced a rigid caste system and sanctified it by the so-called 'divine sanction'. The isolation of the low-caste artisans from the high-caste scholars affected Indian science since the early days. There was in India an ancient tradition that the ritual milking pot was not to be prepared by a low-caste potter! Even metals like gold, copper and newly discovered iron were considered to possess decreasing social status. The concept of hereditary caste permeated into the subject of classification of minerals. According to Rasa-Ratna-Samuccaya (RRS), diamond could have four colours or 'castes'! What was worse, such a classification theory for materials, based on caste, was claimed to have universal applicability:

पदार्थेषु अखिलेषु अपि
(RRS. 4.29-4.31)

We find the excellent compilation *Amarakośa* categorising the materials in four caste-categories, the metallurgical equipments falling under *śūdravarga*. Even the scientifically oriented intellectuals like Caraka, Gotama, Praśastapāda, though not conforming to all the Vedic tenets, tacitly accepted the *varṇāśrama* and the caste-system. The famous Vedantist Śaṁkara not only supported the caste-injunctions of Manu, but also criticised the atomic theory of Kaṇāda (Brahma-Sūtrabhāṣyam, 2.2.17):

तस्मात् एवम् असारतरतर्कसंदृब्धत्वात् ईश्वरकारणश्रुतिविरुद्धत्वात् .....
परमाणुकारणवादे ..... श्रेयोऽर्थिमिरिति वाक्यशेषः ।

Such frontal attacks inhibited free-thinking of scientifically minded but god-fearing intellectuals and diffusion of knowledge across caste barriers.

That the *śūdras* should not be allowed to read the Vedas was almost fully accepted in ancient India. There was a cult of secrecy, of course nurtured by the vested interests. The alchemical and iatrochemical research was marked by professional secrecy, and the astounding claims were made:

रसविद्या दृढं गोप्या मातुर्गुह्यमिव ध्रुवम् ।
भरेत् वीर्यवती गुप्ता निर्वीर्या च प्रकाशनात् ॥
रोगिनां बहुर्मिज्ञातं भवेत् निर्वीर्यम् औषधम् ।
न रोगिविदितं कार्य बहुभिर्विदितं तथा ॥

that 'scientific knowledge remains powerful when kept secret and becomes useless when made public', and that 'the medicine becomes ineffective when its identity is revealed to the patient' (RRS. 6.71-72).

The insular attitude in India resulted in over-compartmentalisation of technology. Only specific tribes like *mundas* practised iron-making (hence the name *mundaloha*), *bhīls* practised zinc-making, *turis* gold and diamond-washing, *Kāñsārīs* bell-metal trading; there were low-caste gem-engravers; the *rasaśāstra* experts dealing with plant and mineral-based medicines were known as *vaidyas* or fallen brāhmins (Roy, 1946). The different castes practising different aspects of trade rarely exchanged informations, only which could result in faster growth of science and technology in India. It has been well said that 'the evils that caste system engendered cannot be over-estimated'.

During Varāhamihira's time there was a healthy respect for foreign scholars (BS.2.4 and 2.32), but in the eleventh century A.D., Al-Beruni observed in India an insular attitude and the vice of vainglory amongst the contemporary Indian scholars (Sachau, 1910, 1983, Vol. 1, pp. 22-23).

At the same time, Al-Beruni noted another reason for the decline of Indian Science: 'Mahmud utterly ruined the prosperity of the country ..... Hindu sciences have retired far away to other places'. Even earlier, the barbarous invasions of the Huns, Scythians and Saracens had stunned the growth of Indian sciences and arts. Ray (1902, 1909) and Chattopadhyaya (1986) have glossed over this aspect of Indian history.

A More Sympathetic View

Having noted the casteist and irrational trends in Indian science, we may also record the postive aspects in our tradition.

In the famous *Nāsadīva Sūkta* (Rv. 10.129), the Ṛgveda provides the earliest example of India's spirit of enquiry. When the post-Vedic culture degenerated into rituals and regimentation in thoughts, a large number of non-conformist intellectuals such as Kapila, Kaṇāda, Buddha appeared on the scene. They brought back the culture of reason and logic (Pande, 1969), and none of them, except the Lokāyata scholars, denied the ethical precepts of the original Āryas.

Pāṇini stressed on proof (*pramāṇa*), original discoveries of knowledge 'which had not existed' (*upajñā*) and their applications (*upakrama*). Kauṭilya stressed on accurate measurements and standardisation in trade and commerce. Whereas the followers of Manu considered philosophy to be a subject subserving the Vedic lores (Arthaśāstra of Kauṭilya, 1.2.2-3), Kauṭilya asserted that *ānvīkṣikī* or investigative philosophy was supposed to investigate, by means of reasoning, not only the hazards of economics and politics, but *also* धर्माधर्मौ त्रय्याम् 'what is spiritual good and evil in the Vedic lore' (AS 1.2.11). It was boldly proclaimed that:

प्रदीप: सर्वविद्यानाम् उपाय: सर्वकर्मनाम् ।
आश्रय: सर्वधर्मानाम्शाश्वत् आन्वीक्षिकी मता ॥

'philosophy is ever thought of as the lamp of all sciences, as the means of all actions (and) as the support of all laws (and duties)' (AS. 1.2.12).

Chattopadhyaya himself has described (1976: 5-7 and 1982: Volume 1, pp. 209-238) how the spirit of debate, enquiry and logic were sustained, during the early Christian era, by Caraka, Gautama and Vātsāyana, who subscribed to the ethical principles of the Vedānta concurrently. Caraka pointed out the merit of scientific debate which 'strengthens conviction by way of removing doubt in the pre-existing stock of knowledge, and breaks the tendency to insularism even of those that would not normally impart their own knowledge to others' (*Caraka Saṁhitā*, 1.25.27., 3.8.15 and 28). In the second century A.D., Gotama or Gautama wrote his famous *Nyāyasūtra* (which was the origin of the science of logic in ancient India) and three centuries later, Vātsāyana wrote the equally famous commentary on it. They emphasized the importances of intellectual doubt (*samasyā*), critical examination of the problem (*parīkṣā*) and final ascertainment (*nirṇaya*) as distinct stages in the scientific and philosophical pursuits.

Āryabhaṭa (476 A.D.) did not subscribe to the traditional notion of five elements. His denial of the fifth element, *ākāśa* or ether, was in accordance with the materialistic philosophy of the Lokāyata (Bongard Levin, 1977). Bhāskara I (629 A.D.) and later Al-Beruni emphasized on Āryabhaṭa's agnostic approach about the space which is not reached by the solar rays, and the notion 'that which is not reached by perception is not knowable'. Āryabhaṭa's theory of the Earth's rotation around its axis and his scientific explanation for the solar and lunar eclipses invited hostility from those who believed in the earlier cosmogenic doctrines, but there is no evidence that he faced any theological inquisition like Galileo.

Al-Beruni rightly accused Brahmagupta (born 598 A.D.) for indulging in half-truths, for criticising Varāhamihira and Āryabhaṭa (who had demolished the Rāhu theory) and yet 'calculating the diameter of the moon in order to explain her eclipsing the sun' (Sachau, 1910, 1983, Volume 2, pp. 110-112).

Chattopadhyaya (1976: 256-270) has discussed why the intellectuals of the ancient India, not only Brahmagupta, but also Kaṇāda, Caraka, Gautama, Vātsāyana, Varāhamihira etc. chose not to go against the Brāhmanic or the Vedic concepts and ideals. According to Chattopadhyaya, theirs was an act of 'subterfuge, a technique of evading inquisition'. He further observed that even in the modern world there are scientists who believe in matter as well as the divine (Newton and Einstein believed in God) and that 'they have split personalities'.

A more sympathetic view may be proposed. The issue whether matter has evolved from consciousness (the spiritual view) or consciousness have evolved from matter (the materialistic view) has not yet been finally settled. This could be the famous 'chicken and egg' problem. Most intellectuals of all ages and countries might have preferred not to put all their eggs in one basket (atheistic material science). It is better to remain more open-minded on this issue (Biswas, 1969: 19-31 and 99-118).

At any rate, the spectacular scientific achievements in the ancient India clearly show that not everything was dead in the intellectual plane. Whereas Śaṁkara ridiculed Kaṇāda's atomic theory, Śubhagupta commented that if atoms are to be doubted on the ground that these are not demonstrable like macroscopic objects, then one should doubt the human mind also!

Ancient India always believed in accurate observations and experiments. Atharvaveda initiated the ancient medical tradition by recommending that the animals in the forest may be closely observed to discover the pharmaceutical properties of the herbs (Av. 8.7.23-26). Caraka has laid emphasis on the importance and utility of scientific experiments (*Caraka Saṁhitā,* 6.8.132, 10.5):

परीक्षायास्तु खलु प्रयोजनम् प्रतिपत्तिज्ञानम् ।
सम्यक् परीक्षा हि बुद्धिमताम कार्यप्रवृत्तिनिवृत्ति कालौ शंसति । परीक्षाकारिणौ हि कुशला भवन्ति ॥

The entire Āyurvedic science of medicine has been based on experimentation; for example, poisoned food and drinks and antidotes were tried on lower animals followed by trial of drugs on human beings. Abundant material on this topic is available in the Saṁhitās of Caraka and Suśruta and in the Aṣṭāngasaṁgraha. For centuries, this approach was pursued by the *Rasaśāstra* scholars.

The author of the 13th century text *Rasakalpa* wrote:

साक्षात् अनुभवैर्दृष्टो ना श्रुतो गुरुदर्शितः
लोकानाम् उपकाराय एतत् सर्वं निवेदिवम् ।

'I have performed the aforesaid experiments with my own hands and have seen them with my own eyes. They are not recorded from mere hearsay or from the dictation of a teacher. These are being promulgated for the benefit of mankind'. (quoted by Ray, 1956: 157 and 370). Yaśodhara made a similar statement in Rasa-Prakāśa-Sudhākara.

The author of the 14th century text *Rasendracintāmaṇi* wrote:

'That which I have heard of learned men and have read in the Śāstras but have not been able to verify by experiment, I have discarded. I am committing to writing only those which I have been able to perform with my own hands' (Ray's translation quoted by Chattopadhyaya, 1982, Volume 1, p. 345).

Rasa-Ratna-Samuccaya (RRS) of the 14th century A.D. defined who could undertake *rasaśāstra* experiments:

सम्यक् साधन सोद्यमा नानाकर्मपरांमुखा

'an ideal dedicated student, a hard working, meticulous research worker in Rasaśāstra' (RRS 6.5-6.7; 6.67-6.68) and described:

अल्यन्तोपवने रम्ये ...... तत्र शाला ......
सम्यक् वातायनोपेता दिव्यचित्रैः विचित्रिता

'the ideal *rasa-śālā* or alchemical laboratory, well-ventilated, well-decorated and set in a pleasing environment (RRS 6.11-6.58). The lay-out and the diversity of equipments were described in detail (Biswas, 1987: 43-44).

CONCLUDING REMARKS

Continuing investigations, both archaeological as well as literary, are likely to augment our knowledge on minerals and metals in ancient India. On the basis of the informations gathered by us so far, we may safely conclude that in the field of material science and technology, ancient India was second to none amongst the contemporary nations in the world.

A veriety of reasons, discussed in this chapter, inhibited the rate of scientific progress in this subcontinent, and India failed to match the spectacular intellectual renaissance in Europe. Nevertheless, when the British investigators of the nineteenth century surveyed our sub-continent, they obtained abundant evidences of ancient mines, working metallurgical outfits in the sectors of iron, copper, zinc, brass and gold, well-informed practices related to gems and non-gem minerals and, briefly stated, of the entire gamut of the ancient and medieval Indian traditions in minerals and metals.

REFERENCES

Biswas, Arun Kumar, 1969. *Science in India,* Firma KLM Private Ltd., Calcutta.

Biswas, Arun Kumar, 1987. Rasa-Ratna-Samuccaya and Mineral Processing State-of-Art in the 13th Century A.D. India. *Indian Journal of History of Science,* 22 (1), 1987, pp. 29-46.

Biswas, Sulekha, Chakravorty, Nayanananda and Biswas, Arun Kumar, 1990. Archaeo-Material Studies in India and Literary Evidences, *Historical Archaeology of India,* edited by Amita Ray and Samir Mukherjee, Books and Books, New Delhi, pp. 49-66. Also 1991, Minerals and Metals in Ancient India, Two Volumes Report submitted to INSA, New Delhi.

Bongard Levin, G.M., 1977. Āryabhaṭa and Lokāyatas, *Indian Journal of History of Science,* No. 2, Vol. 12, November, 1977, pp. 187-193.

Chattopadhyaya, Debiprasad, 1976. *What is Living and What is Dead in Indian Philosophy,* People's Publishing House, New Delhi.

Chattopadhyaya, Debiprasad, 1977. *Science and Society in Ancient India,* Research India Publication, Calcutta.

Chattopadhyaya, Debiprasad (ed.), 1982. *Studies in the History of Science in India,* 2 volumes, Editorial Enterprises, New Delhi.

Chattopadhyaya, Debiprasad, 1986. *History of Science and Technology in Ancient India — The Beginnings,* Firma KLM Private Ltd., Calcutta.

Pande, G.C., 1969. Philosophical Trends and the History of Science in India-Heterodox Trends, *Indian Journal of History of Science,* Nos. 1 & 2, Volume 4, May & November 1969, pp. 42-51.

Ray, Prafulla Chandra, *History of Hindu Chemistry,* (Vol. I 1902-3; Vol. II 1909). Bengal Chemical and Pharmaceutical Works Limited, Calcutta.

Ray, Priyadaranjan (ed.), 1956. *History of Chemistry in Ancient and Medieval India,* Indian Chemical Society, Calcutta.

Roy, S.K., 1946. Glimpses into India's Heritage in Geology, Mining and Metallurgy, *Quarterly Journal of the Geological, Mining and Metallurgical Society of India,* Volume 18, March 1946, pp. 101-123.

Sachau, E.C. (ed.), 1910, 1983. *Alberuni's India* (about 1030 A.D.), Volumes I & II, Munshiram Manoharlal Publishers Private Ltd., New Delhi.

Early Indian Literature as a Source of Archaeometallurgy

P.K. AGRAWALA

For a study of archaeometallurgy one of the two major sources of our information is the vast corpus of early Indian literature itself. Besides archaeological and art evidences surviving to our times from human achievements in past ages, a close scrutiny and thorough analysis of the pieces of informations contained in ancient texts may very well tend to speak themselves through their authors in more or less contemporary vocabulary and diction of those who in all likelihood attained that technical and practical knowledge in their own days with reference to a normative standard then prevailing in that specific stage of cultural advancement. Unfortunately, no specialised texts of any high antiquity have survived to us that profess to talk about metallurgy as a separate branch of human pursuit. However, most of our ancient literary tradition is religious and ritualistic in nature—for that has been in itself the essential historical factor of its continued preservation through the ravages of time and man. This way, the early texts normally represent something of the sort of a highly refined and summary statement of certain technological ideas and practices that had come, for one reason or the other, to be employed and absorbed in the day-to-day religious needs of the community and, therefore, in priestly curriculum of a comprehensive setting.

Another aspect of our ancient literary survivals is their essential poetic composition. This would be in a way found useful, though in a limited application, as including many pieces of cultural evidence which the particular poet in his creative vision employed in the service of his imageries, i.e., his limitless field of analogies and paradigms some times freely drawn from various kinds of material experiences, art practices and involved technological aspects of human achievements in different branches of knowledge.

Even in such cases where a text is not in verse, it was indeed composed in the aphoristic (*sutra*) style professing to give only central ideas in bare minimum words of an otherwise highly developed subject that needed to be comprehended in its fuller perspective and thoroughly explained by the teacher to the disciple from generation to generation. This no doubt immensely helped in evolving technical definitions, phrases and vocabularies of their own kind for each such specialised branch of knowledge and its allied contexts. Obviously, a modern student has essentially to avoid anachronism in drawing his conclusions and has to be extra careful in using and interpreting such texts even in their original diction what to say of their translations in modern languages. Great help can undoubtedly be derived from early glosses, commentaries and such scholiastic tradition itself besides older Sanskrit lexicons, grammars, etymologies and technological semantics involved in a particular context and its proper understanding.

Language can be found to be true carrier of practices and achievements in the society which expressed and understood itself in this potential medium of ever-expansive symbols of names and verb forms. At times, differentiating implications were suggested even through a stress or accent on a syllable or by a modified diction. For example, in his study on ancient Indian life as known to Panini's Grammar, Prof. V.S. Agrawala gives the following piece of an outstanding information as regards the working of a goldsmith: "The phrase *nishtapati suvarnam* 'he heats the gold in the fire only once', (*Nisataptavanasevane*, VIII. 3. 102) belongs to the goldsmith's vocabularly and needs to be explained. The village goldsmith seated before his miniature cupola, has to deal with three kinds of orders. Firstly, new gold or silver in the shape of bar or ingot is brought to him to make ornaments. Secondly, old ornaments are brought in order to be melted and shaped into new forms. In these two cases he subjects gold to repeated heating and expands it by beating for which the expression is *nishtapati survarnam.* In the third case old ornaments are brought to him in order to be repolished or brightened by heating to look like new ones. For this the ornament is heated only once (*anasevane*) and either rubbed or immersed in a solution to be made brighter. This operation was expressed by the cerebralised form *nishtapati* (*suvarnam suvarnakarah*)."[1]

While discussing in detail the position of copper and its alloys in Vedic literature we have ventured elsewhere[2] to conclude that the word *ayas* presumably denoted copper in the earliest hymnal anthology of the Rigveda. If the word *ayas* in the Rigveda meant iron, as several authorities are inclined to believe,[3] we would have to accept the fact that there was no word then prevalent for denoting copper and hence no knowledge of it in that period. This certainly is against all norms of man's archaeological past. However, there are no words normally known in the Rigveda denoting such metals as tin, zinc, lead, or even silver.[4] This last metal of silver is indeed problematic if it were most probably implied by a word which is also a name commonly taken to indicate gold, such as, for example, *hiranya* or *chandra.*

The learned authors of the *Vedic Index* have remarked that "Chandra" appears to denote 'gold' in a certain number of passages from the Rigveda onwards."[5] In the Yajurveda and Atharvaveda,[6] it is quite certain that *Chandra* denotes a metal, but its primary meaning may be better understood as 'silver' even though one finds *hiranya*[7] as given for its symbolical reference or explanation in the Brahmana texts:

chandram hyetach chandrena krinati yatsomam hiranyena, Satapatha 3.3.3.6; *chandram hiranyam,* Taittiriya 1.7.6.3.[8]

An Atharvavedic passage may not be out of place here as a citation which goes to indicate that already several types of '*hiranya*' had come to be used in ritual application: *yo bibharti dakshayanam hiranyam sa jiveshu krinute dirghamayuh,* I. 35.2.[9]

It may be remarked with reference to *ayas* that it had already come to denote iron by the time of grammarian Patanjali who in his Mahabhashya gives a proverbial expression in the language: *ayas kantam ayah samkramati,* on Panini 3.1.7. "iron moves to the loadstone".

The imagery of the iron-stone or magnet stone attracting small pieces of iron towards it, is in fact very common with classical Sanskrit poets.[10] However, this property of *ayas* would not have certainly escaped the notice of the Rigvedic or later Vedic poets if the word had already iron for its earlier meaning. We are not sure in the present state of our knowledge whether Panini also knew the phrase *ayaskanta* for the iron-loving loadstone. Anyway, other earlier references to this effect, if any, are to be looked for and properly analysed in their cultural implications.

Let us take another example. A female demon is mentioned in the Paippalada Samhita by her epithet *ayovara,*[11] which signifies an evil personification to be warded off by *ayas,* i.e., copper. This interpretation may be substantiate by a statement made in the Baudhayana Grihyasutra as follows: *tamrayasenetyatharvanah, karshnayasenabhicharanniti sarvatrikam.* "Doing incantation with the aid of copper (*tamra+ayasa*) is of the Atharvanic kind: with the aid of iron (*karshnayasa*) it is of general usage."[12]

As we very well know, a section appended to the Vedic Kalpa or Srautasutras on the rules of making fire-altars, their diagrams and geometrical shapes, is styled as the *Sulbasutra.*[13] The word *sulba* or *sulva* is generally explained as denoting a string, a cord or a measuring rope, also called *rajju* or *sutra.*[14] It is, however, not quite clear of which material this measuring cord was made, whether of cotton thread or some other fibre. If we were to take the help of later Sanskrit lexicons, the *sutra* or *sulba* would seem to have its reference to the wire of copper. As the *Amarakosa*[15] reads: *atha tamrakam. sulvam mlechochhamukham dvyashta -varishthodumbarani cha,* 2.9.97. It is given here as a synonym for copper, as are the other four words, including *udumbaram.* This seems to be quite in fitness of things as of necessity the sulba cord was meant to be used for the most careful and exact measurements defining complex geometrical shapes and laying of bricks in constructing sacrificial altars where no technical faults or ambiguities of designing were allowed to occur. The word '*sulba*', occurring frequently in Kautilya's Arthasastra (early 3rd century B.C.), has been already recognised quite appropriately to denote copper.[16] In the Sulbasutra context too, a measuring tape or wire of copper should be understood by us to have been called as the "*sutra* of *sulbha*" in its specific connotation indicating its practical use in such fine calculations of the kind needed for constructing the places of sacrifice and fire-altars having definite shapes and fixed layers of numbered brick-units.

REFERENCES

1. Agrawala, V.S., 1953. *India as known to Panini*, Lucknow, pp. 234-35.
2. Agrawala, P.K., 1977. *Early Indian Bronzes*, Varanasi, pp. 40-41; "Ayas and its Alloys in Vedic Literature". *Researches in Indian History, Archaeology, Art and Literature*: Prof. Upendra Thakur Felicitation Volume, Delhi, 1990.
3. MacDonell, A.A. and Keith, A.B., 1958. *Vedic Index of Names and Subjects*, reprint Delhi, two vols. vol. I, p. 31 ff.
4. The word *rajata* occurs at the earliest in the Atharvaveda, 5.28.1: 13.4. 51. On AV. 5.28.1: *"harite trini rajate trinnyayasi trini ....."* the *Vedic Index* observes, "In one passage of the Atharvaveda, the sense of iron (for *ayas*) seems certain" (p. 32). How this conclusion is arrived at by the learned authors is not quite clear. Presumably, this can be seen here a mention of three metals in a specific order of their significance: gold (*harita*), silver (*rajata*), and copper (*ayas*).
5. *Vedic Index*, Im p. 254.
6. See useful references cited, *Vedic Index*, Imp. 254; also, II, p. 197.
7. As noted in the *Vedic Index*, "Rajata as an adjective with *Hiranya* designates 'silver' and ornaments (*Rukṃa*), dishes (*Patra*) and coins (*Niska*) 'made of silver' are mentioned. The word is also used alone as a substantive to denote. 'silver'. Add, *rajatam hiranyam*, Jaiminiya Brahmana I. 63 (cp. Satapatha Br. 12.4.4.7).
8. We have such expressions as early as the Rigveda, *Chandra hiranya*, 9.97.50, that can be instructive if compared with passages referred to above in Note 7, having *chandra* and *hiranya* in allocation. Also, see Kathaka Samhita, 28.4: *chandra-dakshina.*
9. "Whoever wears the gold of Daksha's children hath a long lengthened life among the living." Translation by R.T.H. Griffith, vol. I, Banares, 1916, p. 40.
10. See references given in, for example, in Monnier-Williams, *Sanskrit-English Dictionary*; V.S. Apte, *Sanskrit -English Dictionary*, under *ayaskanta, kantayasa* and allied words.
11. *Paippalada Samhita* 17.15.7.
12. *Baudhayana Grihyasutra* 1.3. 17-18.
13. Datta, Bibhutibhushan, 1932. *The Science of the Sulbas*, Calcutta; Sen, S.N. and Bag, A.K., 1983. *The Sulbasutras*, New Delhi.
14. As observed by B.B. Datta (p. 8), "Thus the terms *sulba* and *rajju* have four meanings:
   (1) mensuration -the act and process of measuring,
   (2) line (or surface) - the result obtained by measuring,
   (3) a measure - the instrument of measuring, and
   (4) geometry - the art of measuring".

15. See Kshirasvami's commentary on these words treating their grammatical consideration and cultural etymologies thereby implied or understood. K.G. Oka, *The Amarakosa*, reprint Delhi, 1981, p. 155.
16. Kangle, R.P., 1960. *The Arthasastra*, Bombay, 2.13.16. 44: 2.14. 20-22, 30-31; etc. Further see, Edgerton, P., 1970. *Buddhist Hybrid Sanskrit Dictionary*, reprint Delhi, p. 531, specifically giving this meaning of *sulba*; his citation, *tamraloham ca sulvam*: and p. 533, *sasulbika* for coppersmith.

Antecedent Stages in the Evolution of Metal Technology

V.S. PATHAK AND PREM SAGAR CHATURVEDI

In the history of human culture, the knowledge of metals and metallurgy marks a very significant stage. The expertise was concerned with identifying and separating the metallic alloys which lay mixed with earth, sand, stone and water and with their processing—forging, smelting and casting for generating the plasticity in metals.

Techniques employed in metallurgy, though novel in certain respect were, in fact, evolved from pre-existing technologies used in wood, leather, lithic and ceramic crafts. They had played a pivotol role in the life of pre-historic man for millennia. These antecedents of the metal technology may precisely be put under two groups, the *takṣ* and the *gharma*-technologies.

The first stage in the technological progress is marked by the *takṣ*. It can appropriately be called the mother of all technologies. Descending from the Indo-European root *tek*, the word for technology found wide-spread acceptance in the shape of Greco-Roman *technos*, Lithunian *tasyati*, Church Slavic *tasati*, Avestan *taš̱* etc. In all these traditions it signifies the manipulative power pre-eminently manifested in the carpenter's work. There is, no doubt, that this is the earliest surviving root denoting the technical skills of man which comprise preponderantly of wood-works. All existing primitive terms for carpenter in the Indo-European tradition, such as Graeco-Roman *tektōn*, Sanskrit *takṣan* and Avestan *tasan* have been derived from this root. In the works of Homer, *tekton* mainly stands for a worker in wood, a carpenter or a joiner (*Illiad* 5.59; 6.315; 13.390; *Odyssey* 9.126; 17.384; 21.430). In the *Avesta* Ahura-Mazda who fashions the earth is frequently spoken as *Geus-tašā*, literally the carpenter of the Mother-Earth (*Yasna* 29.1; 31.11; 44.6; 51.7). The Sanskrit equivalent of *tektōn* and *tašan* is *takṣan* who has his divine counterpart in Tvaṣṭṛ, the creator of all the forms of the world.

In the *Ṛigveda, takṣ* has been employed to express a variety of creative skills. These comprise the composition of hymns and praises (*RV* V.73.1; VI.32.1; IX. 97.22; X.80.7), framing of chariots (*RV* I.111.1; IV.33.8; VII.32.20; VIII.64.5), making of armour for the gods (IV.34.9), carving of *yūpa* (I.162.6; III.8.6 etc.) and wooden vessels (X.53.10; 101.10) and also the fashioning of *vajra* (I.32.3; 52.7; 61.6; 121.3; X.48.3; 99.1). The *takṣ* technology was widely applied in building dwellings and designing means of transport both on surface and water. A number of articles of daily needs and agricultural implements were carved of wood. The text sometimes mentions carpenter's implements, such as *svadhiti* (III.8.6), *vāśī* (X.53.10; 101; 10) and *paraśu* (I.30.4; III.53.22; VI.3.4; VII.104.21; X.28.8) in reference to the creation of these objects. These were made of stone (*aśman*) and also of metal (*ayas*). The techniques practised by the carpenter comprised of cutting, hollowing, strapping the different parts and joinery.

Needless to say, the *takṣ* technology which emerges as the foremost technology in the Vedas and other Indo-European literary traditions, was the primary technology. Based on the manipulative power of hands, this was practised initially with wood in the form of scraping, tearing, cutting and fabricating through addition, substraction, duplication and multiplication. Subsequently, it passed to the leather and came to be applied *mutatis mutandis* to the lithic. With certain limitations this also left its mark in devising the metallic objects as the references imply, to the fabrication of *vajra*.

The next important stage antedating the metal-moulding technology was attained by the *śrī-gharma* technology. Applied with the ceramics, the technology involves two processes. The first is concerned with *śrī* which is required for generating plasticity in the clay and the second i.e. *gharma* involves the firing of the earthen vessel in order to provide it with hardness. Both of these processes being essential, got mingled in the preparation of ceramics.

Interestingly, the very term ceramics which belongs to the Latin *ceramicus*, has linguistically descended from the Greek root *kerannumi* which is equivalent to the Vedic *śṛṇāti* (= *śrī*) and Avestan *sara*. In the Greek tradition, the terms emanating from *kerannumi* denote the diluting of wine with water (*Odyssey* 5.93; 24.364 etc.) and also the mixing and kneading of clay and subsequently several things related to the potter's craft. This is explicitly evident from its several derivatives, such as *kermos* 'potter's earth or clay', *kerāmeia* 'potter's craft', *kerameion* 'potter's workshop' etc. Descending from a tradition of high antiquity, the Greek *kermos* which has its equivalent in Latin *ceramicus*, undoubtedly conveys an earliest sense when kneading the clay with certain ingredients to make it suitable for fashioning the pots, was considered immensely essential.

A parallel tradition associated with the mixing or kneading of clay, is also found existent in the Indo-Iranian literature in the shape of Sanskrit root *śṛṇati* (- *sr*) and Avestan *sara*[1] both meaning to mix. As in the early Greek, the roots express the mixing of liquids, especially *Soma* or cow-products, to be offered as oblation to gods. The notable evolutes of *śṛṇāti-śrī* are *śarma* and *śaraṇa* which denote in the *Ṛigveda* the dwellings made of clay. These are comparable to Avestan *saram* (*Yasna* 41.6) and *saramno* (*Yasna* 49.5; 8; 53.3) which possess almost a similar sense.

This is of utmost significance that the Vedic literature contains a well-evolved tradition of lavigating the clay with the help of certain ingredient. As if the plastic properties endowed in the clay by nature is not enough, the texts of the later Vedic period propose a number of materials to be mixed in the clay required for fashioning pots. The *Śatapatha Brāhmaṇa* XIV.1.2.10-12 suggests the mixing of the soft soil of ants hill (*valmīkavapā*), the soil dug out by the wild boar (*varāhavihatam*) and the plants of *ādāra* (basella cordifolia). All these materials were well-kneaded with the help of water to make the soil soft and sticky. In another context (VI.5.1.1-6) the text refers to the mixing of the resin (*kaṣāya*) of the *parṇa* or *palāśa* (butea frondosa) tree, the hairs of a goat (*ajaloma*), the dust particle of pebble (*śarkarā*), stone (*aśma*) and the iron-rust (*ayorasa*). Yet another tradition, sketched in the *Taittirīya Saṁhitā* (V.1.6.2; 3) and in the *Taittirīya Āraṇyaka* (V.2.13) contains the names of five materials, such as (i) the fragments of the pottery used in the household (*grāmyāṇām pātrāṇām kapālāḥ*) (ii) fragments of the potteries collected from the deserted sites (*armakapālāḥ*) (iii) pebbles of tiny pieces of stones (*śarkarā*), (iv) the hairs of goat (*ajaloma*) and (v) the hairs of a black antilope (*kṛṇṣājinasyaloma*).

Prevalence of a number of traditions of kneading the clay with different ingredients not only affirms the rich knowledge of the Indo-Aryans about the preparation of the clay but also conveys that like the *Graeco-Roman* tradition ceramics were originally concerned with and laid emphasis on mixing and kneading the clay. In this connection, we may also mention about the mixing of hard materials, such as *aśma* (stone) and *ayorasa* (iron-rust) in the clay. About the use of *aśma*, the *Kātyāyana Śrauta-Sūtra* (XVI.3.2) enjoins that the hard stones are first brayed into powder and then mixed in the clay (*tēṣāṁ cūrṇaiḥ saṁsṛjati*). For *ayorasa*, the text avers that the rust, separated from the heated iron and known as *kīṭa*, is to be mixed in the clay. We know from the *Śatapatha Brāhmaṇa* (VI.5.1.1; 4; 6) that mixing of all these materials was attempted in order to equip the vessel with the desired firmness (*sthemne*).

While *srī* (*-śṛṇāti* = Greek *kerannumi*, Latin *ceramicus*) represents only one aspect in the making of ceramics, another one which is no less significant, is marked by the *gharma*. Derived from the Sanskrit root *ghṛ*, to heat, warm, *gharma* has several equivalents in Greco-Latin, Balto-Slavic and Indic languages. Some of them are Church Slavic *grunīcī*, gorēti, meaning burn, Sarbo-Crotian *granac*, Bohemian *hrnek* = *hrnec*, Polish *garnek*, Russian *gorsŏk*, Latin *furnus* 'oven' and Greek *thermos*, meaning hot.[2] Taken cumulatively, these originally indicate heat, burning or oven. The meaning further gets confirmed from the Vedic literature where *gharma* denotes a sacrificial vessel as well as the boiling liquids, heat and furnace.

A number of passages in the Vedic literature refer to *gharma* either in the sense of heat or a fire place or furnace. In the *Ṛigveda* I.112.1, it occurs as an attribute of Agni and indicates a blazing fire. In another passage (III.26.7), the text alludes to fire and employs the terminology *ajasro gharma* to express that it is an inexhaustible source of heat. While describing the preparation of the rice-dish (*odana*), the Atharvaveda XI.3.18 mentions - *caruṁ pañcabilamukhaṁ gharmo'bhīnde* wherein *caru, ukhā* and *gharma* have been used together but in different sense. The *caru* signifies the pot to be employed for cooking the rice-dish, whereas *ukhā* and *gharma* respectively mean the oven and heat which form the essential components to accomplish the cooking. Thus the description provides a picturesque account of a pot (*caru*) lying over an oven (*ukhā*) and getting sufficient heat (*gharma*) to boil its contents.

There are quite a number of descriptions which conceive *gharma* in the sense of fire-place or furnace. The *Ṛigveda* X.16.10 furnishes an interesting account of replacing a flesh-consuming fire by a new one to be used in the sacrifice for the Fathers and refers to *gharma* as a place where the undefiled fire is kept. There is no doubt that *gharma* denotes here a fire-place. A good piece of evidence, affirming the meaning of *gharma* as a fire-place is further provided by the famous anecdote about Atri, narrated in a number of passages (I.119.6; 180.4; V.73.6; VIII. 73.3; X.80.3) of the text. It is said that Atri, the seer, was once thrown by demons (*rākṣasas*) into a *gharma* whence from he was rescued by the Aśvins. Some of the passages aver that, in their attempt to rescue the seer, the Aśvins covered the *gharma* with cold water (I.180.4). If the entire context is properly scrutinised, it seems to be providing a view of a burning fire-place, more appropriately a furnace into which a seer was thrown in a life taking attempt but subsequently rescued and the fire was extinguished through sprinkling the water over the furnace. Almost the same kind of meaning is conveyed by a couple of passages successively occurring in a hymn (XIX. 28) of the *Atharvaveda*. Referring to the binding of an amulet of *darbha* on the occasion of *mahāśānti* ceremony, called *Aindrī*, these invoke the inherent power of the amulet to burn the heart of the foes like a *gharma*. Translating the passages, Whitney takes *gharma* for the heat. It would be, however, more appropriate to render it as a fire-place or furnace because it is in this famous resort of fire that the capacity to burn something lies.

Evidently, *gharma* in the sense of heat or the source of heat in the form of a fire place or furnace is similar to Greek *thermos*, Latin *furnus*. From the standpoint of technology, it represents the thermic principle, the finding of which was a revolutionary step taken in connection with the preparation of ceramics. It was applied for baking the raw vessel in a controlled heat of fire. It was, in fact, the application of this principle that provided the ceramics with its perfection as the unbaked vessels were not sturdy enough and therefore, were of limited use. The thermic principle revolutionised the concept of fabrication and the technique of plasticity. Originating with ceramics, the *gharma* technology depending on furnacing and control of fire soon became a key technology in metals and metallurgy.

Thus, before the advent of metal-moulding technology, two important stages marked by the *takṣ* and *śrī-gharma* technologies were already in vogue. These not only formed the basic background but also gave a fillip to the emergence of metallurgy. While the *takṣ*-technology, based on the manipulative power of hands was of general application, the real key to the growth of metallurgy was presented by the *gharma* or thermic principle, the use of which pushed the knowledge of metal-smelting and metal-casting.

Quite interestingly, the Vedic literature evinces a clear knowledge of metal-smelting and metal-casting respectively by the roots *dhmā* and *sic* which are found mentioned as early as the *Ṛigveda.* There are several references to the root *dhmā* which stands for blowing of fire in order to generate certain amount of heat to smelt the metal. The meaning has eloquently been expressed by the following passage:

अध स्म यस्यार्चयः सम्यक्संयन्ति धूमिनः ।
यदीमह त्रितो दिव्युप ध्मातेव धमति शिशीते ध्मातरी यथा ॥

ṚV, V.9.5.

It alludes to *dhmā* thrice, twice as nouns (*dhmātṛ,* the blower of fire or smelter) and once as a verb (*dhamati,* the act of blowing the fire). The passage envisions an inflamed fire and likens it to one blown by a smelter in course of melting the metal. In the *ṚV,* IV.2.17, men of noble deeds are mentioned purifying their births as the metal is purified by the smelter. In X.72.2, the creative power of Brahmanaspati is compared to that of a smith (*karmāra*) who produces the metallic objects by smelting the metal in the fire. Beside these, there are several other instances wherein the root *dhmā* signifies the power of creation, evidently by smelting.

Like *gharma,* the technology of *dhmā* is directly linked with the control of fire. The heat of the fire was gradually enhanced by passing the current of air with the help of certain contrivance to the extent required for melting the metal. One of these contrivances is *dṛti,* the bellows made of leather which has been mentioned in the *ṚV,* VII.89.2 in connection with blowing of fire by the smelter. The same has again been alluded to in the *ṚV,* IX.1.8 which likens the glittering *Soma* to shining *dṛti* presumably as the latter is always placed in the vicinity of fire. The use of the verbal root *dhmā* in this context affirms this assumption. Apart from *dṛti* or the bellows of leather, wooden pipes were also used as is evident from the *ṚV,* I.85.10.

References to *sic* or metal casting technology based on smelting are found only in embryonic form in the *Ṛigveda. Sic* which means to pour also stands for casting the metal. It is in the previous sense that the root has generally been employed. Nevertheless some of the contexts are pregnant with important implications. Instead of expressing the ordinary pouring of liquids, these depict the placing of creative energies, especially the procreative fluid which results into progeny. This is evident from several descriptions in the text. In the *ṚV,* VI.70.2, Heaven and Earth (*rodasī*) are invoked to cast the procreative energies in the sacrificers for the welfare of mankind.[3] In the succeeding verse (VI.70.3) beings of various forms are assumed to be born because of the invigorating power sprinkled by the Heaven and Earth.[4] In X.61.2 we read -

स इद्दानाय दभ्याय वन्वञ्च्यवानः सूदैरमिमीत वेदिम् ।
तूर्वयाणो गूर्तवचस्तमः क्षोदो न रेत इत ऊति सिञ्चत् ॥

"He (Rudra) bestowing on his worshippers the gift (of wealth) and the defeat (of their foes), casting down (the Rakshasas) with his weapons, has constructed the altar; rapid in movement, most fierce in speech, and shedding the procreative fluid like water around."

Here the sprinkling of the procreative fluid is not only different from the ordinary water but also that it has been done with certain purpose, and as the story goes, Rudra created the Aświns by this fluid.

The most eloquent description of *sic* as a creative energy manifesting itself in various forms, occurs in X.184.1, which mentions:

विष्णुर्योनिं कल्पयतु त्वष्टा रूपाणि पिंशतु ।
आ सिञ्चतु प्रजापतिर्धाता गर्भं दधातु ते ॥

"May Viṣṇu construct the womb, may Tvaṣṭṛ fabricate the forms, may Prajāpati sprinkle (the seed), may Dhātṛ cherish thy embryo."

In the passage, the entire creation is visualised as coming forth from the procreative seed which has been cast by Prajāpati in the womb created by Viṣṇu.

Despite the fact that root *sic* has never been mentioned in the *Ṛigveda* in reference to metal or metallic objects, it is potent with the sense of casting. It is worth to note that in the Sanskrit literature of the later period, the casting of metal in the mould is likened in the Ṛigvedic fashion to casting of human seed in the foetus. A description in the Aśvamedhika Parva of the *Mahābhārata* expresses it thus:

> "As liquified iron, poured (into a mould), takes the form of the mould, know that the entrance of *Jiva* into the foetus is even such. As fire, entering a mass of iron heats it greatly do thou know the manifestation of *Jiva* in the foetus is such."[5]

In a resembling allusion, occurring in the third Chapter of the 'Śarīrasthāna of the *Charak-Saṁhitā*, we read:

> "In whatever womb the impregnating conditions occur, of viviporous and egg-born creatures, in what womb they take shape accordingly, as when gold, silver, copper, tin and lead are poured into various wax-moulds."[6]

When viewed in the light of these anâlogous descriptions, the Ṛigvedic references which employ the root *sic* in the sense of pouring the creative energies become specially meaningful as they also conceive the sense of metal-casting. So far as a clear reference to the metal casting in the Vedic literature is concerned, it is found in the *Atharva-veda* which mentions the casting of the thunderbolt (वज्रं यमसिञ्चत्) by Bṛhaspati.

It is evident, therefore, that before the emergence of the metal smelting and metal casting the technological advancement has taken a big stride in the forms of *takṣ* and *śrī-gharma* technologies which were gradually evolved on the manipulative power of hands and on the knowledge of plasticity and the thermic principle. Practised for millennia, these left their mark on the metals and accelerated the concept of their plasticity. The knowledge of the thermic principle which was initially applied for the ceramics revolutionised the concept of fabrication in the metals by smelting and casting.

REFERENCES

1. Bartholomae as quoted by Taraporewala, Irach, J.S. 1951. *The Divine Songs of Zarathushtra*, Bombay, 258.
2. Buch, C.D. 1965, *A Dictionary of Selected Synonyms in the Principal Indo-European Languages*, The University of Chicago Press, Chicago & London, 5.26.
3. अस्मे रेत: सिञ्चतं यन्मनुर्हितम् ।
4. युवो: सिक्ता विषुरूपाणि सव्रता ।
5. यथा लोहस्य नि:स्यन्दो निषिक्तो बिम्बविग्रहम् ।
   उपैति तद् विजानीहि गर्भे जीवप्रवेशनम् ॥ ९ ॥
   लोहपिण्डं यथा वह्नि: प्रविश्य ह्यतितापयेत् ।
   तथा त्वमपि जानीहि गर्भे जीवोपपादनम् ॥ १० ॥
   Ashwamedhika Parva, *Mahābhārata*, Third Edition V.S.; 2029, Gita Press, Gorakhpur.
6. तत्र जरायुजानाम् अण्डजानाम् च प्राणिनाम् एते गर्भकरा भावा याम्याम् योनिम् आपद्यन्ते, तस्यां तस्यां योनौ तथा तथा रूपा भवन्ति तद तथा कनक रजत ताम्रत्रपुसीसकान्यासिच्यमानानि तेषु तेषु मधूच्छिजृविग्रहेषु .... ।

Literary References on Metals, Metallic Objects of Art and Metal Technology

SWARNA KAMAL BHOWMIK AND MUDRIKA JANI

The art of working in metals is very ancient one in India. Literary references to metal and objects made of metal are quite large in number and variety.

Vedic literature makes mention of *'Ayas'*. The word *'Ayas'* may mean copper or bronze or iron. Presumably copper was represented by the word *'Ayas'*. Ayas was used in making vessels for domestic and ritual purposes.[1]

We get an idea of metal working from the literary evidence of the later Vedic period. In the age of *Kalpasūtras,* a metalsmith's main work was probably restricted to the manufacture of utensils and objects of day-to-day use.[2] Panini mentions in his *Aṣṭādhyāyī* that tools and implements made of copper and other metal were used by a metal smith.[3] Patanjali while commenting on Panini's *Sūtra* "जिविकार्थे चापण्ये। v. 3.", notes in his *Mahābhāṣya* that the Mauryas had made images of Gods for obtaining gold. This most probably refers to metal images. (Kielhorn (editor), *Mahābhāṣya,* vol. II, Bombay 1906, p. 429. Also *Development of Hindu Iconography,* p. 44, by J.N. Banerjee).

The word *Ayas* was used by both Manu and Yajnavalkya. The reference of a golden image of Sita, which was substituted for her, occurs in the *Rāmāyaṇa* during the performance of the *Aśvamedha* sacrifice by Rama in the absence of Sita who was in exile in Valmiki's hermitage.

> कांचनीं मन पत्नीं च दीक्षार्हां यज्ञकर्मणि।
> अग्रतो भरतः कृत्वा गच्छत्वग्रे महामतिः (२५) रामायण, उत्तरकाण्ड सर्ग ९१.

The work of metalsmith in manufacturing agricultural implements, weapons of war in various metals like copper, brass, bronze and iron is described in the *Jātaka*[4] which refer to the eighteen guilds of workers. Moreover the Buddhist *Jātaka* gives numerous references to images made of gold (*Suvarṇa-pratimā*). The Jaina Canonical texts describe in proper technical terms the working processes of a metalsmith, smelting of the ore, forging and casting techniques.[5] The Jaina *Praśṇa Vyākaraṇa* mentions a list of eighteen guilds which includes bronze-smiths (*Kāṁsyakāras*) (*Praśṇa Vyākaraṇa,* pp. 193-194).

The reference comparing the embryonic stage of man with the process of casting of metal images occurs in Aśvamedhika Parva of the Mahābhārata in which horse sacrifice of Yudhishthira has been described.[6] The text is given below:

यथा हि लोहमिष्यन्दो निषिक्तो बिम्बविग्रहम् ।
उपैति तद्वज्जानीहि गर्भे जीवप्रवेशनम् ॥ १८-८ ॥

(अश्वमेधिकपर्व, महाभारत)

The above *śloka* suggests that the entrance of soul into the foetus takes place in the same manner as the liquid iron assumes the form of the image when poured into the mould.

That apart, the *Mahābhārata* mentions of an iron image (*āyasī-pratimā*) of Bhima, which was crushed by Dhritarashtra thinking that he was killing Bhima himself.

तं तु गृह्यैव, पाणिभ्यां भीमसेनमयस्मयम् ।
बभंज बलवान् राजा मन्यमानो वकोदरम् ॥ २ ॥ महाभारत स्त्री पर्व सर्ग १२.

Similar reference occurs in the *Charak Saṁhitā*[7] in the third chapter of the *Śārīrasthāna* which has been dated by scholars to the second or third century A.D. The text is given below:

तत्र जरायुजानाम् अण्डजानाम् च प्राणिनाम् एते गर्भकरा भावा याम्-याम् योनिम् आपद्यन्ते, तस्यां तस्यां योनौ तथा तथा रूपा भवन्ति, तद् यथा कनकरजतताम्रत्रपुसीसकान्यासिच्यमानानि तेषु तेषु मधूच्छिष्टविग्रहेषु, तानि यदा मनुष्यबिम्बम् आपद्यन्ते, तदा मनुष्यविग्रहेण जायन्ते, तस्मात् समुदायप्रभवः सन् गर्भो मनुष्यविग्रहेण जायते, मनुष्यश्च मनुष्यप्रभवः उच्यते, तद् योनित्वात् ॥ २६ ॥ (चरक संहिता-शारीरस्थान-तृतीय अध्याय)

Dr. W. Huchzermeyer translated the above text as follows:

"In whatever womb the impregnating conditions occur, of viviparous and egg-born creatures, in that womb they take shape accordingly, as when gold, silver, copper, tin and lead are poured into various wax-moulds; when it is mould of a human being, they take a human form; therefore, since the foetus has its origin in *Samudāya* and takes a human form, man is considered to be of human origin, since that is his womb."

Kautilya's *Arthaśāstra* mentions about the superintendent of mines and metallurgy under whom a *Lohādhyakṣa* worked. *Lohādhyakṣa* carried on manufacture of copper, bronze (*kāṁsya*), lead, tin, sulphurate of Arsenic (*tāla*), *lodhra* and *ārkuṭa*(?) (Shamasastry, R., 1923; Kautilya's *Arthaśāstra,* Mysore, p. 94 ff.)

The Ratnagotravibhaga of *Maitreyanath* describes the casting of gold images.[8] According to scholars, Maitreyanath lived in the 3rd century A.D. The text is given below:

हेम्नो यथान्तः क्वथितस्य पूर्ण
बिम्बं बहिर् मृण्मयम् एकस्य शान्तम् ।
अन्तर्विशुद्धयै कनकस्य तज्ज्ञाः
संचोदयेद् आवरण बहिर्धा ॥ १२४ ॥ (रत्नगोत्रविभाग)

Dr. W. Huchzermeyer translated the above text as under:

"In the manner as when he sees the full form of liquified gold inside and of earth outside cooled down, the expert should cast (remove) away the outer cover in order to clean the gold inside."

The verse 41 of the 11th chapter of the *Kāmaṇāgama* shortly describes the method of casting of metal image.[9] The text is given below:

लोहजत्वे मधूच्छिष्टम् अग्निवार्दी क्रियन्ते यत् ।
वस्त्रेण शोधयेत् सर्वं दोषं त्यक्त्वा तु शिल्पिना ॥ ४२ ॥
(कामणागम अध्याय ११ श्लोक ४१)

Dr. W. Huchzermeyer translated the above text as under :

"If images are to be cast in metal, the wax has first to be melted in fire and all defects left by the artist are to be removed with a cloth."

The verse 21 of the chapter 34 in the *Suprabhedāgama* shortly describes the art of image making.[10] The text is given below:

मृण्मयं यदि कुर्याच्चेच्छूलन् तत्र प्रकल्पयेत् ।
लोहन्जंच विशेषेण मधूच्छिष्टेन निर्मितम् ॥ २१ ॥
(सुप्रभेदागम अध्याय ३४)

Dr. W. Huchzermeyer translated the above text as below:

"When an image is to be made of earth, one should fix rods inside, if it is to be of metal, it has to be first made to wax."

Metals are included in the list of materials recommended for image making in the *Śaiva Āgama* (Rao, G., 1914, *Elements of Hindu Iconography* I. 1; Madras, p. 48 ff.).

A short description regarding the process of making metal statues is found in the 14th *Paṭala* of the *Viṣṇusaṁhitā.*[11] The text is given below:

लोहे सिक्थमयीम् अर्चां कारयित्वा मृदावताम् ।
सुवर्णादीनि संशोध्य विद्राव्याङ्गारवत् पुनः ॥
कुशलैः कारयेद् यत्नात् संपूर्णां सर्वतोधनम् ॥
(विष्णुसंहिता पटल १४)

Dr. Sivaramamurti translated the text as given below:

"A complete wax image prepared and coated with clay may be cast as a solid one in gold or other metals properly tested and melted in the requisite temperature by experts."

The *Matsya Purāṇa* mentions gold, silver, copper, iron, bronze (*kāṁsya*), stone and wood as materials for use to fashion images of Gods and Goddesses (Vidyasagara Jivananda (editor), 1876, *Matsya Purāṇa,* Calcutta, p. 1102). The *Bhāgavata Purāṇa* mentions eight kinds of images including image of iron (*Bhāgavata Purāṇa* XI, 27-12 quoted in *Pūjaprakāśa,* p. 116. Also see Kane, P.V., 1941, *History of Dharmaśāstra,* 11, part II, Poona, p. 715). The *Agni Purāṇa* also mentions gold and other metals in the list of materials to be employed for image making (*Agni Purāṇa,* Poona, 1900, p. 49). The *Devī Purāṇa* mentions about gold, copper (*tāmra*), bronze (*kāṁsya*), brass (*pittala*), magnetic iron (*vajraloha*), along with images made of other materials (*Devī Purāṇa, Mātṛipratiṣṭhā-vidhi* section quoted in *Śabdakalpadruma Kośa,* III, p. 262).

A Jain canonical *Aṅga* text known as the *Nāyādhammakahāo* refers to a life size image of Princess *Malli* made of gold. The image of Princess *Malli,* who later became a *Tirthaṅkara,* was hollow. The same text further describes a picture-gallery of a banker of *Rājagraha.* The picture-gallery was well-furnished by beautiful pieces of wood-work (*Kāṭṭakāmma*), stuccoes (*Potthakamma*), plaster-work (*leppa*), decorations of flowers and wreathes (*ganṭhima*), dolls made by stuffing, hollow and solid casting techniques (*Purima bharima*) images made of wrapped cloth (*Veṣṭima*), beaten and repouse work. The *Nāyādhammakahāo*[12] has been dated by scholars to a period not later than the 5th century A.D. It may be noted here that the Jain *Agama* texts were edited and written at Valabhi during the 5th century A.D. The *Nāyādhammakahāo* texts were written after the Valabhi and Mathura Councils of early 4th century A.D. and hence they are noteworthy. The text is given below:

तए णं सा मल्ली मणिपेढियाएं उवरिं अप्पणो सरिसियं सरित्तयं सरिण्वयं सरिसलावण्णजोण्वणगुणो
ववेयं कणगमयं मत्थयच्छिंड्डुं पउमप्पलपिहाणं पडिमं करेई २/(८-७३ पृ. १४-१५)
तए णं नंदे पुरत्थिमिल्ले वणसंडे एगं महं चित्तसभं कंरावेइ र अणेगस्वंभसयसंनिविट्ठं यासाइयं ४/ तत्थ
णं बहूणि किण्हाणि य जावं सुक्किलाणि च कट्ठकम्माणि च पोत्थकम्मावि च चित्तलेप्यगंथिमवेढिमपूरिम-
संघाइमाइं उवदसिज्जमाणाइं र चिट्ठंति (१३, पृ. १४२)

The *Madhūcchiṣṭa Vidhāna* describes in detail the process of solid casting of metallic image according to lost form method. It is included in the sixty-eighth chapter of the *Mānasāra*[13] which is dated to the Gupta period by scholars. The *Madhūcchiṣṭa Vidhāna* is considered as the earliest known *Śilpaśāstra.* The text is given below:

बिम्बानां तु मधूच्छिष्टं क्रियां सम्यग् वक्ष्यतेऽधुना ।
शैवं पाशुपतं चैव कालमुखं महाव्रतम् ॥ १ ॥
वामं च भैरभं चैव तन्त्रवत् ज्योतिः षट्क्रमात् ।
अगस्त्यः काश्यपश्चैव भृगुगौतमभार्गवाः ॥ २ ॥

गालव इति ऋषय: प्रोक्ता: कर्षणाद्यर्चनार्थकम् ।
तत्त तन्त्रवशात्सर्वं कुर्यात् षट् संपदास्पदम् ॥ ३ ॥
तत्ततन्त्रविपरीतं चेद् विपत्यं नित्यमावहेत् ।
अनुक्तं तत्र तन्त्रैस्तु ग्राह्यं दोषो न विद्यते ॥ ४ ॥
विश्वकर्मा च विश्वेश: विश्वसार: प्रबोधक: ।
वृतश्चैव मयश्चैव त्वष्टा चैव मनुर्नल: ॥ ५ ॥
मानविन्मानकल्पश्च मानसारो बहुश्रुत: ।
प्रष्टा च मानबोधश्च विश्वबोधो नायश्च तथा ॥ ६ ॥
आदिसारो विशालश्च विश्वकाश्यप एव च ।
वास्तुबोधो महातन्त्रो वास्तुविद्यापतिस्तथा ॥ ७ ॥
पाराशरीयकश्चैव कालयूपो महाऋषि: ।
चैत्याख्य: चित्रक: आवर्य: साधकसारसंहित: ॥ ८ ॥
भानुश्चेन्द्रश्च लोकज्ञ सौराख्य: शिल्पिवित्तम: ।
तदेव ऋषय: प्रोक्ता द्वात्रिंशति: संख्यया ॥ ९ ॥
तान् गोत्रवशाज्ज्ञात्वा शिल्पिन: सर्वत्रयोग्यकम् ।
शुक्लपक्षेऽथवा कृष्णे सर्वमासेषु कारयेत् ॥ १० ॥
शुभयोगे सुनक्षत्रे सुमुहूर्ते सुलग्नके ।
मन्त्रवत्कारयेद गर्धे स्थपतिस्थापकावुभौ ॥ ११ ॥
मधूच्छिष्टं योजयेच्छिल्पी शास्त्रं कारयेत्क्रमात् ।
यथारूपं तथा ध्यात्वा स्थपत्याद्यैश्च वर्धकी ॥ १२ ॥
यन्मधूच्छिष्टबिम्बस्य पंचवर्णैरलङ्कृतम् ।
क्षणिका, बेरमेवोक्तं चलं चाप्यर्चलं तथा ॥ १३ ॥
तस्मात्सर्वप्रयत्नेन तत्रदोषो न विद्यते ।
अक्षै: स्फटिकसंयुक्तं लूतभाजनमेव च ॥ १४ ॥
हस्तेन सर्वबिम्बानां ताम्रपत्रेण योजयेत् ।
अथवा पक्षकानां तु पक्षमूलेन विन्यसेत् ॥ १५ ॥
नखास् सर्वेषु पादौ च संयुक्तं तु गृगे(ताम्रे)णच ।
अथवा शिखशल्यं स्याद् भूषणं च विशेषत: ॥ १६ ॥
स्फटिकै: रत्नसंयुक्तं जीनपात्रेण विन्यसेत् ।
अमू (आम्र) कद्रुमसर्वेषां युक्त्या वर्णं समाश्रितम् ॥ १७ ॥
दारूकीलैश्च बिम्बानां यत्फलार्थे तु योजयेत् ।
आद्रवद्रव्यस्माद् दोषेमेवं न विद्यते ॥ १८ ॥
सर्वालंकारसंयुक्तं वेरं ग्रामप्रदक्षिणम् ।
यजमानादि संवेक्ष्य सैव्यं तत्रार्चनं भवेत् ॥ १९ ॥

तस्माद् दैवतार्थे संग्राह्य चान्यग्रामे विलासयेत् ।
पश्चातु शिल्पिशालाश् च बेरमेवं प्रशस्यते ॥ २० ॥
मानोन्मानप्रमाणेन शोभयेच्छिल्पि वित्तमः ।
द्वित्र्यङ्गुलाधिकं वाथ मधूच्छिष्टेन लेपयेत् ॥ २१ ॥
तदूर्ध्वे मृत्तिकां लेप्य शोषयेत्तद् विचक्षणैः
तत् पिण्डमुतापयेद्वा मधूच्छिष्टोद्गमैः पुनः ॥ २२ ॥
कर्तुरिच्छा यथा लोहैर्धृतः एतत्प्रविश्यति ।
पूर्णयेज्जलं संप्रोक्ष्य त्यक्त्वा तदर्धमृत्तिकः ॥ २३ ॥
बिम्बं सर्वाङ्ग संशोध्य श्वेतगन्धानुलेपयेत् ।
तत्पीठोपरि संस्थाप्य पुष्पमालैरलङ्कृतम् ॥ २४ ॥
ग्रामं प्रदक्षिणं कुर्यात्सर्वमङ्गलघोषणैः ।
शिल्पिशालां प्रविश्य कारयेदुक्तवत्क्रमात् ॥ २५ ॥
तस्योपाङ्गविहीनं चेत्पुनः सन्तानकं भवेत् ।
शिरो वा मध्यकाये वा हीनं चेद् विसृजेत्पुनः ॥ २६ ॥
कर्तुशिल्प्यचलं चित्तं कारयेत्पूर्ववद् भवेत् ।
चलं चित्तं न कुर्याच्चेत्स्थाननाशं धनक्षचम् ॥ २७ ॥
अनेकजन्मकृत पुण्यं क्षिप्रमेव विनश्यति ।
तस्मात्सर्वप्रयत्नेन पुतश (पूतं) च कारयेत्सुधीः ॥ २८ ॥

T.K. Palaniappan[14] translated the above Sanskrit original text of the *Madhūcchiṣṭavidhāna*, 68th chapter of the *Mānasāra*, and the translation is quoted below:

The bees wax method used in the production of (metal) images is clearly described below:

Agastya, Kashyapa, Bhrigu, Gautama, Bhargava and Galava are the six sages who have explained from the ploughing of the temple site (*karṣaṇam*) to the worship of images (Arcanā) in the finished temple, the science describing the process in the books, namely, *Viśvakarmā, Śaiva, Paśupatya, Kāla-mukha, Mahāvrata, Vāmana* and *Bhairava.* If with the help of, and in accordance with, the methods here set forth in respect to the construction of temples and making of images is faithfully adhered to, it will lead to wealth and prosperity, but if deviated therefrom will cause poverty and distress. However, if certain useful formulae and practices not described here are found in similar texts dealing with the process, there is no objection to the *Sthapati* adopting them for his own improvement (Sloka, 1 to 4).

The 32 Śilpaśāstras, viz., *Viśvakarmā, Viśveśa, Viśvasāra, Prabodhaka, Vṛita, Maya, Traṣṭā, Manu, Nala, Mānavin, Mānakalpa, Mānasāra, Praṣṭa, Mānabodha, Viśvabobha, Nāya, Ādisāra, Viśāla, Viśvakāśyapa, Vāstubodha, Mahātantra, Vāstuvidyapati, Pārāśariyaka, Kālayūpa, Caitya, Citraka, Avarya, Sādhakasāra, Bhānu, Indra, Lokajna* and *Saura*, were revealed by the great and famous *Viśvakarmā* (Lord of the Arts and Architect of the gods from whom all *Sthapati* claim descent); therefore, only the most skilful sculptors (*Sthapati*), who abide by the traditional practices laid down in these texts, should be chosen for executing works of metal-art employed in the building of temples (*Śloka* 5 to 9).

Basing a decision on the conjunction of the planets, at the most auspicious time of any day during either the bright or the dark half of the month, the *Sthapati* and the *sthapaka* (guru, chief architect or chief *Sthapati*) may prayerfully commence prepairing the casting furnace and the (wax) image; As laid down in the formula for it in this treatise, the bees-wax and dammar (the resinous sap of the sal tree) must first be correctly mixed to make the 'prepared wax'. The proposed image must first be completely visualised and realised in the mind of the *Sthapati* through contemplation until it is finally to be modelled in the 'prepared wax' according to the instructions of the chief *Sthapati* (Śloka 10 to 12).

When the wax image is completed it should be purified with five powdered pigments - (*Panchavarṇa*). Although there are two distinct kinds of *mūrti* (idols), viz., *cala* (embulatory, for processional use) and *acala* (stationary, for remaining inside the temple), both are classified as *Kṣaṇikā mūrtis* (transitory types), Hence, these should both be made with equal concentration and concern. In preparing wax models of these *mūrtis*, at the jointures of the component parts of the body they should be reinforced with copper rods or nails; and although the wax model will melt away, there should be no objection to using those supports. (*Śloka* 13 to 19).

The finished wax model of the idol should then be taken in procession to the *Sthapati's* foundry or workshop to be cast. The chief *sthapati* must first calculate what allowances should be made for the contraction or expansion of the wax model and enlarge it accordingly by adding wax to the extent necessary (*Śloka* 20 and 21).

The wax model is then ready to be covered with layers of mud and (when dry) must be heated and thereby the wax is expelled. After casting, the mould should be sprinkled with water to cool it so that it (the cast icon) may be removed after it is broken open (*Śloka* 22 and 23).

When the *Sthapati* is satisfied that the cast image is a perfect as he can make it, he should besmear it with sandal-wood paste, adorn it with flowers and, accompanied by musicians, take it out in procession so that the villagers may approve and worship it. Afterwards, the image must be taken back to the workshop of the *Sthapati* so that in case there is some slight defect or distortion in the minor component parts of the image's body, such as the limbs (*upāṅga*), they may be rectified by recasting those particular portions. But if the defect be a major one affecting, for example, the torso (*pratyaṅga*), then the whole casting will have to be condemned, a fresh wax model be made and the image cast a new. In this event, the *Sthapati* and the *Sthāpaka* must retain a firm belief in the successful outcome and show an undeviating determination, failing which the village will face destruction. These will follow loss of prestige and wreath for both the *Sthapati* and *Yajamana*, and the merits and virtue (*Puṇyam*) they have acquired through many births will be wiped away. Therefore, in each phase in the production of the image every care should be taken to carry out principles and instructions laid down in the *Śilpaśāstra.* (*Śloka* 24 to 28).

As most of the early known, *Śilpaśāstra,* describe only the technique of solid casting, it may be surmised that the solid casting technique is earlier and easier to execute than the technique of hollow casting. Owing to religious faith rather than technical reasons, all metal images were, however, made by the technique of solid cast.

The *uttarabhāga* of *Śilparatna* also dates back to the Gupta period and unlike *Madhūcchiṣṭavidhāna,* it describes both the techniques of solid casting and hollow casting. The Chapter I deals with the need for the metal craftsman to collect different kinds of clay and the method of preparing the clay for modelling a clay image for hollow casting. The Chapter II deals with the technique of casting of metal images. The text is given below :

अथ मृदः ।

तीर्थक्षेत्राचलारामपुण्यदेशोद्भवां मृदम् ॥ ४४ ॥
शुक्लां रक्तां पीतां कृष्णामादाय चूर्णयेत् ।
यवगोधूममाषाणां चूर्णानां गुग्गुलोरपि ॥ ४५ ॥

प्रथमोऽध्यायः । उत्तरभाग शिल्परत्न ।

मधूच्छिष्टेन निर्माय सकलं निष्कलं तु वा ।
बद्धवा मृदा दृढं शुष्कमधूच्छिष्टं बहिः सृजेत् ॥ ३२ ॥
तत्र लोहार्द्रमापूर्य कारयेल्लौहमुत्तमम् ।
अथ तत्प्रक्रियां वक्ष्ये कर्मकारहिताय तु ॥ ३३ ॥
कठिनां मन्दकठिना मृद्वी मृदुतरा तथा ।
मूषाकरणयोग्येति पंचधा मृत्तिका स्मृता ॥ ३४ ॥
पूर्वोक्तां नाकुजां वाथ मृत्स्नामादाय यत्नतः
मृल्लोष्टचूर्णसंयुक्तां यथायुक्ति विमर्दयेत् ॥ ३५ ॥
शुद्धाम्भसा पूगचर्मसारं युक्त्या सुयोजयेत् ।
कारयेत् कठिनामेवं शिलामुसलताडिताम् ॥ ३६ ॥
तस्मिन् गोमयसंयुक्तो स्यान्मन्दकठिना पुनः ।
मृद्भाण्डचूर्णसंयुक्तां तत्पादांशकमृत्तिकाम् ॥ ३७ ॥
पेषण्यां पेषयेद् यां सा मृद्वीति कथिता पुरा ।
तदेव गोमययुता मृत्स्ना मृदुतरा स्मृता ॥ ३८ ॥

तुषाद्यङ्गारचूर्णेन संयुक्ता घटमृत्तिका ।
कार्पासपटचूर्णेन समं मुसलपीडिता ॥ ३९ ॥
एषा मूषामृदाख्याता कर्तव्या सानया दृढम् ।
यत्किचिंदीप्सितं तन्तु किचिंन्नयूनप्रमाणतः ॥ ४० ॥
कृत्वा द्वितीयया शुष्के तष्ट्वा लिप्त्वा पुनः पुनः ।
अन्ते गोमयसारेण लिप्त्वा श्लक्ष्णीकृते पुनः ॥ ४१ ॥
कृत्वा तस्मिन् मधूच्छिष्टपट्टं युक्तधनान्वितम् ।
इष्टप्रमाणं तत् सर्वं कृत्वा तेन सलक्षणम् ॥ ४२ ॥
भूषणादीनि सर्वाणि मधूच्छिष्टस्य तन्तुना ।
कृत्वा तिन्त्रिणिसारेण वेणुसारेण वा तथा ॥ ४३ ॥
कृतेन युक्तशस्त्रेण सुस्निव्धं कारयेदथ ।
पुनर्विलेपयेन्मृत्स्नामार्द्रां मृदुतरां क्रमात् ॥ ४४ ॥
शुष्के तस्मिन् पुनर्लिप्त्वा मृदूया संशोषयेत् पुनः ।
कठिनेन पुनस्तत्र कृष्णायसपटादिभिः ॥ ४५ ॥
बद्ध्वा यथाबलं पश्चात् कठिनेन विलेपयेत् ।
आवृत्त्यालेपयेत् तेन यथातोषं यथाबलम् ॥ ४६ ॥
अप्रधानावयवत्रस्थाने स्यात्लोहपद्धतिः ।
एवं शुष्के पुनर्यन्त्रवशादातप्य वहिन्ना ॥ ४७ ॥
मधूच्छिष्टं बहिस्त्यक्त्वा तं दहेत् काष्ठवेष्टितम् ।
यदा कूश्माण्डपुष्पाभं वर्णमन्तर्गतं पुनः ॥ ४८ ॥
लोहमार्गादिरन्ध्रेण दृश्यते तत् तदैव हि ।
अवरूह्य भुवः श्वभ्रे विन्यस्योर्ध्वमुखं यथा ॥ ४९ ॥
मूषायामार्द्रितं लोहं शीघ्रमादाय यत्नतः ।
निक्षिप्य तन्मुखे मन्दमन्तर्भागं सुपूरयेत् ॥ ५० ॥
पांसुना बहिराच्छाद्य कालाच्छीतलिते पुनः ।
तीक्ष्णायसकृतैः शस्त्रैर्मृदं त्यक्त्वा सुधीः क्रमात् ॥ ५१ ॥
सुश्लक्ष्णं कारयेत् तैस्तैरूचितैः शस्त्रभेदकैः ।
एवं तत्कर्मकुशलैर्लोहजं कारयेद् बुधः ॥ ५२ ॥
धनं चेल्लोहजं बिम्बं मधुच्छिष्टेन केवलः ।
कृत्वा मृल्लेपनादीनि पूर्ववत् क्रमतश्चरेत् ॥ ५३ ॥
द्वितीयोऽध्यायः । उत्तरभागः । शिल्परत्नम् ।

The above text has been translated into English by T.K. Palaniappan.[15] The translation is quoted below:

Chapter 1. This chapter deals with the need for the craftsman to produce white, red, yellow and black sand (clay) from tanks, sacred temple waters, and from gardens. It also describes the method of preparing this clay.

Chapter 2. This chapter deals with the technique of casting as described below.

Although it is always best to make *Śiva Liṅgas,* images etc., out of wax, cover them with layers of clay, expel the wax and then fill with molten metal, in order to help and benefit the Sthapati, I also submit the following technique of casting (*Śloka* 32, 33).

The various grades of clay required to fall into the following five categories.

Very hard clay : Clay plus powdered pottery tiles and nut water.

Medium clay : The above mixture should first be dried, then pounded on a pounding stone to fine powder and mixed with dung.

Soft clay : Three parts of clay plus one part of powdered pottery, mixed together on a grinding stone (*Śloka,* 34 to 37).

Fine clay : The above mixture is again pounded on the grinding stone with an admixture of dung (*Śloka,* 38).

Crucible making clay : husks which have been burned to powder and cotton fabric pounded to powder, are mixed together in equal parts, and in turn, pounded on the grinding stone. The crucible is to be made of the above mixture and arranged by means of thin wire and thread (*Śloka* 39, 40).

Modelling the Core : An image is first modelled, but to a slightly smaller dimension, out of the 'medium clay' described in the second category. This must then be covered with the 'Fine Clay' described in the fourth category. Over this is put a layer of wax, the detailed modelling of which is done with the tools made of bamboo or tamarind wood (*Śloka,* 41 to 43).

Over this, when finished is put on application of the 'soft clay' described in the third category, into which is inserted little metal strips, or nails. An orifice is made at the bottom of the clay mould, the latter being allowed to dry and with a slight application of heat, the wax is expelled through the orifice.

The mould is heated red hot and put in position, orifice on up. Into this the molten metal is poured; flush to the top, by skilled and experienced hands (*Sloka* 44 to 48).

If the image to be cast has to be solid and heavy, then the wax model of it must be prepared out of a solid piece of wax. This is covered with layers of mud, the wax expelled and the cavity filled with molten metal. This results in a solid metal image (*Śloka* 49 to 53).

The *Śilpaśāstra* which is known as *Abhilaṣitārthacintāmaṇi*[16] provides the most important source on the art of metal casting. The *Abhilaṣitārthacintāmaṇi* is also known as the *Mānasollāsa.* The king Someshvara Bhulokamalla of the western *Cālukya* dynasty of the *Kalyāṇi* line is generally believed to be the author of this text.

According to scholars, the author had completed this *Śilpaśāstra* in 1131 A.D. There is only the description of the technique of solid casting and the process of hollow casting is not described. The author has devoted twenty-one verses for describing the method of making bronze figures by solid casting technique. S.K. Sarasvati has discussed the technique of casting by lost wax process with relevant text from *Mānasollāsa* which was also translated into English by him. (S.K. Sarasvati, "An ancient text on the casting of metal images', *Journal of the Indian Society of Oriental Art* IV (1936), p. 139 ff.).

The text of the *Mānasollāsa* is given below:

नवतालप्रमाणेन लक्षणेन समावृतम् ।
प्रतिमा कारयेत् पूर्वं उदितेन विचक्षणः ॥ ७७ ॥
सर्वावयवसम्पूर्णाम् किंचित् पीतां दृशोः प्रियाम् ।
यथोक्तैर्आयुधैर् युक्तां बाहुभिश्च यथोदितैः ॥ ७८ ॥
तत्पृष्ठे स्कन्धदेशे च कृकटयां मुकुटेऽथ वा ।
काशपुष्पनिभं दीर्घं नालकं मदनोद्भवम् ॥ ७९ ॥
स्थापयित्वा ततश्चार्यां लिम्पेत संस्कृतया मृदा ।
मर्षीं तुषमयीं घृष्ट्वा कार्पासं शतशः क्षतम् ॥ ८० ॥
लवणं चूर्णितं श्लक्ष्णं स्वल्पं संयोजयेन्मृदा ।
पेषयेत् सर्वमेकत्रे सुश्लक्ष्णे शिलातले ॥ ८१ ॥
वारत्रयं तदावर्त्य तेन लिम्पेत समन्ततः ।
अच्छस् स्यात् प्रथमो लेपः छायायां कृतशोषणः ॥ ८२ ॥
दिनद्वये व्यतीते तु द्वितीयस्स्यात् ततः पुनः ।
तस्मिन् शुष्के तृतीयस्तु निबिडो लेप इष्यते ॥ ८३ ॥
नालक्ष्य मुखं त्यक्त्वा सर्वमालेपयेन्मृदा ।
शोषयेत् तं प्रयत्नेन युक्तिभिर् बुद्धिमान् नरः ॥ ८४ ॥
सिक्थकं तोलयेदादावर्चालग्नं विचक्षणः ।
रीत्य ताम्रेण रौप्येण हेम्ना वा कारयेत तु तम् ॥ ८५ ॥
सिक्थाद् अष्ठगुणम् ताम्रम् रीतिद्रव्यम् च कल्पयेत् ।
रजतं द्वादशगुणम् हेम स्यात् शोडशोत्तरम् ॥ ८६ ॥
मृदा संवेष्टयेद् द्रव्यं यदिष्टं कनकादिकम् ।
नालिकेशकृतिं मूषां पूर्ववत् परिशोषयेत् ॥ ८७ ॥
वह्नौ प्रतापितमर्चासिक्थं निस्सारयेत् ततः ।
मूषां प्रतपयेत् पश्चात् पावकोच्छिष्टवह्निना ॥ ८८ ॥
रीतिस्ताम्रं च रसताम् नवाङ्गारैः व्रजेद् ध्रुवम् ।
तप्ताङ्गारैर्विनिक्षिप्तेः रजतम् रसताम् व्रजेत् ॥ ८९ ॥

सुवर्णम् रसताम् याति पञ्चकृत्वः प्रदीपितैः ।
मूषामूर्धनि निर्माय रन्ध्रं लोहशलाकया ॥ ९० ॥
संदंशेन दृढं धृत्वा तप्तं मूषां समुद्धरेत् ।
तप्तार्चानालकस्यास्य वर्तिं प्रज्वलिताम् न्यसेत् ॥ ९१ ॥
संदंशेन धृतां मूषां तापयित्वा प्रयत्नतः ।
रसं तु नालकस्यास्ये क्षिपेद् अच्छिन्नधारया ॥ ९२ ॥
नालकाननपर्यन्तम् संपूर्णं विरमेत् ततः ।
स्फोटयेत् तत्समीपस्थं पावकं तापशान्तये ॥ ९३ ॥
शीतलत्वं च यातायां प्रतिमायां स्वभावतः ।
स्फोटयेन्मृत्तिकां दग्धां विदग्धो लघुहस्तकः ॥ ९४ ॥
ततो द्रव्यमयी सार्था यथा मदननिर्मिता ।
जायते तादृशी साक्षाद् अङ्गोपाङ्गोपशोभिता ॥ ९५ ॥
यत्र क्वाप्यधिकं पश्येच्चारणैस्तत् प्रशान्तयेत् ।
नालकं छेदयेच्चापि पश्चादुज्ज्वलतां नयेत् ॥ ९६ ॥
अनेन विधिना सम्यग् विधायर्चां शुभे तिथौ ।
विधिवत् ताम् प्रतिष्ठाप्य पूजयेत् प्रत्यहं नृपः ॥ ९७ ॥

मानसोल्लासः अभिलषितार्थथचिन्तामणि । (श्लोक ७७-९७)

C. Sivaramamurti translated the text into English. The translation is quoted below.[17]

The image should be prepared of wax, according to the *navatāla* (nine *tāla*) measurement, light yellow in colour and beautiful to behold, with proper disposition of limbs including arms and weapons according to the iconographic texts. (*Śloka* 77-78).

Long tubes, each with a flared mouth resembling a *kāśa* flower, should be added on the back, at the shoulder, on the nape, or on the crown (79).

The figure had then to be coated lightly with clay well prepared by adding to it charred husk, tiny fits of cotton and salt, all finely ground on a stone. The coating has to be repeated three times, each after a interval of two days, and dried in the shade. The final coating is the most liberal one. The mouth and the cannel of the tube, wherever fixed, should be left free when the coating is applied (*Śloka,* 80 to 84).

The weight and quantity of brass, copper, silver or gold for preparing the image is determined by the weight of the wax in the mould; brass and copper are to be eight times the weight of wax, silver twelve and gold sixteen times (*Śloka,* 85, 86).

The chosen metal should be encased in clay to form a crucible of the shape of a large cocount, and then dried and heated in the fire. The mould is then heated to allow the wax in it to melt and run out (*Śloka,* 87, 88).

The crucible is then heated again; the heat for melting copper, silver and gold in progressive order - mild embers, flaming embers, and blazing embers to five times the normal heat. A hole made with an iron rod in the crucible, which is to be lifted up from the fire with the help of iron tongs, the molten metal is poured into the mouth left open on the mould, after cleaning them with a burning wick so that it would run in a hot stream to completely fill up the entire space within, and reach upto the mouth of the tubes (*Śloka,* 89 to 93).

The fire is extinguished and the mould allowed to cool down, when the burnt earth composing the mould is carefully broken to reveal the image as originally prepared in wax. The tubes and other superfluous projections are now cut and the image finished and furnished (*Śloka,* 94 to 96).

After making an image by this method, the king should install it on an auspicious day according to the usual rites and should daily worship it (*Śloka,* 97).

Numerous texts of the Mediaeval period (7th century to 13th century A.D.) have recorded traditions and techniques of metal casting in India.

The Jain *Anuyogadvāra cūrṇi*[18] describes the technique of hollow casting. Both solid and hollow casting techniques by the lost-wax process are described in the *Viṣṇudharmottara Purāṇa.*[19]

The *Viṣṇu Saṁhitā*[20] gives a reference regarding the image making. A verse states that if an image is to be made of metal, an image must be made in wax and then coated with clay. The text is given below:

लोहे सिक्थमयीमर्चांकारयित्वा मृदावृताम् ।
सुवर्णादीनि संशोध्य विद्राव्याङ्गारवत् पुनः ॥ ६५ ॥
चतुदर्श पटलः । विष्णुसंहिता ।

The *Āgama* literature of the *Pāñcarātra* branch of the *Srīvaiṣṇava* religion is not only vast but also has an interesting past. The *Pāñcarātrāgama* literature are primarily bhakti scriptures and devoted to the study of Vaiṣṇavism. The number of the *Pāñcarātra* scriptures (called variously *Āgama, Saṁhitā* or *Tantra*) is given as 108. They are dated earlier to 5th/6th century A.D. The *Ahirbudhnya Saṁhitā* is one of the oldest *Pāñcarātra* texts. The *Pāñcarātra* scriptures deals with the five-fold manifestation of Lord Vasudeva viz., *Para, Vyūha, Vibhara, Antaryamin* and *Arcā.*

The *Pāñcarātrāgama* literature provides a few passages relating to the famous Indian 'lost-wax' method of casting metal icons. The two quoted below - one from *Vasiṣṭha-Saṁhitā* and the other from the *Padma-Saṁhitā* which is the most popular text among the *Saṁhitā* of the *Pāñcarātrāgama* - concern with materials to be employed and methodology of casting metal icons under the title *Lohasaṁgraha.*

In the *Vasiṣṭha-Saṁhitā* (137) gold, silver and copper are given as options with regard to acceptable metals to be sought depending upon the guiding factors in their section including financial feasibility, purity of the ore available and caste-affiliation of the person financing the enterprise. In these options, copper is favoured above other metals or alloys probably by reason of its economy and workability. Metal icons are considered the best of all other kinds of icons. The text is given below:

१३७ ... लोहं तु त्रिविधं प्रोक्तं तन्त्रेऽस्मिन् प्रतिमाविधौ ।
सुवर्णं रजतं ताम्रं यथा वित्तानुसारतः ।
सुवर्णं द्विविधं प्रोक्तं जातरूपं च हाटकम् ॥
जातरूप प्रधानं स्यात् तद्लाभे तु हाटकम् ।
रूप्यं च द्विविधं प्रोक्तं रूप्यं रजतमेव च ॥
रूप्यमत्र प्रशस्तं स्यात्तद्लाभे तु राजतम् ।
ताम्रं च द्विविधं प्रोक्तं ताम्रमौदुम्बरं तथा ॥
ताम्रमत्र प्रशस्तं स्यादौदुम्बरमलाभतः ।
अत्र च द्विविधं प्रोक्तं शुद्धं मिश्रितमेव च ॥
शुद्धमेतत् प्रशस्तं स्यात् मिश्रणं तु विवर्जयेत् ।
अर्धचित्रं तथालेख्यं लोहैः कुर्यात् कदाचन ॥
सुवर्णं रजतं ताम्रं ब्राह्मणानां विशेषतः ।
सुवर्णं रजतं चैव क्षत्रियस्य विशेषतः ॥
रजतं चापि ताम्रं च वैश्यस्य च विशेषतः ।
ताम्रमेव तु शूद्राणां सर्वेषां ताम्रमेव वा ॥
लोहजाः प्रतिमाः सर्वा निर्दोषास्समुदाहृताः

Vasiṣṭha IV: 836-910.

In the *Padma Saṁhitā* 138, methodology of casting image has been outlined. Bees wax (*Madhūcchiṣṭa*) is softened and from it an icon is modelled. During the process of modelling of wax image, the ācārya - director attends to certain visual. Ghee is applied to the wax figure and measurements are verified. The metals to be used in the casting are melted. At this stage, it is still possible to attend to certain alterations in the wax figure. Once the wax model is packed in plaster and the bees wax melted away, additions and alterations are not possible. At this stage, the molten metal is poured into the cast. After a suitable waiting time, the plaster is chipped away. The rough-metal image is ready at this stage. The details are articulated at the finishing stage. After details are rendered, the icon is made ready for use. It is also pointed out that the pedestal (i.e., *āsana*— base) of the icon as well as its encircling *prabhā*— arc must also be made of the same kind of metal. The text is given below:

१३८ ...... बिम्बं प्रभामासनं वा लोहजं परिकल्पयेत् ॥
द्रावयित्वा मधूच्छिष्टं प्रतिमां तेन कल्पयेत् ।
धान्यराशि विनिक्षिप्य तद्मध्ये तिलतण्डुलम् ॥

चक्रजं तु लिखेत्तस्मिन् प्रागग्रात्रिक्षिपेत्कुशन् ।
वस्त्रं च नूतनं तस्मिन् विस्तीर्य प्रतिमां क्षिपेत् ॥
प्राचीनशिरसं पश्चाज्जुहुयादनले गुरूः ।
पुष्पै समिद्भिलोजैश्च सपिषाष्टोत्तरं शतम् ॥
जुहुयान्मूलमन्त्रेण नृसूक्तेन चरूः पुनः ।
हुत्वैवं विधिनाचर्यः स्त्रव्वी कनकभूषणः ॥
संपाताज्येन शंसिस्चेत्प्रतिमां तत्वपद्धतिम् ।
ध्ययेत्तत्र शुभेलग्ने पूजयेन्मूलविद्यया ॥
यजमानो गुरूं पूज्य बहुमिस्तोषयेद्घनैः ।
ब्राह्मणान् विदुषो वितैर्बहुभिः परितोषयेत् ॥
स्थपतिः प्रतिमां कुर्याद्रथकाराभ्यनुराया ।
मानोन्मानप्रमाणैस्तां प्रतिमां कारयेच्शुभाम् ॥
मृदालेपं ततः कुर्यात्शिल्पशास्त्रनुसारतः ।
सुवर्णं रजतं ताम्रं विशुद्धं लोहमुत्तमम् ॥
आदाय धान्यराश्यां तु पूर्ववन्न्यासनम् ।
पूर्ववच्चैव गुह्यादाचार्यश्रोदितक्रमात् ॥
द्रावयित्वा ततः शुद्धं लोहं न्यायसमाजितम् ।
स्थपतिः प्रतिमां तेन कुर्याच्छास्त्रोक्त वर्त्मनि ।
लोहजां प्रतिमां चैवं कल्पयेत्परमात्मनः ।
स याति परमं स्थानं पुनरावृत्तिवर्जितम् ॥
प्रतिमादस्तपादादिवैकल्य शान्तिमाचरेत् ।
आचार्यो देवहोमाद्यैस्ततः शान्तिर्भविष्यति ॥
समाद्धयादवयवान् विकलान् स्थपतिः पुनः ।
पदमसंहिता । क्रियापाद् । अध्याय १९ श्लोका - ३३६-४६ अ ।

(Smith, Daniel H. and Venkatachari, K.K., 1969, *A Source Book of Vaiṣṇava Iconography* according to *Pāñcarātrāgama* texts, Pāñcarātra Pariśodhna Pariṣad, 93, T.P. Koil Street, Triplicane, Madras - 5, pp. 45-47).

The techniques of making metallic images have been mentioned in other later texts, such, for example, as the *Samarāṅgaṇa Sūtradhāra,* and thc *Yuktikalpataru.* The mediaeval Jain texts such, for example, as the *Ācāra-Dindkara* (14th century A.D.) provides a long list of materials to be employed for making images. The list of materials recommends the preparation of image of gold, silver or copper; but disallows the preparation of images of bronze (*kaṁsya*), lead (*sisa*) or tin (*vaṅga).*[21]

Vasunandi (c. 12 century A.D.) in his *Śrāvakācāra* recommends gold, silver, brass as materials along with other materials for use in making images of *Tīrthaṅkara* and *ācārya.*[22]

A Digambara writer, *Āśādhara* (12th century A.D.) recommends the use of gold, silver, brass or bronze for image making along with other materials like stones and wood.[23]

There are Sanskrit literary references regarding the chemistry of metals.

The *Rasaratnākara*[24] of Nagarjuna (c. 8th century A.D.) mentions that copper could be converted into brass (which was regarded at that time as artificially prepared gold) by heating it with calamine (a zinc ore) and carbonaceous matter. Brass prepared from copper and calamine (zinc ore) was mistaken for gold. The text is given below:

गन्धकशुद्धिः ।
क्रमेण भूत्वा तुरगेण रंजितः ।
करौति शुल्वं त्रिपुटेन कांचनम् ॥

The English translation of the above text is given below:

What wonder is it that, calamine .... roasted thrice with copper converts the later into gold? Calamine, a zinc mineral, when roasted with copper in the presence of reducing organic matter is likely to give rise to brass, which possibly passed as artificially prepared gold.

The method of preparation of brass, as mentioned in the *Rasaratnākara* has been repeated in *Rasārṇavatantra* (12th century A.D.) which also mentions the method of extraction of zinc from calamine (zinc ore) and colouring of copper by means of calamine to turn it into gold.[25] The text is given below:

मृत्तिका गुडपाषाण मेदतो रसकस्मिधा ॥
किमत्र चित्रं रसको रसेन भावितः ।
क्रमेण भूत्वा तुरगेण रञ्जितः ।
करोति शुल्वं त्रिपुटेन कांचनम् ॥

The translation of the above text by a scholar is given below:

Rasaka (Calamine): There are three kinds of it, namely, of the colour of earth, of the appearance of treacle and of the colour of stones, what wonder is that *Rasaka* mixed with (certain organic matters) and roasted three times with copper converts the latter into gold.

The *Rasendracūḍāmaṇī* of Somadeva (12th century A.D.) mentions a process for making a brass like alloy of copper, tin, zinc, and lead.[26] This most probably refers to the alloy known as 'lead-brass'. This literary evidence shows that during the period between 9th century and 10th century or early part of the 11th century A.D., Indian chemists (Alchemists) were engaged in the production of an alloy which is composed of copper, tin and zinc by roasting 'bronze alloy' with 'calamine' (zinc ore). "Zinc-bronze" should be the scientific name of such an alloy. This 'zinc-bronze' alloy may be regarded as the transitional stage between the bronze-alloy and the real brass-alloy.

The fact that the alloy brass is an alloy of copper and zinc is known from the statement in Bhāvamishra's *Bhavaprakāśa* that the brass alloy is a semi-metal (*upadhātu*) of zinc and copper; because, it is derived from those two metals. The author of the famous work *Bhāvaprakāśa* is the renowned Bhavamishra who lived about the time of the emperor Akbar. Bhavamishra devoted one complete chapter on the subject of *Dhātu-Upadhātu* (i.e., metal and metal alloys) in the first part of his treatise *Bhāvaprakāśa.* He states that there are seven *upadhātu* and included bronze and brass amongst them. The text is given below:

सद्योपधातवः स्वर्णमक्षिकं तारमक्षिकम् ।
तुत्थं कांस्यं च रीतिश्च सिन्दूरं च शिलाजतु ॥ ५१ ॥

Bhavamishra mentions several terms synonymous with brass, and two different types of brass referred to in the couplet quoted below:

पित्तलं त्वारकूटं स्यादारो रीतिश्च कथ्यते ।
राजरीतिब्रह्मरीतिः कपिला पिङ्गलापि च ॥ ६९ ॥

In the first line of the above couplet, Bhavamishra shows that brass has four Sanskrit names, viz., *Pittala, Ārkuta, Āra* and *Rīti.* In the same line, the author clarifies that when the heated alloy in the crucible turns red in colour, the brass is known as *'Rājarīti'*; and when it assumes a yellow colour, the metal is known as 'Brahmarīti'.

In the investigation of non-ferrous metal technology of Gujarat,[27] it has been found that the medieval period in Gujarat witnessed a gradual change in the use of non-ferrous metalic alloys from bronze to zinc-bronze, then to lead-brass alloy and finally to the stage of real brass-alloy. Lead-brass alloy is conspicuous by the presence of lead in acknowledgeable quantity along with metals like copper, tin and zinc. In lead-brass alloy, lead occurs hardly more than 13 per cent, and zinc occurs hardly more than 20 per cent, while tin occurs hardly 2 per cent. Towards the middle of the 11th century metal working communities of Gujarat were able to master the extraction of metallic zinc from the calamine ore (zinc ore); and they were successful in evolving and producing an alloy derived from a combination of tin, copper, zinc and lead metals. This fact is evident from the text of the *'Rasendracūḍāmaṇī'* of Somadeva (12th century A.D.) as stated above. The metal working communities were, however, do not know, how to make real brass alloy viz., an alloy of metallic copper and metallic zinc only. The literary evidence of real brass technology is recorded in the Bhavamishra's *Bhāvaprakāśa.*[28] In the first line of the couplet No. 70, Bhavamishra explains the properties of brass alloy and categorically states that this metal is only an alloy of copper and zinc. The text is quoted below:

रीतिरप्युपधातुः स्यात्ताम्रस्य यशदस्य च ।
पित्तलस्य गुणा ज्ञेयाः स्वयोनिसदृशा जनैः ॥
संयोगजप्रभावेण तस्याप्यन्ये गुणाः स्मृताः ॥ ७० ॥

These references show very clearly that real brass was known to be an alloy of zinc and copper as early as the 15th century A.D. and during that period this brass was a popular alloy which was commercially produced.

Even in China, the metallic zinc was unknown in the 14th century A.D. It was known in China later than India. Probably, in 16th century A.D., the metallic zinc was known to China. According to Browne (1961, S. 576) several blocks of metallic zinc (95% Zn) have been found in the Chinese Province of Kwantung bearing Chinese letters which indicated that this zinc was manufactured in 1585. It may be surmised that metallic zinc was available in China towards the later part of the 16th century A.D. As such, the real brass technology must have developed later on.

In Europe, on the other hand, the metallic zinc was, however, not known before the second part of the 18th century A.D. The Romans made considerable brass from metallic copper and zinc oxide which is an ore of zinc. The metallic zinc was, however, not known to them at that time. In the later years of the republic, brass was made from metallic copper and calamine (zinc carbonate) which is another ore of zinc. Metallic zinc was, however, not known at that time also. Until the 19th century A.D., calamine brass was made.[29]

The preparation of brass from metallic copper and metallic zinc was unknown to Europe before 18th century A.D. Brass alloy was, however, prepared by the fusion of metallic zinc and copper in a very small scale in the late 18th century A.D.[30] and until 19th century A.D., calamine brass was prepared.[31] The real brass technology developed in Europe after 19th century A.D. In India, on the other hand, particularly in Gujarat, metallic zinc was extracted from zinc ores as early as 8th-9th century A.D. and the lead-brass technology was known as early as 10th-11th century A.D. The real brass technology was known as early as 15th century A.D. Lead-brass alloy was used right upto the 14th century A.D. and with the progress of time with the improvement of furnace design, acquisition of greater skill in the use of metallurgical techniques, and realisation of the composition of real brass as an alloy of copper and zinc only, the importance of lead was minimised to the extent of negligible quantity and the real brass technology developed. The advancement in real brass metallurgy was accompanied by the absence of lead or by the presence of negligible quantity of lead.

The *Ambā-mātā* area of the North Gujarat is rich in lead-zinc-copper ores and this region extends over a length of 2 kms. towards the north-westernly direction from the outskirts of Ambaji village (24° 20 ' : 72° 51'), falling on the northern side of the Ambaji-Abu road.

Majumdar[32] mentions that large ships loaded with copper were regularly sent from Broach, the ancient port of Gujarat, to the Arabian and Persian coast. Gujarat had become a prominent international trade centre owing to her wealth and her position at the North-Eastern corner of the Arabian sea since olden times. The gulf of Cambay was an ideal approach to river ports in the age of small sailing ships.

Ancient historical literatures like *'Puratan Prabhandha Sangraha'* (1030 A.D.), *'Vicharashreni'* (1137 A.D.) *'Tirthamala'* (1447 A.D.), have made interesting references concerning the occurrence, mining and smelting of copper and other base-metals in the particular region of Ambā-mātā, around Ambaji in Banaskantha District of North Gujarat.

There are inscribed images of Tīrthaṅkara in the temples situated in Kumbharia—a place which is nearly two km. east of Ambaji. Kumbharia was a famous Jain pilgrimage centre in the medieval period of Gujarat. From the inscriptions written on the images, which are worshipped in the temples devoted to the *Shantinatha* and *Mahavira*, it is learnt that temple building activities began in this area towards 11th century A.D. and continued upto 12th/13th centuries. During the temple building activities, metal industries developed in Kumbharia for the extraction and purification of metallic ores. Kumbharia is situated near the mountain known as *Arasur* which is full of mines of non-ferrous metals. Metal workers of Kumbharia exploited mines of the Arasur mountain. These facts are mentioned in the old Sanskrit treatise entitled *Puratan Prabhandha Sangraha* (1030 A.D.) and the *Upadesh Saptati* (A.D. 1477).

Thus, we find from the Sanskrit literary references considerable amount of evidences regarding metals, metallic objects of art and non-ferrous metal technology of India from the Vedic period to the post-medieval period of Indian History. The literary references suggest continuous activities and progress of metallurgists and chemists of Ancient India.

REFERENCES

1. Coomaraswamy, A.K., 1927, *History of Indian and Indonesian Art*, London, p. 8; Majumdar, R.C. and Pusalkar, A.D. (editors), *The Vedic Age* (History and Culture of the Indian People, vol. 1), Bombay 1951, p. 397.
2. Ram Gopal, 1959. *India of Vedic Kaplasutras*, Delhi, p. 136.
3. Panini's *Aṣṭādhyāyī*, 2nd edition, Varanasi, 1963, p. 234.
4. Mehta, R.L., 1939. *Pre-Buddhist India*; Bombay, p. 199. And see Cowell, F.B. (Editor), *Jatakas*, I, p. 343; III, p. 93; IV, p. 105; V, p. 282.
5. Sikdar, J.C., 1964. *Studies in the Bhagavati Sutra*, Muzaffarpur, p. 268; Sikdar, J.C., 1947. *Jaina canon*, Bombay, p. 187.
6. *Mahābhārata*, Poona critical edition, Āśvamedhika Parva XIV. 18.8.
7. Agrawala, P.K., 1977. *Early Indian Bronze*, p. 135.

8. *Ibid.*, p. 136.
9. Rao, G., 1914. *Elements of Hindu Iconography*, vol. I, p. 50.
10. *Ibid.*
11. Sivaramamurti, C., 1963. *South Indian Bronzes*, p. 14.
12. *Nāyādhammakahāo* (N.V. Vaidya's ed., Poona, 1940), 13, p. 142.
13. *Mānasāra* (Madhūcchistavidhāna).
14. Translation by T.K. Palaniappan in Reeves, R., *Perdue Casting in India*, 1962, pp. 29-31.
15. *Ibid.*, pp. 15-16.
16. *Manasollasa* (Abhilaṣitārthacintāmaṇi, verses 77 to 97). Also, see R. Shamasastry (Editor), *Abhilaṣitārthacintāmani* of Someshvara Deva, Part I, *Prakarana* 1-3, Mysore 1926. And G.K. Shrigondekar (Editor); *Manasollasa*, 2 vols. Baroda, 1939, *Prakarana* 1, W. 77-97.
17. Sivaramamurti, C., 1963. *South Indian Bronzes*, pp. 16-17.
18. *Aṇuyogadvāra Cūrṇi* with Vritti, Ratlam, 1928, pp. 7-8.
19. *Viṣṇudharmottara Purāṇa*, 111, 43-44.
20. *Viṣṇu Saṁhitā* (Chap. XIV).
21. *Acāra Dinakara* II, p. 142 ff.
22. Vasunandi's *Śrāvakācāra*, V. 390, p. 123.
23. Jinayajñakalpa, quoted in *Jaina Siddhanta* 6, *Bhaskara*, II, No. 1, 1935, p. 12.
24. Ray, P. (editor), 1958. *History of Chemistry of Ancient and Medieval India*, Calcutta, p. 129.
25. *Ibid.*, pp. 138-39.
26. *Ibid.*, p. 152.
27. Swarnakamal, 1980. *Studies in Metallic art and technology of Gujarat*, Baroda, pp. 166-169.
28. Bhavamishra's *Bhāvaprakāśa Pūrva Khaṇda*; translated by Shri Girija Shankar Maya Shankar Shastri, 55, Published by Sastu Sahitya Vardhaka Karyalaya, Ahmedabad and Bombay, pp. 409 and 112.
29. Copper Development Association, 1934. *Copper through the ages*; C.D.A. Publication No. 3, p. 35 and p. 38. Copper Development Association, 1968, *Copper, brass, bronze, wrought mill products*; *tolerances* Part I, Handbook.
30. Bridge Port Brass Co., 1920. *Seven centuries of brass making*, Bridgeport, p. 15.
31. Gowland, W., 1912. Copper and its alloys in early times. *The Journal of the Institute of Metals*, VII, p. 44; Michael, H.V. and Asaro, F., 'Chemical study of the Plate of Brass", *Archaeometry*, vol. 21, Part I, February, 1979, pp. 3-14.
32. Majumdar, R.C., 1970. Tales of Maritime Adventures, *Sunday Bharat Jyoti*, April 12, Bombay.

Iron in Early Buddhist Texts

PRIYA SRIVASTAVA AND VIBHA TRIPATHI

Ancient Indian literature is a mine of information on Social, Cultural and Economic life. Early Buddhist texts are a reflection of life during early historic period, particularly in middle and parts of lower Ganga Valley. An attempt is being made here to examine these texts for data on growth of iron technology in ancient India, the variety of objects in use and the degree of its usage in vogue in day-to-day life as well as the level of metallurgical skill achieved by smiths.

The early Buddhist texts are datable to 5th-4th cent. B.C. The earliest text is said to be composed immediately after 'Mahaparinibban' of Buddha in the beginning of the 5th cent. B.C. in the first Buddhist Council.[1] It is claimed that this council recorded spiritual preachings of Lord Buddha who had delivered his sermons at several places. The subsequent two Councils (second[2] and third[3] Buddhist Councils) were held at interval of hundred years each during 4th-3rd cent. B.C. respectively. The discussions held during these Councils are said to be recorded in the 'Tripitakas'.

Present paper is based on material contained in the Tripitaka and therefore, should represent the socio-economic and cultural material pertaining to 5th-3rd cent. B.C. with a provision for inclusion of earlier material and practices continuing upto 5th cent. B.C. The three original early texts of Buddhism (Tripitaka) are Sutta, Vinay and Abhidhammapitaka. All the three of them have sections and subsections; it may briefly be divided thus - Suttapitaka has five sections or Nikayas known as Dighikāya-, Majjhim, Saṁyukta, Anguttar and Khuddak Nikāya. Khuddak Nikāya has a group of 15 texts in it viz., Khuddak pāth, Dhammapada, Suttanipāta, Udān, Etiuttak, Vimānvatthu, Pettvatthu, Thergāthā - Therigāthā, Jātaka, Niddes, Apdanpāli, Pattisambhidāmagga, Buddhvamsa and Chariyāpitaka. Vinaypitaka has well defined three sections i.e., Suttavibhanga, Kandak (Cullavagga and Mahavagga) and Parivāra. The third Abhidhamapitaka has been devided in 7 sections i.e. Dhamasamgaṇi Vibhaṅg, Kathāvatthu, Dhātukathā, Puggalapaññayatti, Yamak and Paṭṭhān etc. These texts provide a rich information of the socio-economic condition of contemporary society.

A close study of these texts will lead to a proper assessment of level of technological advancement and whether it had any role to play in life. A detailed itinerary of types of iron objects prevalent in contemporary society, is being made which is followed by examination of techniques of iron smelting and forging and manufacturing of objects. We can also gain an insight in tools, implements and actual workshops of iron workers as well as their socio-economic status.

Tool Types

It is of foremost importance to list the variety of objects of iron mentioned in the early Buddhist texts.

Objects of Commerce	*Weapons*	*Ornaments & Toiletteries*	*Carpentry/ masonary implements*	*Agricultural implements*	*Household object*	*Building material*
Weight	Arrow	Iron ear-ring	Hammer	Plough share	Dish	Chain
Weighing Scale	Sword	Gold plated ear-ring	Chisel		Dias (Iron stool or Bench)	Rod
Coin of iron	Spear-head	Nail-Parer	Axe	Showel	Couldron	Nail
	Trident		Shears	Hoe	Basin	Lock
	Macehead		Adze	Spade	Jar	Key
	Shield and Helmet			Trowel	Vessels	
	Chariot			Hatchet	Pincers	
	Goad			Sickle	Backing-sheet Knife Scissors Niddle Probe	

The above table shows clearly that iron objects were commonly being used in early phase of life by 5th-4th cent. B.C. The frequency of these references leads to the assumption that they were not monopolised by a social strata but was a commodity of common man used in day-to-day life.

The Pali texts suggest a profuse use of iron in structural activity. In 'Samyukta Nikaya' a reference is made to a huge cello at Sravasti which measured one yojana in length, one yojana in width and one yojana in height. In contained mustard seeds-

> "Āysaṁ Nagaraṁ yojaṁ vitthāreṇa, yojaṁ Ubbedhena Puṇṇa sāspān guḷikābaddhaṁ"

There are other reference to structures of huge dimensions like fortified city of Varanasi having iron gates/towers 'Prakara' around the city -

> "Ayopākara Pariyanto Ayasā Patikujjito. Tassa ayomayā bhumi Jalitā tejasāyutā samantā yojanaṁ sataṁ karitavā tiṭṭhati sabbadā.[5]

The use of iron in fortification is indicated by several other references in Pali texts. The 'Nāgrāj Jataka' describes a fortification having iron rods and deep moats all around in Karambiyapattan.[6]

> "Assa Nagaraṁ Mahantaṁ, attharakha aysaṁ bhaddasālam. Sāmanta khāta parikhā upeteṁ avaṁ pi me to edha guhyamantā."

Similarly, a high walled palace with profuse use of iron has been mentioned in connection with Varanasi.[7]

The excessive height did not permit the sunrays to enter the fortified enclosure, the text says.[8]

Evidence of Mining and Metallurgy

Although the data on mineral deposits in Pali texts is not exhaustive, there is sufficient evidence to deduct that most of the metals were known during that period in their mineral forms.

"Abhidhammpitaka" recognises the presence of various mineral deposits deep down the surface of the earth. In precise expression it alludes the hardness of the earth to those deposits.[9] According to 'Digha Nikaya' and 'Jaradpan Jataka'[10] the region lying between Sravasti and Varanasi[11] was rich in mineral deposits and was tapped for several valuable minerals. Although modern geological records don't indicate presence of ores in this area but we have to bear in mind that the modern sources keep a record only of economically viable mineral deposits in the present context.

A well developed metal industry at several big centres is attested to by literary texts, viz. a 'Lohargrama' or 'Kammargama' in Magadh (South Bihar).[12] It may safely be argued that these smelting—forging industries had a regular supply of raw material. Mineral deposits must have been located in a manageable proximity. This automatically leads to the conclusion that mineral deposits whether small or big were conveniently located in North India and were being exploited by the smiths of 6th-5th cent. B.C.

Technology of iron working has been described, though incidentally in Pali texts. The furnace used in iron smelting is said to be charged for long hours, at times for the whole day long and the temperature in them was sufficiently high. A provision for draught was made through bellows. The bellows were operated continuously. The operation produced a strong (high) sound which has been compared with regular breathing of a living being -

"Kammār gagganyā dhammānay saddohoti."

This exercise was concluded after a long and tedious working resulting into a pure glowing lump of iron. 'Digh-Nikaya'[13] and 'Samyukta Nikaya' discuss at length how iron bloom was being produced as a result of whole day working. The bloom had impurities in a red hot condition with high heat and gaseous concomitants. It was beaten up, cooled down and then weighed again several times. The weight of bloom thus got reduced after repeated beating and heatings in the forge. Iron produced in this fashion is described to be lighter, softer and more elastic -

> "Puriso diwsaṁ santatto ayoguḷo tejo sahgato hoti vāyo sahgato vā āditto sampajjalito sajotibhuto tadā lahuttaro cā hoti mṛiduttaro cā kammaññattaro cā. Yadāpan so ayoguḷo nevaṁ tejo sahgato nev vāyo sahgato sito nibhuto tadā gurutaro ca hoti patithañtaro ca akammaññattaro ca ti. Puriso divasaṁ santtaṁ ayogulaṁ addittaṁ sammpajjalitaṁ sajotibhutaṁ tulayāṁ tuleyya tamenaṁ aparen samayen sitaṁ bibbuta tulaya tuleyya. Yadā vā aditto sampajjalito sajotibhuto, tamen ayoguḷo lahuttaro ca hoti mṛiduttaro ca kammaññattaro ca. Yadāpana so ayoguḷo sito nibbuto ti tamen guruttaro ca hoti patithantaro ca akammaññattaro ca ti."[14]

Roughly translated the passage means,

'The bloom of iron becomes light, malleable and shining when it stays in direct contact with fire and air otherwise it turns heavy and blackish. The iron worker beats the bloom then weighs it again after heating then only it acquires the high quality of lightness, softness and smoothness, not otherwise.'

This appears to be referring to production of wrought iron which imbibed some carbon too because of repeated contact with charcoal. The knowledge of quenching has been referred to in Pali texts. The 'Samyutta Nikaya', 'Suttanipata', 'Samkicca Jataka' and 'Mahavagga' elaborately describe the method of production of agricultural implement. The plough-share were heated for whole day and then immersed in cold water producing a hissing sound.

"Phaledivasaṁ santatte udake pakkhipatto citcitāyati sandhūpāyati."

It is a clear reference to quenching. The other category of iron objects were quenched or made steely for better effectiveness. 'Nikayas'[15] explains that an iron bowl 'Ayokapāla' (i) (Helmet or an object shaped like skull) and pan were being heated for the whole day. The pan (ii) was being douched with cold water intermittently. The water thus sprinkled used to evaporate immediately.

(i) "Diwasaṁ santatte ayokapāle haññamāne papatikā uppātitavā nibbāteyya."

(ii) "Balwā puriso santatte ayokatāhedwe vā tinikā udake usitāni nipāteyya dandho."

This also hints at another method of quenching being practised. Perhaps in case of bigger objects water was sprinkled or poured over them than dipping them in pot or water bowl.

'Dhammapada' mentioned that special treatment was given in manufacturing of edges of arrow heads which were subjected to excessive heat and then beaten up and shaped into a very straight and sharp edged object.[16]

> "Ujuṁ karoti usukāro dvisu vā tejanaṁ."
> (An expert arrow maker heats and shapes the arrow is a way that it became straight and sharp)

It might be suggestive of case hardening. It is mentioned in 'Samyukta Nikaya' in connection with knives, nail parer, jeveline, rods, nails and spearheads.[17]

(a) "Tiṇheṇa khuren āditten sampajjaliten sajotibhuten."
(a knife starts shining like glowing sun)

(b) "Tatten nakhaccheden aditten sampajjaliten sajotibhutāya."
(a nail parer with a sharp glowing edge)

(c) "Tiṇhāya sattiyā āddittāya sampajjalitāya sajotibhutāya."
(a weapon having sharpness and glistening qualities of glowing flame)

(d) "Tattāya ayosalākāya addittāya sampajjalitāya sajotibhutāya."
(a sharp needle glistening like sun)

(e) "Tattan ayokhilaṁ."
[a nail of iron which was heated up (to make)].

Hammer (of iron or beating iron) have been mentioned in 'Udanpali.[18]

"Ayoghanen hatasseva jalito jātavedaso."
(The spark resulting from a beating iron hammer dies promptly).

The master artisan supervised the working. This is quite significant as in absence of measuring instruments, the experience accummulated over years of working made them indispensible to flawless production of objects.

Types of Iron

In addition to the above techniques, and terms in connection with iron, we came across other terms like 'Sikkayas' which was considered to be an exceptionally high quality iron having a whitish hue above it. According to 'Mahāumagga Jataka' this was being fashioned at Kampilya for manufacturing weapon grade objects. 'Sikkayas' was resistant to corrosion. This was prepared with special technique removing all possible impurities from the matrix of the bloom imparting it a shine comparable to sun rays. According to this reference, the King of Kampilya used weapons manufactured with 'Sikkayas' against the King of Videh and his victory was assured because of the superior quality of his weapons. The other qualities associated with this kind of 'Ayas' in the text are that its hue was like a glistening fish; yet another adjective compares it with shining star.[19]

"Veḷḷāḷino vitamalā *sikkāyasa* daḷahā. *Pāṭhinavaṇṇa* (fish) nettiṅsi teladhotāyo pabhassarā."

It may be decided on the basis of the above that iron was not only largely being used in daily life during the above period but also different types of iron was being produced for diverse usages.

Methods of preservation of iron objects against corrosion have also been suggested in Pali texts. The 'Mahāummagga Jataka' describes the use of oil for preservation of 'Sikkayas' weapons like swords, jevelines etc. against corrosion.[20]

It is well known even today that oil is the most effective agent against corrosion. 'Cullavagga' also explains several modes for safe keeping of small object like iron needles. The lord has suggested that they should be stored in tubular containers. If the tubular container is not effective enough in warding off corrosion to insulate it further, yeast, barley meal, powdered stone, beewax and gum could be used,[21] as is clear from the passage below -

"Sūciyo kaṇṇakitāyo honti anujanāmi .... sūcināḷikaṁ ti.
Sūcinaḷike pi kaṇṇakitāyo honti ... kiṇṇen puretuṁ ti.
Kiṇṇen pi kaṇṇakitāyo honti ... sattuā puretuṁ ti.
Sattuā pi kaṇṇakitāyo honti ... saritakaṁ ti.
Saritake pi kaṇṇakitāyo honti .....Madhusitthakeo sāretuṁ ti.
Saritake pi saritakeṁ paribhijjati saritakasipātikaṁ ti."

It is evident, from the above passages that not only complex techniques of iron manufacturing was known to the contemporary society, even the preservation and warding off corrosion was well understood. Techniques of quenching and tempering were known and fully being practised by the smiths. 'Suci Jataka'[22] mentions of 'Lohargram' or village for the smiths, leads to the assumption that large number of families were engaged in the work and it had assumed the dimension of fully developed profession. The entire class of these artisans occupied territories of villages earmarked for them.

"Etodāni pātayanti sūciko balisāni ca.
Sūcim kammargāmasamim bikketabbā pajānatā.
Ācāriyāva jānanti kāmmam sukat dukkat."

The archaeological data in this context is not very rich and, therefore, it has led to the natural deduction of a scarcity of iron in the society of 5th-6th cent. B.C. We frequently debate the impact of iron in socio-economic milieu. Without entering into the controversy here, suffice is to say that iron was playing a key role in the life of common people. Not only a large variety of weapons with different degrees of strength and sharpness were being used for waging wars but carpenters and stone workers had iron implements capable of fashioning better commodities. Agricultural implements even if not attested to by excavations (for a variety of reasons) were used frequently. All this speaks for the effluence in contemporary society that is reflected through a study of these Buddhist texts.

REFERENCES

1. Kashyap, Jagdish (General Editor), 1937. *Cullavagga, Vinaypitaka,* Vol. II, Pt. I, p. 411, Nalanda, Bihar.
2. *Ibid.*
3. Kausalyayan, Anand, 1908. *Deepvarisa,* Chap. 30, *Mahavarisa,* Chap. 20.
4. Kashyap, J., 1959. *Samyutta Nikaya,* Vol. II, p. 157, Nalanda, Bihar.
5. *Ibid.*, 1937, *Jatak Pali,* Khuddaka Nikaya, Vol. III, p. 66. Nalanda, Bihar (N.B.)
6. *Ibid.*, p. 391.
7. Kashyap, J., 1959, N.B. *Cariyapitaka, Khuddaka Nikaya,* Vol. VII, p. 66.
8. *Ibid.*, Vibhangpali, Abhidhammapitaka, p. 62.
9. Kausalyayan, Anand, 1941. Jataka Atthakatha, Hindi Translation, p. 268. Hindi Sahitya Sammelan, Prayag.
10. Kashyap, J., 1958. N.B. *Digha Nikaya,* Vol. II, p. 260-61.
11. Kashyap, J., 1932. *Jataka,* Pt. I, p. 138.
12. Kashyap, J., 1959. *Samyutta Nikaya,* Pt. I, p. 106.
13. Kashyap, J., 1958. *Digha Nikaya,* Vol. II, p. 249.
14. Kashyap, J., *Samyutta Nikaya,* p. 168, *Sutta Nipata, Khuddak Nikaya,* Vol. I, p. 282, *Jataka,* Pt. I, p. 601, *Mahavagga,* Vinaypitaka, p. 60.
    Kosambi, D.D., 1975. The Beginnings of Iron Age in ancient India, *Journal of the Economic and Social History of the Orient,* Pt. II, p. 309.
15. (i) Kashyap, J., 1960, N.B. *Anguttara Nikaya,* Vol. III, p. 207.
    (ii) *Ibid.*, Bapat, B.P., 1956, N.B. *Samyutta Nikaya,* Vol. IV, p. 170.
16. Buddhist Legent, *Dhammapada,* 29.2, p. 186.
17. Kashyap, J., *Samyutta Nikaya,* Vol. IV, pp. 152-153. *Majjhim Nikaya,* Vol. III, p. 236. *Anguttar Nikaya,* Vol. I, p. 130. *Mahaniddes Pali,* pp. 346-47. *Callaniddes Pali,* p. 251.
18. Kashyap, J., 1959. *Udanpali,* p. 177, Nalanda, Bihar.
19. Kashyap, J., *Jataka Pali,* Pt. II, p. 224.
20. *Ibid.*
21. *Cullavagga,* p. 204.
22. *Jataka,* Pt. p. 138.

'Ayas' in the Rigveda—A Note

D.N. TRIPATHI

The word *ayas* occurs at a number of places in the Rigveda. Though it is definitely a metal, its exact nature is not known beyond doubt. It has been variously interpreted to connote copper, bronze, iron or metal in general. Gordon Childe thought it referred to copper.[1] According to Zimmer it was bronze.[2] Monier Williams translate *ayas* as iron, metal, an iron weapon, gold and steel.[3] Scholars like M.N. Banerjee, L. Gopal and N.R. Banerjee have taken ayas as iron.[4] Mahidhar while explaining a passage of *Vajasaneyi Samhita* (XVIII. 13) takes ayas as a general terms for metal denoting both iron and copper. The present note aims at a fresh analysis of the Rigvedic data in the light of the recent archaeological discoveries.

A study of the Rigvedic hymns reveals that *ayas* was a baser metal used for making weapons, implements and vessels. Thus a hymn of Rigveda dealing with the blessings of weapons refers to arrows tipped with *ayas.*[5] Likewise another hymn informs us about the vāsi made of *ayas.*[6] Many other weapons are also referred to have been made of ayas. In this context sharp-edged weapons like *asi, svadhiti, parasu, pavi* and *kshura* deserve our attention. In a Rigvedic hymn *asi* has been referred to as a sharp weapon used for cutting the bull into bits.[7] *Svadhiti,* the axe, was used for cutting trees. *Agni* has been called as *'svadhitir-vanānām'*[8] svadhiti along with *asi* has been referred in the context of horse-sacrifice.[9] These implements could kill an animal with one stroke. Like *svadhiti parasu* was also used for cutting woods and clearing forests.[10] *Pavi,* a weapon of *Maruts,*[11] has been compared with *vajra.* It was also used as a metallic tyre of the wheels of the chariots.[12] *Kshura* was also a sharp-edged weapon. Indra has been requested to make the intellect of man sharp like *kshura,* the razor.[13] Thus it is obvious that strong and sharp weapons could be manufactured from *ayas.* The hardness and sharpness of this metal have been alluded in other contexts as well. The epithets such as *ayahsipra,*[14] *ayohanuh,*[15] *ayahsirsa*[16] are symbolic expressions for the hardness of the metal. The sharpness of the metallic objects and weapons is suggested by the expression *'ayaso-dhara',*[17] meaning the sharp blade of the objects made by *ayas.* The Rigvedic references about the hardness and sharpness of the objects made from *ayas* led N.R. Banerjee to conclude that it was no other metal but iron. But his suggestion does not seem to be convincing. It is to be noted that in a hymn of *Atharvaveda, svadhiti,* the axe, has been described as *lohita.*[18] In the later Vedic literature *lohayasa* or *lohitayas* stands for copper. Thus one can easily conclude that *svadhiti* described as *lohita* in *Atharvaveda* was a weapon made of copper. As such the possibility of the use of copper weapons and implements during the Rigvedic period cannot be ruled out. Moreover, the available archaeological data goes against the theory propounded by N.R. Banerjee. We shall, however, discuss the archaeological evidence on the subject later. In certain hymns of the Rigveda, there are allusions about the colour of *ayas* which may help us in identifying the metal. The Agni has been called *'ayodanstra'.*[19] According to Zimmer it refers to the colour of the flames which resembles *ayas.* As the colour of the flames is reddish it may be presumed that the *ayas* was also reddish in colour. Likewise there is another reference to the car seat of *Mitra* and *Varuna* which is made of *ayas* but shines like gold in the rays of the morning sun.[20] Here also the reddish colour of the *ayas* is alluded. Now, some passage of Atharvaveda throw further light on the problem. In Atharvaveda[21] *ayas* is divided into two species, the *shyama* and *lohita.* While the first is a black coloured metal the second stands for the red-coloured metal. Almost all the vedic scholars have identified *shyama* with iron and *lohita* with copper. Loha is mentioned in *Vajasaneyi Samhita*[22] and *Taittiriya Samhita*[23] as distinguished from *shyama. Lohayasa* has been mentioned in the *Jaiminiya Upanisada Brahmana*[24] in contrast with *karsnayas,* and in *Taittiriya Brahmana*[25] with *krisṇayasa.* Thus in vedic literature *loha, lohayas* or *lohitayasa* suggesting a red-coloured metal stands for copper and is distinguished from iron which is black in colour. As the Rigvedic *ayas* also seems to be red in colour it may be identified with copper. Though some scholars have tried to identify Rigvedic *ayas* with bronze[26] but the difficulty is that the tin, an essential component of bronze, has not been mentioned at all in the Rigveda.

Now the question arises that if *ayas* stood for copper in Rigvedic times why it was called *lohayas* during the later vedic age. The answer lies in the semantic changes of the word ayas resulting from the technological developments in the society. In early Vedic age metallurgy was in its primitive stage and varieties of baser metal were not known. The word *hiraṇya* and *rajata* denoted precious metals and *ayasa* denoted baser metals. Thus theoretically *ayas* was a general name for baser metal. But as copper was the popular metal for use, *ayas* practically meant copper. By later Vedic age when other varieties of baser metal such as iron, tin, and lead came to be known, the copper was called *lohayas* in order to distinguish it from other metals. As bronze was a mixed metal, combining both copper and tin and having a distinct colour, it was probably simply called *ayas* during the later vedic times. This is suggested by a passage of *Vajasaneyi Samhita*[27] where *ayas* has been mentioned in contrast with *hiranya, loha, shyama, sisa* and *trapu.* Such semantic changes are not strange because we know that the word *loha* which definitely stood for copper during the later Vedic age became popular name for iron in historical times.

The identification of Rigvedic *ayas* with copper, proposed above, is corroborated by the archaeological evidences. In view of the archaeological findings it can be safely asserted that iron was not known in India before 1000 B.C. Iron implements have been discovered from Gandhara Grave Culture in Pakistan and from cairn burials of Baluchistan. The earliest possible dates of these cultures is *c.* 900 B.C. to 1000 B.C. In Saraswati-Ganga valley iron is mainly associated with Painted Grey Ware Culture. The period covered by Painted Grey Ware Culture is roughly 800 B.C. to 400 B.C. Thus archaeologically the introduction of iron in northern India cannot be dated earlier than 1000 B.C. Among the early iron using people of northern India, the authors of Painted Grey Ware Culture covered comparatively a larger geographical area roughly extending from Punjab to eastern U.P. The Gangetic valley was the cradle of this culture. B.B. Lal associated this culture with Mahabharat people, while N.R. Banerjee and some other scholars tried to associate it with Rigvedic Aryans. But the geographical horizons and the material aspects of this culture correspond to the later Vedic culture. It was noted by A. Ghosh that in Saraswati valley the remains of the Painted Grey Ware Culture were found above the dry bed of Ghaggar. It suggests that the Painted Grey Ware Culture flourished in northern India only after the disappearance of Saraswati into deserts. As it is evident from the hymns of the Rigveda, Saraswati was flowing majestically during the later Vedic period. It has been mentioned in the *Panchvinsa Brahmana* that Saraswati was lost in the sands of desert near *Vinasana.* In the light of this evidence Painted Grey Ware Culture cannot be associated with Rigvedic people. In fact the Rigvedic culture seems to have flourished much before 1000 B.C. It may be recalled that one of the factors that has influenced the dating of the Rigveda is the date of Boghaz Keui inscription fixed firmly at 1365 B.C. This inscription records the names of such Vedic deities who are typically Indian. In the light of this well known evidence it is difficult to fix the date of Rigveda after 1400 B.C. Picture known from archaeology reveals that the cultures flourishing in the Sapta-Sindhu region, the cradle of Rigvedic people, during the second millennium B.C. were copper age cultures. Copper ores available in abundance in Rajasthan helped the early growth of copper technology in that area.

Thus through the comparative study of literature and archaeological material it may be concluded that *ayas,* though carrying the sense of baser metal in the Rigveda, was actually copper.

REFERENCES

1. Childe, V.G., 1926. *The Aryans,* p. 29n.
2. *Vedic Index,* Vol. I, p. 31.
3. Williams, M., 1960. *A Sanskrit-English Dictionary,* p. 85.

4. Banerjee, M.N., 1927. On Metals and Metallurgy in India, *I.H.Q.*, Vol. III, p. 16. ff.; Gopal, L., *Uttar Bharati*, Vol. IX, No. 3, pp. 71-78: Banerjee, N.R., 1965. *Iron Age in India*, pp. 158-60.
5. *RV*, VI.75.15.
6. *RV*, VIII.29.3.
7. *RV*, X.79.6.
8. *RV*, IX.96.6.
9. *RV*, I.162.20.
10. *RV*, VI.3.4.
11. *RV*, I.66.10.
12. *RV*, V.52.9.
13. *RV*, VIII.4.16.
14. *RV*, IV.37.4.
15. *RV*, VI.71.4.
16. *RV*, VIII.101.3.
17. *RV*, V.30.15; VI.47.10.
18. VI141.2.
19. *RV*, I.88.5; X.87.2.
20. *RV*, V.62.8; y. V.62.7.
21. *Ath. V.* XT.3.1.7. *Maitrayini Samhita*, 18.2.9.
22. XVIII.13.
23. IV.7.5.1.
24. III.17.3.
25. III.62.6.5.
26. *Vedic Index* (Zimmer's opinion), Vol. I, p. 31.
27. XVIII.13.

The Search of Metals and Techniques for Minting Earliest Coins in India

B.R. MANI

Gold, silver and copper were frequently used in India from Chalcolithic times and coins were minted later with these metals. Archaeological evidence of regular use of gold for manufacturing earliest coins is absent. Silver and copper were used for minting punch-marked and uninscribed copper-bronze cast coins. Although references, of gold coins have been traced in the Vedic literature and gold was procured for making ornaments and other objects from a very early period, in the absence of archaeological evidence of any early regular gold currency system it is not desirable to go into details of antiquity of gold coins.

The three techniques applied for minting earliest coins have been discussed into details by scholars (Thakur, 1961). They are casting technique with the help of clay moulds, striking technique with the help of die and punching technique with the help of several punches (bimbaṭaṅka). Without going into details of the methods of each of these techniques, as this is not required here, we may discuss about the chronological position of application of these techniques from the available indications from literary as well as archaeological sources which combined together give a clear view of the earliest use of metal and technique of minting coins.

Native silver has always been rare in India and silver has not been reported to occur in India as a primary mineral. Sources of silver have been attributed to secondary sources either associated with gold from Kolar gold mines of Karnataka or with galena(argentiferous) (Bhardwaj, 1979 : 132). In small quantity it is obtained from lead ores in Bihar, and Orissa. Silver was obtained from the zinc blended argentiferous galena from Zawar mines near Udaipur in Rajasthan in ancient times. Badwin mines of Burma and Faranjal mines of Afghanistan seem to be perennial sources of silver in ancient times alongwith mines of Bactria and Transoxiana. Thus it seems that silver was imported in India in large quantity at a later date between *c.* 600 B.C. and *c.* 300 B.C. onwards.

Copper has been abundant in India. Mining and smelting of copper ores have a remote antiquity at Singhana near Khetri (Rajasthan) and Sikkim (Bhardwaj, 1979: 91). Besides Rajasthan, a flourishing copper mining and smelting is believed to have flourished in different areas of Bihar. "Important copper mines of Rajasthan being those at Khetri and Singhana in the Jhunjhunu district and Kho-Dariba in Alwar District. An organic sample from Dariba has been dated by C-14 to Ca. 360 B.C. In Bihar there are important mines of copper at Rakha and Mosabani in Singhbhum district. The copper belt is about 100 miles in length and is marked by many ancient workings" (Bhardwaj, 1979: 94). Evidences of ancient copper mining and smelting have been found at the copper mines of Baragunda (Hazaribagh district), Barachakri and Chander Pahar in Singhbhum district. Evidences of a number of old copper workings are present in Almora and Chamoli districts of Uttar Pradesh. A list of 42 old mines have been appended by Bhardwaj (1979). Cassiterite which is the main ore of tin is reported from a large number of sites from Hazaribagh, Ranchi and Gaya in Bihar. Thus it is clear that there has been abundance of copper and tin in India for minting copper and bronze coins which are found in cast variety and also in punch-marked type. Copper in association with lead and zinc has been found in a large quantity at different sites of Tamilnadu (Kuppuram, 1989: 27-28). Copper was abundantly available in India and was one of the chief export items as mentioned by Pliny and in the *Periplus of the Erythraean Sea* (Jaggi, 1981: 239).

Stamping and casting of metals were known to Indians from Vedic period as far as literary evidences are concerned and at least from Harappan civilization in the third millennium B.C. so far as the archaeological evidences are concerned. Attempts have been made to trace the origin of the words *mudrā* or *muhar* and *sikkā* denoting die-struck and (cast) coins respectively from *Ṛigvedic* compound word *Lopāmudrā* and *Atharvavedic* root *sich* in the sense of casting metal, like casting of thunderbolt (*Vajraṁ Yamasincata*) by Bṛhaspati of the Āṅgirasa race. Variants of *sikta*, a past participle of the root in Pāli also (*Sitthaka*) denote a bee-wax mould. The word travelled westward and in Syrian and Aramaic *sykta* denoted a coin-mould. Arabic and Persian languages adopted it in the forms of *sikkatun* and *sikkā* and the latter returned to India in the medieval times and till then it is used commonly for a coin (Pathak, 1981: 1-18).

The first element *lopā* of the compound word can be derived from Sanskrit root *lup* meaning to pierce or to strike and Lithuanian *lanpyati*, Russian *lupit* and Serbo-Croatian *lupiti* are all connected with the action of striking. Roots *lup* and *rup* have been regarded as variants of the same and *rūpa* has been regarded as a stamped object from which word rupayā (rupee) has been derived. *Mudrā* can be related with Sumerian *Musarum* or Akkadian *Muzra* for denoting written object or writing of Assyrian musrau (seal, writing). The Persian word *muhr* used for coin or seal or signet ring has its origin in this. Thus both the elements of the compound point towards a stamped object, if not to object with legends or die-struck coin.

Words like *rukma* or *nishka* in connection with a particular form in gold (*suvarṇa* or *hiraṇya*) have been repeatedly mentioned in the Vedic and later literature. But since there is no archaeological evidence of a regular gold money, under the present circumstances it is not possible to discuss about the gold coins of earliest numismatic phase. *Rukma* and *nishka* are supposed to be gold discoid pendants or circular flat ornaments which in course of time seem to have assumed a particular form with design and weight standard and were used in money-transactions. *Rukma* is also mentioned in the Aryan documents discovered in the Near East at Nuzi (Pathak, 1981: 6). On the basis of a reference from *Śatapatha Brāhmaṇa*, Prof. V.S. Pathak has described it as an ornament worn on the neck and as a symbol of sun shaped accordingly as a round plate with twenty-one small projections radiating around.

On the evidences corelated from *Atharvaveda* and interpreting an imagery from *Kumāra-Sambhava* of Kālidāsa, it has been proposed that *nishka* should be a torque — a hollow rounded with spokes projecting from it (Pathak, 1981: 7). It would be our humble submission that perhaps the same form of a disc or sun with spokes or rays which are found on some early copper and silver coins may be representing the *nishka*. Merely on the basis of a few interpretations and commentories it is not possible to think that *nishka* was only a gold object and not a silver or copper one. In this context it would not be out of place to mention that origin and derivation of the word mishka in Sanskrit is not known. "In the Semetic group, Aramaic *mishka* (cast metal) from √*nsk* to pour, comparable to Skt. √*sich* which also means both to pour and to cast metal, comes closest to the Vedic *nishka*. The Akkadian √*nsk* = cast metal and √*nsq* = precious stones seem to belong to the same group" (Pathak, 1981: 7-8). These evidences pieced together may at least point towards an inference that the *nishka* was originally a cast piece of metal used for ornaments, as well as for money transactions and perhaps the early copper cast coins with single sun or disk symbol formed a type of *nishka*.

While discussing about *jātarūpa* of later Vedic literature, *rūpya* of Pāṇini's *Ashṭādhyāyī* and *rūpiya* of Pāli literature, Prof. Pathak quotes the *Pārājika* when monks used to beg - *'āvuso, Kahāpaṇa dehī'ti'* i.e., 'Gentleman, give us *Kahāpaṇa* (*Kārshāpaṇa*)' and the shocked people commented *'ime samaṇā sakyaputtiyā rupiyāṁ paṭigaṇhantiti'*, i.e. 'Oh Buddhist monks, accept rūpiyas (coins)' (1981: 12). This gives difference between *Kārshāpaṇa* and *rūpiya* and it is possible that the original meaning of *Kārshāpaṇa* was a copper coin and *rūpiya*, a coin or later a silver coin. *Kārsha* might have been used to denote weight also.

Further, interpreting the word *āhata* Prof. Pathak concludes that the word *āhata* in the sense of beaten or stamped must have been used later as found in *Vinaya-Piṭaka* (lakkhaṇāhataṁ) and *Kāśika* (nighaṭika ṭāḍana), but its original meaning even in the *Ashṭādhyāyī* of Pāṇini were 'rūpādāhata' is referred in the sense of moulding coin (1981: 13-15; 1982, 97 ff.) and not in the sense of stamping or punching. In Ṛgveda (V. 12.13), the Rain god is mentioned as *āhanas*, because he pours in the womb of the earth the creative fluid, the rains and creates various forms. In the context of embryo the word *āhata* is used with the meaning 'filled within, enveloped in' in the *Atharva-Veda* (I. 11-4). Derivatives of ā √han in the sense occurs in the *Taittirīya Saṁhitā, Taittirīya Brāhmaṇa* and *Śatapatha Brāhmaṇa* and *āhana* does not convey the idea of striking, but is associated with the concept of mould. Prof. Pathak also states that the root *han* of which *āhata* is the past participle is traced to the Indo-European *gwhen* (> han) meaning 'to fill in' which radiates ideas to mould or to fix within. The old Persian inscription of Darius I from Susa (F) contains the word *ājata* (participle of ā + √ jan = ā √han = past participle *āhata*) which is the old Iranian form of *āhata* for the bricks which are moulded (*ājata*). "In short, the Vedic word *āhata* in the *Ashṭādhyāyī* means 'moulded' and thus *āhata* coin may mean a cast coin and not a punch-marked coin as generally believed.

On the basis of stratigraphic evidences from some of the important excavated sites of north India, it has already been proposed that uninscribed copper cast coins were the earliest monetary issues of the country which precede punch-marked silver coins alongwith which they remained in circulation for a pretty long time and the copper punch-marked money was a later appendage (Mani, 1990).

The matter was further examined and 35 excavation reports of important archaeological sites throughout the country were consulted. Findings have been very similar to the earlier study by the author, although in the cases of excavation reports from Deccan and South, it has been found that coinage started at a comparatively later stage, some times around third century B.C. or so in most of the cases and uninscribed copper cast coins are in many cases absent while punch-marked silver and copper coins are found in the earliest strata of the Early Historical Period, closely followed by Sātavāhana coins, as in the cases of Salihundam, Dhatva and Bhokardan. The lower dates of punch-marked coins are said to be 3rd century B.C. at Kaundinyapura, Nagara (Gujrat), 2nd century B.C. at Nevasa, Salihundam, Vaddamanu and 1st century B.C. at Bhokardan. Although uninscribed copper cast coins are found in association with punch-marked coins from Nevasa, Kaundinyapura, Nagara and some times they out-number the punch-marked issues as in the case of north India. Thus it seems that the punch-marked coins in Deccan were introduced in Maurya and succeeding period. The abundance of uninscribed copper cast coins and punch-marked coins in the Pre-Mauryan levels at the important excavated sites of north India is because of the growth of urban centres in the Uttarāpatha in 7th-6th centuries B.C. particularly the sixteen great states (shodaśa-mahājanapadas).

Stratigraphical evidences from archaeological sites like Kausambi, Taxila (Bhir mound), Mathura, Sravasti, Atranjikhera, Hastinapur, Kayatha, Maheshwar, Sonpur, Tilaurakot and Piprahwa-Ganwaria complex indicate the coexistence of punch-marked and uninscribed copper cast coins together in the Pre-Mauryan periods. It is also to be mentioned that the uninscribed copper cast coins out-number the punch-marked issues at most of these sites. There are further indications from sites such as Kausambi, Sravaṣti, Atranjikhera, Kayatha, Navdatoli and Sonpur etc. that uninscribed copper cast coins were the earliest issues of India as they are found in the lower levels of stratification than those in which uninscribed copper cast coins and punch-marked coins are found together. Die-struck and cast copper coins numbering 37 have been found in the lower most deposit (Phase A, Period I) datable around *c.* 6th century B.C. at Tilaurakot. They bear four-spoked wheel symbol and have been supposed to be the earliest coins of the region (Mitra, 1972: 89 ff). Interestingly, in the 1949-50 excavations at Kausambi 23 uninscribed copper cast coins were recovered from the lower levels but no punch-marked coin was found there. While the earliest punch-marked coins at Kausambi are datable around 535 B.C. to 465 B.C., the antiquity of uninscribed copper cast coins have been suggested in the excavation report of 1957-59 as 885 B.C. to 815 B.C. These dates may or may not be accepted, but it is very clear that copper cast issues, found much below the stratified levels from where punch-marked and other coins are found, came into existence in the earliest numismatic phase.

Finally, on the basis of stratigraphical evidences our humble submission is that copper coins, particularly the uninscribed copper cast coins were first to be issued in India as also the metal was easily accessible and silver coins, found only in the punch-marked variety in the earliest numismatic phase, seems to have been issued in the period when trade developed with outside the bounds of the country when silver was imported some times after the 6th century B.C. Words frequently used in early literature like *nishka, āhata* and roots like '*sich* (from which the word *sikkā* got derived at a later date) seem to have denoted the meaning of moulded or cast coins in their original usage. In this connection it may be pointed out that in general, easily obtainable base metals were used for coinage in its earliest stage throughout the world and precious metals were substituted afterwards. Among the three main techniques adopted for minting coins, the casting technique is a stage of minting having its own drawbacks like breaking of the piece because of air bubbles and fluctuation in weight standard and disturbed device. As such it was a primitive method. It seems that with the introduction of new methods of minting, i.e., striking with die or punching, the casting technique was gradually given up and it was later adopted for production of forged or sham issues as a number of clay-moulds for making punch-marked and Kushan coins have been reported from various sites. Although some very early die-struck coins are available in copper, the technique seems to have gone out of vogue when punch-marked coins were issued and the die-struck coins were reintroduced with Indo-Greek contacts and the technique continued afterwards.

REFERENCES

Ansari, Z.D. and M.K. Dhavalikar., 1975. *Excavations at Kayatha,* Pune.

Bhardwaj, H.C., 1979. *Aspects of Ancient Indian Technology.* Delhi.

Deo, S.B. et al., 1974. *Excavatins at Bhokardan (Bhogavardahana)* 1973. Nagpur.

Dikshit, M.G. et al., 1968. *Excavations at Kanndinyapura.* Bombay.

—— 1983. *Excavations of Vaddamanu and Explorations in the Krishna Valley Basin,* Birla Archaeological and Cultural Research Institute, Hyderabad.

Gaur, R.C., 1983. *Excavations at Atranjikhera, Early Civilization of the Upper Ganga Basin.* Delhi.

Indian Archaeology A Review, 1954-55. (15 - Mathura).

Jaggi, O.P., 1981. *Technology in Ancient India.* Delhi.

Joshi, M.C., 1989. Mathura as an Ancient Settlement. *Mathura: The Cultural Heritage,* AIIS, New Delhi.

Kuppuram, G., 1989. *Ancient Indian Mining, Metallurgy and Metal Industries.* I. Delhi.

Lal, B.B., 1954-55. Excavations at Hastinapur and other Explorations in the upper Ganga and Sutlej Basins 1950-52. *Ancient India* 10 & 11. New Delhi.

Mani, B.R., 1990. 'Chronology of Earliest Indian Coins, Stratigraphic Evidence and Scope of Investigation', Paper presented in the Seminar on 'Archaeological Perspective of U.P. and Future Prospects' organised by the U.P. State Archaeological Organisation in October, 1990 at Lucknow.

Marshall, J., 1951. Taxila, 3 vols. Cambridge, 1951 (Reprinted New Delhi, 1975).

Mehta, R.N. and D.R. Shah, 1968. *Excavation at Nagara.* Baroda.

Mehta, R.N. et al., 1975. *Excavations at Dhatva.* Baroda.

Mitra, Debala, 1972. *Excavations at Tilaura-Kot and Kodan and Explorations in the Nepalese Tarai.* Calcutta.

Pathak, V.S., 1981. 'A Semantic Study of Numismatic Terms'. Presidential Address at the 68th Annual Conference. *JNSI,* XLIII, I.

Pathak, V.S., 1982. 'Ahata : A Semantic Study', *JNSI,* XLIV.

Ray, S.C., 1989. Stratigraphic Evidence of Coins from Excavations at Mathura. *Mathura: The Cultural Heritage,* AIIS, New Delhi.

Sankalia, H.D. et al., 1958. *The Excavations at Maheswar and Navdatoli* 1952-53, Poona.

Sankalia, H.D. et al., 1960. *From History to Pre-History at Nevasa* (1954-56), Poona.

Sharma, G.R., 1960. *The Excavations at Kausambi* (1957-59), Allahabad.

Sharma, G.R., 1969. Excavations at Kausambi 1949-50, *Memoirs of the A.S.I.* 74, Delhi.

Sinha, B.P. and B.S. Verma, 1977. *Sonpur Excavations* (1956 and 1959-1962), Patna.

Sinha, K.K., 1967. *Excavations at Sravasti,* 1959. Summary of results. B.H.U., Varanasi.

Srivastava, K.M., 1986. *Discovery of Kapilavastu,* New Delhi.

Subrahmanyam, R., 1964. *Salihundam: A Buddhist Site in Andhra Pradesh,* Hyderabad.

Thakur, Upendra, 1961. 'Mints and Minting in India', *Journal of the Numismatic Society of India,* XXIII: 175 ff.

Verma, T.P., 1985. A Note and a copper coin from Narhan. Narhan, 1983-85: A Preliminary Report, *Bharati.* New Series 3. BHU. Varanasi.

Section IV
Socio-economic Implications of Metals

Social Implications of Technology : A Study of Iron in pre - c. 200 B.C. India

VIJAY KUMAR THAKUR

Social institutions are closely bound up with the contemporary techno-economic apparatus. Moral problems do not change rapidly, but the behaviour of the society is fashioned in accordance with the instruments that the material world offers it. In other words, the available technology decisively shapes the society, lends its form to emerging institutions and normally breaks the traditional normatives of the group. Technology, therefore, directly interacts with the society, shapes it and provides a sustenance-base as well, a point aptly made out by Leroi-Gourhan. "With man, the relationship between the individual and society varies as a direct function of the evolution of techno-economic structures, and in order to understand certain characteristics of the social body in the various stages of evolution, these techno-economic structures must be defined. The most direct impact of the level of technology on the social group is related to the very density of this group. From the moment that intellectual evolution created values characteristic of *homo sapiens*, the relationship between technical level and social density became the principal factor of progress" (Gille, 1990). Thus, the nature of technological know-how and the level of social development constitute the two variables of a civilisation and it is this interaction that has to be comprehended in any study pertaining to the role of technology in social change. It is with this perspective that the present paper aims at analysing the role of iron technology in shaping the society of early India.

The study of iron in early India, both in its technical and social manifestations, suffers from a serious limitation. These are essentially uni-dimensional in approach. The society as a prime technological variable gets ignored with the result that society is always at the receiving end, almost as a helpless entity incapable of any dynamism of its own. Such an understanding of the social role of technology sees techniques as independent of the *milieu* in which it evolves. This tantamounts to divorcing the process of the origin and growth of a technology from the social system, a line of argument that can lead to funny conclusions. The study of technology has to keep in mind the social climate. It is common knowledge that a technological innovation is a product of a social system and in turn, it gets accepted if the society is ready for it. The social readiness depends upon a variety of factors

- the form of mode of production, the nature of class formation, the ecological factor, etc. This may explain the changing role of a technology in succeeding historical periods.

Iron in early India has been an extremely popular topic of debate in recent Indian historiography, but the focus statically remains on its antiquity and technicalities. The earlier writings were mostly confined to the use of literary data, the two representative scholars of this trend being Neogi (1914) and Banerjee. They suggested that iron was introduced in India by the Ṛgvedic Āryans and the key to their conclusion was the interpretation of the Ṛgvedic word *ayas* as iron, a deduction that has eluded a consensus of opinion amongst historians (Varma, 1982). Lallanji Gopal and N.R. Banerjee (1965), represent the second trend which is not much different from the first. The Āryan obsession persists but with a change. They associate the beginnings of iron with the later Vedic culture by corroborating literary reference with archaeological data. The problem with these two and other similar studies has been an erroneous perspective that confines all enquiries regarding the use of iron in the Ṛgvedic period to the interpretation of the term *ayas,* least realising that evidence to this effect may come from elsewhere also. It is pertinent to cite M.A. Mehendale's arguments in this context, "The problem presented by *'ayasa'* is no doubt somewhat complicated. But let us see the facts as they are known to us. Cognates for the word 'iron' are not available in the Indo-European languages and, hence, knowledge of iron for Indo-European period cannot be established. On the other hand *'ayasa'* has cognates in other Indo-European languages where it means 'copper-bronze'. Add to this that until recently, iron age in India could not be pushed back to the Ṛgvedic period. Hence, although in later Saṁskṛta *'āyasa'* means 'iron' it was reasonable for more modern scholars, Thieme, for example, to assume that *'āyasa'* in the *Ṛgveda* meant copper or bronze but not iron. But now we know that iron was certainly known in the late *Ṛgvedic* period (*'asi'* = black hence made of iron knife), and archaeological evidence pushes back the knowledge of iron to a much earlier period ".... I would ..... interpret *'āyasa'* in the early parts of *Ṛgveda* to mean copper" (Varma, 1982). It is fairly certain that the Āryans were not the vehicle of the diffusion of iron in India, but it was here that they encountered the metal for the first time.

A different approach to the problem is noticed in the analysis of Kosambi (1973). Relying primarily on Buddhist literary data, he moves away from the trend of Ārya-centred theorisations. He takes the reference to an iron ploughshare in the *Suttanipāta* as an obvious evidence for a widespread use of iron in agriculture thereby suggesting an earlier date, somewhere around the 8th century B.C., for its beginning in the middle Ganga valley. Kosambi's analytical tool was extremely refreshing for he attempted to underline the social roots of this technological development, a point which had hitherto remained ignored.

Archaeological studies of iron in early India have widened the perspective by pointing out the regional use-patterns of this metal. But in these studies also some of the older conceptual determinants keep on reappearing with stoic regularity and reinforced data-base. Thus, Tripathi (1973), who has comprehensively surveyed the recent archaeological data, accepts the basic diffusionary postulate of N.R. Banerjee and suggests the introduction of iron in the northwest by 1000 B.C. and its arrival in the Ganga valley by 800 B.C. In a subsequent paper (Tripathi, 1990), she has revised her earlier views and proposed a multicentred, indigenous origin of iron in India. Another notable contributor to the debate, Chakrabarti (1973), while rejecting the diffusionist stance, strongly advocates for a polycentric origin of iron in the sub-continent. His delineation of the different iron-bearing foci, at times, seems to be very tenuous and dangerously generalised. His assertion that by *c.* 1000 B.C. "the use of iron came of age in India", can hardly be accepted for the Ganga valley. He, however, pointedly refers to this region as falling within this chronological scheme just on the basis of the limited evidence forthcoming from Jakhera and Hulas (Chakrabarti, 1988: 119). His basic problem is a contradiction in terms of a sober realisation of a polycentric origin of iron in India on the one hand, and a forced attempt to work out a standardised chronology for the beginning of iron age in the region on the other. It seems that most of these scholars do not make a clear distinction between stray and isolated iron-yielding site/sites of a cultural phase and the currency of iron know-how throughout a culture when they try to delineate the beginnings of iron age in India. The confusion is further confounded due to the discovery of some early iron objects that chronologically go beyond the available dates of iron-use in other areas of the sub-continent. Thus, the date for iron at Ahar is *c.* 1300 B.C., (Sahi, 1976: 367-69) for Nagda *c.* 1200 B.C., for Noh and Atranjikhera 12th-11th centuries B.C. and for Hallur *c.* 1100 B.C. (Varma, 1982). Chirand, in middle Ganga valley, has also yielded iron in an early datable context. How does one explain the occurrence of iron at these sites in a context much earlier than its use at other sites of the same region ? Vibha Tripathi, addressing herself to this problem exclusively in the context of Ahar, suggests, "where copper smelting was quite well known there might be an accidental stumbling on some form of a metallic iron through copper smelting ...." This is a suggestion with far-reaching consequences in the context of understanding the very process of the beginnings of the use of iron in India. This is not only a rejection of her earlier diffusionary stance, but a coherent advocacy for the independent beginnings of iron-use in different parts of India. In other words, the social climate played a crucialrole in the evolution of this technology, a point, which has not been considered so far. Settlements like Ahar, Nagda, Noh, Atranjikhera, Chirand, etc., were materially the most developed centres of their respective regions and thus, the seminal role played by these foci in the initial stages of this process.

The delineation of the Iron Age in the Indian context has two crucial issues : (i) the earliest iron centres have to be taken into cognisance and (ii) an element of standardisation in terms of use-pattern has to be a major criterion. Recently Snodgrass (1980: 271) has suggested that the Iron Age is distinguished by "working iron", that is, iron used to make the functional parts of the real cutting and piercing instruments that constitute the basis of early technology. This postulate which is valid for the Mediterranean envisages "three broad stages in the development of iron technology," each successive stage increasingly replacing bronze implements with iron till the latter predominates. In western Iran the three successive stages are: I.; 1450/1350-1100 B.C.; II - 1100-800, III 800-500 B.C.) The Indian situation certainly admits of some basic similarities. As the beginning of Iron Age is normally synchronised with the date of the first appearance of the metal in the given geographical zone, the genesis of the Indian Iron Age can be put around 1300 B.C. on the basis of the evidence forthcoming from Ahar, Noh, Prakash and Bahal. By 200 B.C. the metal had come of age in the sense that its impact was being felt all-around. This impact was manifest both in terms of the material culture of the Ganga valley and the dissemination of these developed elements in outlying regions. Thus, the time-frame for the evolution of iron technology in India *vis-a-vis* its social context, one may suggest, had the following stages:

(i) The beginnings, *c.* 1300-1000 B.C.
(ii) The phase of adaptation *c.* 1000-600 B.C.
(iii) The phase of percolation, production and expansion, *c.* 600-200 B.C.

This proposed periodisation is not solely based on the changing patterns of iron-use in the sub-continent; it also keeps into focus the social dimensions of the technological development. Further, this exercise of stage-based analysis does not suggest any point of break/gap in terms of the evolution of this technology, but it is simply an attempt to identify the basic changes by taking certain specific points of reference. Thus, *c.* 1000 B.C. does not mark a break in the tradition of iron-use inaugurated around 1300 B.C., but it indicates the beginning of a phase of a relatively more meaningful interaction between this technology and the society of the times.

(i) **The Beginnings,** *c.* 1300-1000 B.C.

The knowledge of iron was acquired in the sub-continent sometime towards the middle of the second millennium B.C. The metal makes its earliest appearance at Ahar near Udaipur in *c.* 1300 B.C. and then it occurs at Nagda, Prakash and Bahal in *c.* 1200 B.C. This evidence from Malwa and northern Deccan, alongwith the evidence of Atranjikhera and Hallur, lends support to our proposed chronological sequence. Though it is difficult to talk in terms of any homogeneity in the pattern of iron-use at these sites, yet it is obvious that by *c.* 1000 B.C. iron was known in a vast geographical area from the western coast in Karnataka to western parts of Uttar Pradesh. By supplementing this archaeological data with the Ṛgvedic references (I, 162-20, X 79-6) suggesting iron knife or *'asi'* being used for animal sacrifice as the Indo-European *ṇsis* means black and thus iron. But the word *ṇsis* does not occur in family book of Ṛgveda (Cf. Varma, 1982). The geographical horizon of the iron zone can be further extended to the Saptasindhu region. In spite of this wide geographical expanse, the pattern of iron-use in the pre-1000 B.C. phase is severely constrained. The iron repository includes simple huntng tools with a relative preponderance of arrow-heads and points. The material *milieu* of the *Ṛgveda* and the contemporary chalcolithic sites hardly had a greater scope for the use of this metal. In an age when food collection was still crucial for the sustenance of the group, the best use to which the new technology could be put to was hunting, and that too as a supplement to the elder and time-tested tools of copper and bronze. The social climate for a greater and better use of iron was lacking. It seems that this new technology was not a product of social requirements, rather it was a chance find. The study of Ahar iron by Vibha Tripathi (1985) clearly underlines this accidental stumbling on iron in the process of copper smelting. She is equally convincing about similar patterns of development at other early iron-using sites. One cannot rule out the possibility of even the ignorance of this metal on the part of the users in this early phase, a point hinted at by Tripathi also. Iron, therefore, does not seem to be socially meaningful during the first phase of its occurrence due to a gap between the social climate and the emerging body of technological know how. It is in terms of this incompatibility that the relationships between the technical and social systems need to be analysed in order to gain an insight into the processes of mutual adaptations of the two systems.

(ii) **The Phase of Adaptation**, *c.* 1000-600 B.C.

The second phase of iron technology in India is marked by a process of adaptation. By *c.* 1000 B.C. iron had become a social reality. Its potentialities came to be realised and the emerging social system attempted to harness it for achieving its immediate objectives. One encounters an interesting pattern of social climate controlled exploitation of technological knowledge. A perusal of the available data suggests that the iron technology was not occupying the driver's seat in the context of on-going process of change, it was rather the other way round. The existing social system, a system that had just started undergoing a transition from a semi-nomadic to a settled life pattern, was not congenial for an expansive use of this metal. The use of iron remained limited in scope as the period marks the beginning of the process of the gradual social acceptability of this technology.

The period from *c.* 1000-600 B.C. marks a transition from a tribal/pastoral to a settled/agrarian mode of production. Despite the frequent occurrence of prayers for the increase of cattle in the later Vedic texts, agriculture was emerging as the chief means of livelihood. The role of iron technology in hastening this process of transition has been very often emphasised, notably by R.S. Sharma (1968: 59-60; 1983: 117-118) in his explanations of the origins of Buddhism. The archaeological picture, however, is different. By *c.* 1000 B.C. iron implements are reported from the whole of upper Ganga basin. Sites like Noh, Atranjikhera and Kauśāmbī have yielded a very impressive array of iron object. A careful look at the repository of iron finds of the PGW phase makes interesting reading. These include, besides slags and shapeless bits, knives, arrow-heads, spear-heads, knife-blades, spades, nails, daggers, hoes, tongs, adzes, fish-hooks, chisels and scissors (Tripathi, 1976). M.D.N. Sahi (1982) talks in terms of the following classification of the contemporary iron objects:

(i) Weapons used for warfare or hunting;
(ii) Household objects;
(iii) Craft tools;
(iv) Objects used in agricultural operations.

Vibha Tripathi's (1990) classification is slightly different:

(i) hunting tools;
(ii) agricultural tools;
(iii) household objects;
(iv) building material.

Tripathi, while referring to the agricultural tools, admits their paucity and notes the availability of only a sickle from Atranjikhera and an axe from Rajghat. Sahi (1982), however, tries to project an impression of a relative abundance of iron agricultural tools during the period. He refers to the broken pieces of socketed axes from Hastināpura and simple type axes from Atranjikhera, Jakhera and Noh, which "were used by the PGW people for cutting the stumps of the burnt down trees of the dense forests facilitating reclamation of the land for agricultural purposes." Besides raising genuine doubts about the stratigraphical position and dubious identification of some of these finds, one may ask Sahi a couple of questions. In a *milieu* when iron was virtually confined to the manufacture of weapons, is it not more plausible to suggest that iron axes initially came in use more as a weapon than an agricultural aid ? If iron-axes were available for forest-clearance, why were the forests being burnt for the same purpose ? Obviously the agricultural usages of axe is under serious doubt during this period. Sahi also refers to two ploughshares, one each from Alamgirpur and Jakhera. He, however, ignores the extremely doubtful identification of the Alamgirpur piece (Banerjee, 1965) and the post-600 B.C. datum of the Jakhera share (Sahi), a situation that demolishes his hypothesis of a meaningful use of iron in agriculture in the PGW phase (Sahi, 1982).

It needs restatement that the available body of archaeological data does not permit a hypothesis of iron-based agriculture in the pre-600 B.C. period. Iron agricultural implements are very scarce despite a gradual increase in the use of the metal. The discovery of two furnaces for iron smelting and forging of iron objects from the excavated site of Suneri village in Rajasthan supplements this picture. The furnaces, associated with the PGW culture, have the open type hearths provided with bellows (Sharma, 1983). The word *bhāstra*, which came to denote a bellow in post-Vedic times, is mentioned in the *Śatapatha Brāhmaṇa* (I, 1,2,7; VI, 3.16) in the sense of a leather container, a fact that may suggest that it was only towards the end of the Vedic period that bellows came in use. Thus, the impact of this new technique was felt not by the PGW people but their successors.

The later Vedic texts portray a similar picture of a growing knowledge of iron. The contemporary literature has many terms for iron - *śyāma* in the *Vājasaneyī samhitā* (XVIII. 13; a text of *c.* 800 B.C.), and *śyāma ayas* in the *Atharvaveda* (XI. 3.1.7; a text of *c.* 800 B.C.), *śyāmena* in the *Atharvaveda,* IX. 4, *kṛṣṇāyas* in the *Chhāndogya Upaniṣad* (K.C. Varma) and the *Jaiminīya Upaniṣad Brāhmaṇa,* (II, 90), the latter also using the word *karṣṇāyas.* The texts also refer to a variety of iron objects, indicating its growing popularity as a working-metal. The *Atharvaveda* refers to an iron-knife (*asi*) used for cutting animal flesh (IX, 5.4) and the *Jaiminīya Upaniṣad Brāhmaṇa* mentions an iron-shield (IV. 1). The texts also allude to their day-to-day familiarity with iron-lances/spears (Singh 1965). Like archaeology, the literature of the period attests to the growing knowledge and use of iron in contexts other than agriculture, a conclusion that has been rejected by D.K. Chakrabarti. He has culled four pieces of evidence from the contemporary texts to postulate the use of iron in agriculture in the Ganga valley and the Indo-Gangetic divide by *c.* 800 B.C. (Chakrabarti, 1979: 24). His first two pieces of evidence, cited from the *Śatapatha Brāhmaṇa* (XIII, 2.2.16-19; XIII 3.4.5), clearly bring out the association of the peasantry masses with iron. The conclusion that the association of iron with the common people means its association with agriculture as well, however, is open to doubt. The passages under discussion have associated different metals with different status groups; the group at the top is associated with gold and the group at the bottom is associated with iron while silver takes an intermediate status. Thus metals do not represent any professional or economic association, they reflect the social standing of their associates in a hierarchical order corresponding with their own respective values. Thus, the *Śatapatha Brāhmaṇa* hardly refers to the use of iron in agriculture as "an accomplished fact". The third piece of evidence, forthcoming from the *Atharvaveda* (X. 6.2.3), refers to a metal ploughshare, which has been interpreted to suggest the prevalence of iron ploughshare (Macdonell & Keith 1958: 509). It has been argued that the *Atharvaveda* knew of iron, and therefore, the metal of the ploughshare was iron. This argument deliberately ignores the possibility of the use of a non-iron metal ploughshare, a point which becomes relevant if one keeps in mind the archaeological sequence of the Ganga valley. References to ploughs drawn by six (*Atharvaveda*, VIII), eight (*Taittiriya Samhitā*, I, 8.7; V 2.5.2), twelve (*Maitrayani Samhitā* II 6.2) or even twenty four (*Kathaka Samhitā*, 15.2) oxen hardly make much positive sense in the context of the use of iron in agriculture. It has been suggested that the ploughs had to be drawn by so many oxen at a time because the ploughshare was made of iron (Chakrabarti, above). This interpretation obviously implies the increase in the weight of a plough as a result of the use of iron ploughshare so much so that it had to be drawn by more than a pair of oxen. Anyone who is familiar with the countryside might realise the unrealistic nature of this argument. Even today when the villagers are using iron ploughshare they hardly need more than a pair of oxen for their ploughs, a fact which must have been equally applicable for the agriculturists of even the earlier periods. These references to multi-oxen yoked ploughs, instead of suggesting the existence of iron ploughshare, indicate the use of heavy ploughs. With the expansion of agriculture and the growing emphasis on the cultivation of rice, sugarcane and cotton, the agrarian scenario started changing rapidly. It was becoming increasingly difficult to break the hard soil with a simple wooden plough. Moreover, the new crops required deep ploughing, an objective that could hardly be achieved with the available plough. A multi-oxen yoked plough/multi plough technique was adapted to break the hard soil and prepare it for the cultivation of the new crops. As iron ploughshare was not known, the wooden ploughs had to be made heavy in order to achieve the desired results. Once iron ploughshare came in vogue in the post-600 B.C. period this cumbersome technique was given up in favour of two-oxen ploughs with iron shares. This PGW practice has interesting historical parallels. The ancient Germans had also evolved a unique system of tillage for the heavy clay lands of north European forests. They went for deep ploughing in the region by yoking eight oxen to a plough (Childe, 1942: 242) obviously to overcome the limitations of the pre-iron technology. Thus, there is hardly a reasonable case for putting the beginning of the use of iron in agriculture in north India to *c.* 800 B.C. This is also true of iron's role in the process of the expansion of agrarian area. A spurt in the PGW settlements undoubtedly suggests the clearance of forests for habitation and agriculture. The primary need of the period was the breaking of new grounds, and history provides innumerable instances of this development in pre-iron days. The best primitive method of getting rid of jungles was by burning it. The famous story in the *Śatapatha Brāhmaṇa*, detailing how Videgha Māthava reached the banks of Sadānīrā (modern Gandaka) by following fire-god, Agni (I, 4.1.10-17), illustrates the currency of this practice in the later Vedic times. We also come across instances of the clearance of jungle only from land to be cultivated, otherwise it was not disturbed (*Taittiriya Samhitā* V. 2.5). The iron repository of the PGW phase contains too few iron axes to permit the postulation of any meaningful role played by this metal in the process of agrarian expansion. One can not, however, rule out a limited supplementary role played by copper and bronze axes in this process.

In spite of the absence of any relationship between iron and agrarian production, the post-1000 B.C. witnessed the full-fledged beginnings of plough based cultivation. The *Śatapatha Brāhmaṇa* contains an exclusive *Brāhmaṇa* concerning rites in connection with ploughing (VII. 2.2). In addition to barley, we have now wheat, sugarcane and rice. Varieties of paddy were known — *kṛṣṇavrīhi*[1] (black paddy), *śuklavrīhi*[2] (white paddy), *āśudhānya* (a swift growing paddy harvested within two months),[3] *hāyana* (red husk paddy),[4] *nīvāra* (which grew in shallow tanks)[5] and *mahāvrīhi.*[6] A host of other crops are referred to in the contemporary texts (Ray). The practice of irrigation and the knowledge of manures strengthened this pattern of economic transformation have been highlighted by Ray. This inaugurated a process of the restructuring of social relations. The peasants started producing a surplus,[7] a development that paved the way for social stratification in the form of castes. Iron played a crucial role in this context. The social climate was not yet congenial for the acceptability of iron as a technology of production, but it did need it in order to pattern the on-going process of change. This involved the attempt to work out a social structure to sustain these developments. Obviously the ruling group wanted to control the system of production to derive the maximum benefit from it. Marx, while outlining this mechanism of adaptation, writes, "The ideas of the ruling class are in every epoch the ruling ideas: i.e. the class which is the ruling *material* force of society, is at the same time its ruling *intellectual* force. The class which has the means of material production at its disposal, has control at the same time over the means of mental production, so that thereby, generally speaking, the ideas of those who lack the means of mental production are subject to it. The ruling ideas are nothing more than the ideal expressions of the dominant material relationships, the dominant material relationships grasped as ideas; hence of the relationships which make the one class the ruling one, therefore the ideas of its dominance" (Marks and Engels 1965: 417).

The new technology, therefore, became an expression of the dominant group. The peasants, no doubt, would have benefited from it, but it came to be confined to the service of the class which was trying to perpetuate its hold over the means of production. The process of class formation had a primitive beginning in the later Vedic society (Thakur: 246,) and by monopolising iron technology for the manufacture of weapons of war in order to strengthen itself, the dominant group tried to intensify this formation. The mode of production did not feel any appreciable impact of iron but the emergence of the new production relations was facilitated by this technology. Even this process was not very smooth. As this was the formative age, the idea of a somewhat homogeneous ruling class was yet to crystallise. The process of class formation itself illustrates this situation in the form of an interesting attendant development. The later Vedic texts hint at the beginnings of a brāhmaṇa-kṣatriya conflict, ostensibly for shares in the surplus. (Macdonell & Keith, 1958). As the conflict intensified towards the end of the Vedic period, the contemporary texts found it imperative to stress unity and co-operation between the two components of the ruling group. This was an intra-class struggle; a struggle that followed the iron-aided rudimentary class formation. It is a conflict at the top between ideology (the ritual control) and physical power and the latter assumes a challenging status due to the help of iron weapons. Iron then is helping the journey towards a highly structured society.

The elitist exploitation of the new technology is further illustrated by the process of state formation in the later Vedic times. The process, which was inaugurated by the generation of a social surplus, received a fillip from the nascent iron technology. The emerging dominant group exclusivised the use of iron for increasing its own coercive/striking powers by limiting it to war weapons. Thus, the use pattern of iron technology came to be conditioned by the compulsions of the social relations. The process of state formation seems to have started towards the beginning of the 10th century B.C. with smaller communities coalescing together into larger units. The Ṛgvedic Krivis and Purus after merging into a homogeneous Kuru group went on to align with the Pañchālas to rule over western Uttar Pradesh. Possibly the magnified version of the *Mahābhārata* story echoes the beginnings of the complementary processes of territorial political formation and formation of larger communities around 900 B.C. The use of iron, therefore, seems to have been largely instrumental in the dissolution of the Ṛgvedic tribal political organisation and the consequential evolution of fixed territorial political units. This scenario is clearly spelt out by the evidence collected from the contemporary political regions. Iron weapons are forthcoming from the kingdom of Kuru in the upper part of the Ganga-Yamuna *doāb* (the area corresponding to modern Thaneshwar, Delhi and the greater part of the upper Ganga *doāb*), (Raychaudhuri, 1972: 21) of the Pañchālas in Bareilly, Badaun, Farrukhabad, the adjoining districts of Rohilkhand and the Central *doāb*, of the Matsyas in Bharatpur, Jaipur and Alwar, and then of the Madras in the Punjab. It is, however, a matter of conjecture as to which of these political units possessed the largest number of iron weapons. But as Atranjikhera, in the Pañchāla area, has yielded the largest deposit of iron weapons discovered so far (Raychaudhuri, 1958), it is worthwhile to argue that these people had an edge over their contemporaries in this respect. It is also pertinent to note that the Kuru-Pañchāla area was the epicentre of the Painted Grey Ware (Tripathi, 1976: 21-24), which represents the contemporary iron age culture of India. The formation and the political superiority of the larger kingdoms of the Kuru-Pañchālas, therefore, seems to have been largely influenced by the use of iron-tipped arrows and spears.

The strengthening of the coercive power of the state due to the use of iron weapons further intensified the process of state formation by aiding the emergence of two important organs of government—taxation system and bureaucracy. The use of the adjective *viśamattā* (devourer of the people) (*Aitareya Brāhmaṇa*, VIII, 17) for the king indicates the beginning of regular taxes as well as his new found capacity to extract the same. The officer called the *bhāgadūgha* collected the royal share. The emergence of a bureaucracy is attested by the list of the *ratnins* maintained in the later Vedic texts; at least 12 of them seem to be officials.[8] Contemporary references to the *sthapati* and the *śatapati* may suggest the beginnings of a regular system of provincial administration (Raychaudhuri, 1958: 44) and the *adhikṛta* may be taken as a village officer appointed by the king as stated in *Prasna Upanishad* (cited by Raychaudhuri, 1958: 45). The prime coercive agency of the state - a standing army—gained in importance towards the end of the Vedic period. In the list of the *ratnins*, given in the earlier texts of the later Vedic period, the *senānī*, or commander-in-chief, invariably comes after four or five other officers,[9] but in the *Śatapatha Brāhmaṇa* (V. 3.1) he heads the list. The growing use of iron weapons may have been largely instrumental for the emergence of a standing army which, in turn, heightened the status of the *senānī* by *c.* 700 B.C. The state, despite these changes, was yet to correspond to its definitive paradigm, but the material *milieu* and the iron weapons did help it to undergo the initial phase of transition from tribal polity to the later day *saptānga* state model.

The social role of iron during the second phase of its evolution was rather limited. The two obvious constraints were limitations of source and primitive nature of this technology. The people of the upper Ganga and Sutlej basins were not familiar with the rich iron sources of south Bihar. The iron deposits of Mandi in Himachal Pradesh, Patiala in the Punjab and the Kumaon hills in Uttar Pradesh, that these people exploited, were neither rich nor easily accessible (Bharadwaj, 1979: 154). The technological poverty of iron metallurgy is also evident (Bharadwaj, 1979: 158). The limitation is illustrated by the wasteful rich metallic ferrous slags. Normally small objects of iron with slag inclusions were manufactured. It appears that wrought iron was extracted from solid state reduction of iron oxide with charcoal (Tripathi, 1990). Evidence of carburisation is not forthcoming. The nature of the limitation of this technology has led to the suggestion that even weapons were made in small quantities and probably lay in the sole possession of chiefs and princes (Sharma). Besides these physical constraints, the basic disability from which this technology suffered during this period was the lack of a congenial social climate. The economic transition was well under way and it had its own dynamism and rationale for continuance. In the process a new set of social relations was inaugurated and iron played an important role in crystallising these relations. When the productive relations are changing rapidly, as when a new class is rising into a position of power, there is a particular incentive to harness the newly acquired scientific and technological power for the intensification of this process. Iron technology had class content in terms of its application during this period. It did not determine the process of social change, rather it established a foothold as a socially dependent/conditioned technology.[10]

(iii) The Phase of Percolation, Production and Diffusion, c. 600-200 B.C.

Towards the beginning of the third phase, iron technology is marked by an essential continuity of the earlier tradition. This point can be meaningfully illustrated by redefining the cultural sequences of the Ganga valley, especially for the first 100 years of this technological continuity. It is commonly held that the two cultures — the PGW and the NBP — besides providing safe chronological criteria, represent two differing traditions—temporally and spatially. Despite an established chronological hierarchy, a significant overlap can be noticed between the earlier PGW and the succeeding NBP cultures (Agrawal & Kusumgar, 1979). This overlap has been virtually ignored (Agrawal, 1971: 88). Their primary distribution zones are taken to be different and they are said to be representing different socio-economic formations. PGW culture seems to have a village base while NBP with plentiful iron objects attained an urban bias (Agrawal, 1971: 244). This alleged dichotomy between the two cultures, however, does not stand, for "when looking at both NBP and PGW in relation to their associated cultural assemblages, and within the scope of a regional analysis, no significant qualitative differences are apparent" (Anna King, 1984: 110). Anna King, after analysing the relevant cultural sequences of Hastināpura, Allahapur, Ahichchhatra, Atranjikhera, Noh, Śrāvasti, Kauśāmbī, Rajghat, Prahladpur, Vaiśālī, Rajgir and Chirand, concludes, "It seems difficult to accept PGW and NBP ware as isolated phases representing different cultures ... spatial distributions, associated wares, all the material assemblages, similarities in morphology, decoration, etc., all these elements tend to confirm the notion that both wares are coeval and form part of one single, integral culture. Furthermore, it can safely be stated that the appearance of NBP is not connected with the appearance of new ceramic shapes or types, nor as ushering a new and different cultural environment." Excavations at Atranjikhera confirm the element of cultural continuity. The excavator of the site, R.C. Gaur, writes that during the period of overlap (*c.* 600-500 B.C.) "the general characteristic features of the material culture more or less remained the same as they were towards the end of the preceding PGW period." (Anna King, 1984).

The 6th century B.C., therefore, did not witness any basic change either in the material life of the people or the use pattern of iron implements. No agricultural tool was noticed at Atranjikhera during *c.* 600-500 B.C. In this light, it is difficult to accept the suggestion of a direct linkage between iron technology and the origin of urbanisation in the Ganga valley (Thakur, 1981: 61-65). One of the initial advocates of this theory R.S. Sharma (1983: 72), has subsequently revised his stand, though indirectly, as far as the date of the introduction of iron in the field of agrarian production is concerned. Recently he has suggested that only from *c.* 500 B.C. one encounters "more" agricultural implements. He clearly implies that before the 5th century B.C. iron did not play any significant role in the field of production, an argument which can be stretched backwards to contradict his theory of the use of iron *vis-a-vis* the origins of urban centres in the 6th century B.C. If, as seen from Sharma's own account, enough iron agricultural implements were not available to the pre-500 B.C. people, then how can one talk in terms of any perceptible link between iron aided agriculture and beginnings of urbanisation in the Ganga valley? There is yet another dimension of the problem already touched upon. For arguments sake it may be suggested that if the use of iron tools in agriculture became more frequent from *c.* 500 B.C., it may have affected the economy of at least one century earlier as well. But the PGW and the early NBP archaeology fails to identify any worthwhile assemblage of iron agricultural implements, thereby discounting any possibility of the production of an iron-aided surplus (Thakur, 1981).

In fact, as shown by Rau (1976: 7-50), the beginnings of an urban life dates back to the closing decades of the later Vedic period. An analysis of the contemporary textual evidence suggests that the term 'pura' occurring frequently in early literature stands for temporary emergency defence establishment and not an urban centre as interpreted by many scholars. It is significant to note that the term *nagara*, a usual word for town in epic and classical Sanskrit, gained currency during this period. The original meaning of the word is hardly definitive. It occurs first in the *Jaiminīya Brāhmaṇa* (I. 125) in the context of a dwelling not a town. The precise term is *naunagaraṁ* which can be taken as a house-boat. The *nagara*, as a town, gets the earliest reference in the *Taittirīya Āraṇyaka* (I. 11.7; I.31.2). The *Samavidhāna Brāhmaṇa* indicates its familiarity with a hierarchical order of settlements by referring to a *nagara* as an upgrading of *grāma* and *nigama* (II. 4.2) and valuewise lying between *grāma* and *janapada* (II. 5.6). These citations indubitably refer to a new type of settlement identifiable with a town. It is, however, clear that the term *nagara* came to denote a town only in two of the latest Vedic texts implying the emergence of a town life towards the end of the 7th century B.C. This conclusion is also supported by the contemporary archaeological finds. The PGW - early NBP society had not reached the stage of abundant surplus production and, therefore, qualitatively superior institutional formations could hardly gain in strength. Nevertheless, the beginnings of a nascent urban process towards the end of the seventh century B.C. need not be doubted. The number of such towns increased in the 6th century B.C. The capitals of the *mahājanapadas* assumed urban status but towards the beginning their material make-up was not different from the preceeding period. Even in the early West the situation was not different. Analysing Iron Age cities, Bernal (1965: 153) writes, "Early iron age cities rarely had populations of more than a few thousands, as against the hundreds of thousands of bronze age cities."

Attempts to explain the origins of Buddhism in terms of an iron-aided agriculture-sponsored urbanisation in the Ganga valley (Sharma, 1968: 58-65; 1983: 116-134), suffers from a similar misconstrued perspective. At the time when the Buddha was formulating his ideas iron technology had hardly penetrated the domain of agricultural production (Sarao). The Buddha was born, brought up and attained mental maturity in the pre-500 B.C. urban *milieu*. The urban links of early Buddhism is undeniable but it was not a link with towns sustained by the surplus produced by an iron-aided agriculture (Thakur, 1984: 206-208). The emergence of the territorial states, *i.e.* the sixteen *mahājanapadas*, also represents a similar pattern of socio-economic change. Iron, no doubt, had helped this process by decisively enhancing the coercive power of the ruling group, but in terms of the generation of a surplus for sustaining the system the technology was yet to become relevant.

Iron continued to aid the process of social stratification and state formation, the two roles assigned to it by the emerging social relations in the previous phase. By the beginning of the 6th century B.C. the crystallising social structure was getting formalised and attempts to standardise the new social relations followed. The *Dharmasūtras* manifest this tendency. The ideological upheaval of the 6th century B.C. was seemingly a reaction against the formal crystallisation of a stratified society and the growing complexities of the political system. Even the brāhmaṇa-kṣatriya conflict, a struggle for gaining ascendancy in the social system, was resolved by working out a compromise to ensure that this tension would never develop to a breaking point because that would harm the interests of both the groups (Bhattacharya). Thus, iron seemingly served the interests of the group that had monopolised and exploited it for gaining dominance in society and control over means of production. The technology was now on the verge of coming out of the bounds of dependence. When the productive relations, undergoing a process of rapid change tend to get a definitive shape, as when a new class, rising into a position power, succeeds in establishing itself, "there is a particular incentive to improvement in production which will enhance the wealth and power of this class" (Barnal, 1965: 48). The iron technology, therefore, was prepared for a break-through from the phase of social dependence to assuming the status of prime mover in the society by *c.* 500 B.C.

During the second phase of the NBP (NBP Culture is no longer believed to be homogeneous in time or space), *i.e.* c. 500-350 B.C., changes become perceptible. New trends begin to appear in the form of new pottery types alongwith better specimens of NBP. The NBP of this sub-phase as reported by Gaur (1983: 456), "is excellent in finish and has many colour variations, including those with silvery and golden tinge." The beginning of the use of kiln-burnt bricks at Atranjikhera as well as at Rajghat during this sub-phase (Crain, 1984, Narain & Roy, 1976) adds another facet to this process of change in the material life of the people. It is in this context that the role of iron technology assumed a new dimension. Instead of being confined to the process of the crystallisation of a stratified society, iron now came to play a meaningful role in the process of production. This was the period that marked a decisive beginning in terms of the penetration and transformation of barbarian clan societies through the influence of class economies of the early iron age cities with their emphasis on private property, a monarchical state-system and production of weapon.

This sub-phase (*c.* 500 B.C. - 350 B.C.) witnesses the expanding horizon of iron use, both spatially and technologically. The people of almost the whole of northern India, except the eastern exteriors, grew familiar with this metal before the beginning of Mauryan rule. Its use in agriculture becomes fairly certain, though not all-pervasive. The Jakhera ploughshare of *c.* 500 B.C. and the two ploughshares from middle NBP phase, one each from Kauśāmbī (Sharma, 1983: 95) and Atranjikhera (Gaur, 1983: 425), illustrate the changing scenario. Iron-axes, sickles and chisels, though not profuse, keep on appearing in the middle NBP context. The situation, however, does not warrant any conclusion in favour of an "unprecedented growth" of the crafts, industries and some occupations (Tripathi, 1986: 33-34). It is pertinent to note that most of these agrarian tools are forthcoming from the upper Ganga valley and a deduction on that basis in terms of a parallel use-pattern for the contemporary middle Ganga valley, especially for the area of modern Bihar, does not seem sound. Sharma's (*op. cit.*) arguement about absence of agricultural implements in this area is not fully acceptable, as the climate in this area is not favourable for preservation of iron. He himself has stated this fact (pp. 93-94). Even literature does not support the picture of a widespread use of iron in agriculture during this sub-phase. The term *ayanangala* or iron ploughshare occurs in a later Pāli text (Pali-English Dictionary). It is again in the prose introduction to a *sutta* of the *Suttanipāta* that a *phāla* or ploughshare possibly of iron is mentioned. (The immersion of plough into water here should not be taken for iron alone, but for any metal). The contemporary texts, however, do provide a picture of wide familiarity with this metal. This proliferation of iron technology must have been aided by the access of the people to the richest iron mines found in Singhbhum. The evidence for a limited exploitation of these mines at the local level is forthcoming from an earlier period also. In a context of *c.* 1000 B.C. an iron "sickle-like" object coexisting with iron slag and a profusion of neolithic celts and microliths have been noticed at Barudih in Singhbhum (Chakrabarti : 119). But now the mines seem to have been opened up for the outside world. The analysis of the iron-ore relationship in the context of tools discovered from Rajghat indicates the use of Singhbhum and Mayurbhanj ores at the site (Bharadwaj : 397). This alongwith the known technology of wrought iron and use of bellows proved to be a decisive factor in the dissemination of this technology in the Ganga valley.

Thus, many of the functions assigned to iron technology in the pre-500 B.C. period seem relevant for this period. The earlier cultural trends now get strengthened. Iron now starts playing a vital role in this transformation. The most significant development of the period is the standardisation of cultural elements over a vast geographical area. Thus, the NBP finds of the pre-Mauryan times have been reported from all over the Ganga valley as well as the region of Malwa from sites like Ujjain, Besnagar etc. The emergence of this NBP cultural complex owed a lot to iron-aided agriculture. The use of iron tools helped the expansion of agriculture, a development illustrated by a significant spurt in the number of NBP settlements primarily in eastern U.P. and Bihar where over five hundred sites have so far been located. The use of iron ploughshare facilitated agriculture which was further boosted by the beginning of transplantation of paddy. The available surplus was used for rearing up more urban centres and developed trade. A proliferation of crafts, the growth of guilds and the beginning of metallic currency illustrate the nature of the new economic pattern (Thakur 1981: 66-74). The impact of iron technology, in a slow moving society, was gradually becoming perceptible.

The process of state formation was continuously sustained throughout the pre-Mauryan times. This was the period when Magadha rose to prominence by breaking the political hegemony of the small and primitive states of northern India. The process marked a transition from small and mutually incompatible political units to a large unified kingdom. The role of iron was decisive in this context on two counts. First, the iron technology and the fertile soil of the region combined to produce a sizeable amount of surplus that strengthened the material base of the political institutions of the region. Second, Magadha had a virtual monopoly over the main source of iron metal in the Dhalbhum and Singhbhum districts with quick access to iron encrustations on the Barabar hills (Kosambi, 1975: 155). This enhanced both the military might and the material strength of Magadha in relation to other contemporary states.

Thus, the period from *c.* 500 B.C. marks a point of transition when iron technology breaks the bounds of its exploitation for protecting and perpetuating the emerging production relations and begins to influence the means of production. It also begins to spread in areas hitherto unfamiliar with iron and iron based agrarian technologies. The process of state formation is further intensified. The emergence of a large homogeneous cultural unit comes to be sustained by the broadening base of this technology. The linkage of NBP sites with sites of permanent settlement underlines the growing strength of the material culture of the period.

Although the expansion and refinement of iron technology had been considerable since *c.* 500 B.C., it was not until the second half of the 4th century B.C. that its full impact was felt on society. It is with the arrival of the Mauryas on the scene that dominant economic and political institutions came to be formalised. The period witnessed a further strengthening of iron technology both spatially and technically. A study of the NBP sites suggests that many new regions, where the precise beginning of the Iron Age was uncertain till the pre-Mauryan times, now came to acquire its developed traits (Roy, 1986: 193). Both at Hastināpura (Lal, 1954: 97-99) and Śrāvasti (Sinha, 1967: 67-68) this phase is richer in iron objects compared to the preceding NBP phase. Similar is the case with Kauśāmbī (Sharma, 1960: 45-48). Out of eleven types of iron arrow-heads and five types of spears and javelins reported at the site, as many as eight types of the former and all the five types of the latter occur throughout this phase. This pattern is repeated at a number of sites. At a second set of sites which have recorded cultural antecedents—Ropar, Sonepur and Sohagaura, for instance — iron appears in the 4th century B.C. A third set of sites, where this phase of the NBP is not preceded by any earlier cultural phase, illustrate the physical and spatial progress of iron more poignantly. Excavations at Taxila underline this pattern of technological expansion. In fact as one moves towards the Mauryan times, the site starts yielding a larger number of iron objects, as is best revealed at Taxila (Marshall, 1951: 104, 107) a fact that gives strength to the suggestion that the Iron Age in India almost reaches a stage of culmination in the 3rd century B.C. Iron objects are forthcoming, for the first time, from Śiśupālgarh, Nasik and a host of other sites in Maharashtra and central India (Roy, 1986). Except a few sites, the settlements of the period are decidedly marked by a profusion of iron objects. Iron-yielding sites even in the south, especially in Karnataka and Andhra, suddenly increase in number in the 4th-3rd centuries B.C., but it has been suggested that this had nothing to do with the northern developments (Nagaraju & Gururaja Rao, 1979: 321-329). It is true that a direct linkage between the two cannot be established, but it is difficult to ignore the discovery of some of the typical idioms of Mauryan culture from the region. Notable examples are Mauryan inscriptions and NBP sherds at Amravati. A sudden spurt in iron-working in the area may possibly be explained in terms of the intrusion of the developed elements of late NBP culture, but with the admission that the Mauryan stamp on the cultures of this area is not easily discernible from archaeological evidence.

A greater exploitation of iron mines fulfilled the growing demand for the metal on the one hand, and helped its further spread on the other. The literary evidence is very specific on this point (*Arthaśāstra* II, 12). The Mauryan state exercised a monopoly over mines and trade in mineral products. Kauṭilya, realising the importance of mining, provides for an *ākarādhyakṣa* (superintendent of mines), a *khānyādhyakṣa* (superintendent of digging), a *lakṣaṇādhyakṣa* (superintendent of elements) as well as a *lohādhyakṣa* (superintendent of iron working). Archaeology, however, attests to the working in gold, silver and copper only (Murray, 1983). The location of numerous small heaps of iron slags scattered all over the iron belt in south Bihar, however, suggests substantial iron working. The discovery of a number of pits, containing charcoal, iron slag and sand, from Period I at Saradkel near Ranchi (*Indian Archaeology: A Review* 1964-65: 6), lends credence to this view. The excavator refers to these pits as iron-smelting ovens. This was possibly a factory site. A better organisation of iron-working, therefore, seems to have become the hallmark of the period. Kauṭilya was so conscious of the importance of metals and mining that he went to the extent of asserting that mining was the source of all power (*Arthaśāstra* II. 2). This concern of the state and the growing social demands on iron helped the refinement of its techniques. Now we have better evidence, both in terms of quality and quantity, of smelting and forging of iron objects. The excavations at Dhatwa bear out this development. The smelters here succeeded in extracting 99.76% pure iron. They also devised the technique of manufacturing hard implements, like hoe, in two stages involving the process of "forging welding" (Hegde, 1973). The best set of evidence of smelting and manufacture of iron objects is forthcoming from Ujjainī where lime, representing calcium compounds, was being used as a flux. Analysing the archaeological data, N.R. Banerjee writes, "The remains of a forge with a groove for the introduction of the working end (or nozzle) of blower or bellows, an improvised stand made from the study and large neck of a broken vessel to support a water jar to store water for quenching, a small or miniature jar to collect small quantities of water according to necessity, and a shallow but large enough bowl to contain water near at hand for quenching help to reconstruct the process of a blacksmith's working. The use of an anvil also of iron and iron tools like pincers for handling the red-hot iron bar or tool, is clearly indicated. The iron nail and chisel lying near the forge, which is lined with bricks and clay, burnt hard, and contains ash and charcoal, suggest the variety of objects produced, sharpened or remade at the forge. This method is even now in vogue and would by itself indicate an advanced stage of development of the process of iron working, with a long background." (Banerjee, 1965: 179).

The post *c.* 350 B.C. period, therefore, has two dimensions of the social implications of iron technology—manifestation of its full impact on society and furtherance of this trend through technological expansion. The Mauryan state, which is the best expression of the material *milieu* of the times, was essentially a culmination of the socio-economic process inaugurated during the later Vedic times, which led to the transition from a pastoral/agrarian economy to an established and full-fledged agriculture based economy. The widening and strengthening of the base of agriculture could become possible due to the widespread use of iron ploughshares and extension of agricultural area through deliberate deforestation. Additional foci of agricultural production developed beyond the Ganga valley, especially in the fertile areas of the Indus and parts of the peninsula. It is needless to add that the political authority, due to both enhanced agrarian production and acquisition of coercive powers over the centuries, was in a better position to extract an unprecedented quantity of surplus. This may give an idea as to bow the Mauryan state was able to sustain a huge army and a large bureaucracy. In fact, the Mauryan state, with the help of its coercive arms, seems to be an attempt to organise the earlier tendencies to give it a definite direction. This was sought to be achieved through the formation of a centralised kingdom controlling extensive geographical area, an experiment which was first of its kind in Indian history. It is relevant to note that the distribution of Aśokan edicts, which gives an idea of the extent of the Mauryan empire, synchronises with the distribution of ancient sites of permanent settlement, the sole exception being a group of inscriptions in Karnataka (Thapar, 1966: 81). Thus, the Mauryan conquests need to be viewed as a process of consolidating the achievements of pre-Mauryan material culture. These conquests were attempts to follow and unify a network that had been expanding for hundreds of years (Heitzman, 1984). The emergence of the empire became possible only when the urban tradition had matured to a certain level. The concentration of population in urban centres was due to the production of a sizeable amount of surplus, its appropriation and its commercialisation. The pre-Mauryan times had three major nucleus of communications — the north to south-west route (from Śrāvasti to Pratiṣṭhāna), the north to south-east route (from Śrāvasti to Rājagṛha), and the east-west route (along the river courses of the Ganga and the Yamuna) (Rhys Davids, 1903: 103). The Mauryas capitalised on it by providing better communication facilities to create an additional social basis for empire — the commercial groups.

The Mauryan state with the help of iron technology played the role of a homogenizer of cultures.[11] An improved network of communication alongwith a vast army and a large bureaucracy provided it with the necessary perquisites to achieve this end. The existence of a common script must have helped the process. The role of ideology also needs emphasis in this context. An attempt at bringing about homogenity is reflected at the ideational level in the form of the policy of *dhamma* enshrined in the Aśokan inscriptions scattered all over the country. It was an attempt, based on current religious disputations, to provide a common factor. It was sought to be a vehicle of persuasive assimilation, not military confrontation, "in which conforming to the broad ethical ideals of *dhamma* was central" (Thapar, 1986: 22). The military strength of the Mauryas was so substantial that Aśoka was not faced with the compulsion of elaborating upon the traditional ideology for the legitimisation of the political system. This, coupled with the emergence of a new pattern of political authority, provided an opportune moment for redefining the "non-imperialistic" political normatives. The *dhamma* of Aśoka needs to be viewed as an alternative ideology armed with new requirements of the political system.

The new political system, represented by a vast bureaucratic complex aimed at complete administrative centralisation, forced the pace of agrarian growth with its new found monopoly over a developed iron technology. Their compulsion to tap fresh resources for sustaining the overgrown political structure prompted them to found new settlements and to rehabilitate the decaying ones. Kauṭilya devises an extremely interesting mechanism to bring the virgin soil under cultivation (*Arthaśāstra* II, 1). The state also acquired its own farms worked under the supervision of the *sītādhyakṣa* (*Arthaśāstra* II. 24). The agrarian expansion was greatly helped by the initiative taken by the state in the field of irrigation. Iron technology became instrumental in the completion of massive irrigational projects. Archaeology also attests the process of planned deforestation undertaken by the Mauryan state. A recent study of 74 excavated NBP sites identifies 32 sites with early NBP culture and 57 sites with late NBP culture (Roy, *op. cit.*). The frequent occurrence of ring-wells during this period now made the founding of settlements away from the river-streams less problematic. A natural consequence of this tremendous spurt in agrarian activities was an increase in the number of agricultural taxes. The cultivators had to bear the burden of additional levies - the *piṇḍakara*, the *senābhakta* and an irrigation cess, (*Arthaśāstra* II. 4) besides the payment of an emergency tax, the *praṇaya* (*Arthaśāstra* V. 4). The state's concern for protecting and supporting the pattern of agrarian expansion, illustrated by Aśoka's repeated exhortions in favour of non-injury to animals,[12] was an essential component of the system.

This large-scale agrarianisation of the Ganga valley as well as the outlying regions generated such an amount of surplus that a number of new towns came to be established and the material roots of the older urban settlements strengthened. The transformation of Pāṭaliputra into a cosmopolitan centre conforms to this new reality of the material life. The Mauryan towns functioned as centres of state-control over crafts and commerce as there are numerous references to roads constructed by Aśoka in his inscriptions. Consequently the organisation of trade became smoother and "the crafts gradually assumed the shape of small-scale industries" (Thapar, 1973: 72). It is significant to note that the earliest discernible well-organised network of long-distance trade belonging to the 3rd century B.C., connected only those areas that had been permeated by stable settlements (Schwartzbeg, 1978: 19). The iron technology based rural settlements were undoubtedly providing support, and occasional direction, to the growing urban economic forms of the Mauryan times.

The strong material make-up of the Ganga valley became instrumental in affecting far-reaching changes in the peripheral zones. The process admitted of two distinct patterns—diffusion of iron based agrarian technology in the areas and helping the organisation of the local elements of material culture by assimilating these into the mainstream. Places like Mahasthan[13] in Bangladesh and Khanamihirer Dhibi[14] in West Bengal, which have yielded iron implements and indubitable evidence of their emergence as stable settlements in the 3rd century B.C., may have acted as nuclei of the transition to settled agriculture in their respective zones. Areas like Kalinga seem to have received the new elements in a greater measure. The region came into contact with Magadha in the 4th century B.C., but the transition was possibly affected in the 3rd century B.C. due to an excessive concern of Aśoka for this area after its assimilation within the empire.[15] The evidence comes in the form of the earliest iron implements unearthed from five sites in the area - Śiśupālgarh, Jaugada, Asurgarh, Kharliagarh and Gudbhela. Iron came to be used from the 3rd century B.C. and the objects included agricultural implements. The process of agrarian transformation may have been intensified due to the spread of the knowledge of steel or semi-steel making in the area due to Mauryan contacts. Significantly the growth of material culture in Orissa was not confined to the coastal districts, as is often suggested (Lal: 79). The presence of the developed elements of material culture—iron implements and coins — from widely scattered areas points to a process of hospitable reception of iron technology almost in the whole of the province. The Chedi kingdom, was a product of this agrarian scenario and the Hathigumpha inscription of Khāravela, which shows the king's concern and initiative for providing irrigational facility to the cultivators, amply testifies to the effects of the penetration of iron technology in the area.

The role of the Ganga valley culture in the dessimination of its elements in the Deccan, however, cannot be determined with clarity. The region does have a history of iron use right from the Hallur days but unlike the north, it failed to evolve an urban culture, at least, till *c.* 300 B.C. This requires an explanation especially from those who advocate for an exclusively autonomous pattern of social formations in that region (Parashar, 1981). It is a common knowledge that the area had a strong tradition of iron use, but this point needs to be stated with the admission that the historical transformation in the area could not be brought about before the 3rd century B.C. It is here that the Mauryas seem to have played a vital role. The earliest documented evidence of Mauryan intrusive element comes in the form of a concentration of Aśokan inscriptions from Mysore and a fragmentary one from Amravati in Andhra. It is true that the excavations at Brahmagiri (Wheeler), where two Aśokan inscriptions were found, did not yield any identifiable material remains of the Mauryas, but the beginning of a new ceramic type and the first appearance of iron at the site in Period II (contemporaneous with the Mauryas) are of consequence. This ceramic type also occurs at Śiśupālgarh (Lal), Kesarapalli (Sarkar) (Andhra) and Korkai (Champaklakshmi) (Tamilnadu). One wonders whether the four sites are providing for a common intrusive element of the Mauryan culture. The Amravati evidence (Rea, 1912) is, however, more decisive. About one km. from this Buddhist establishment we have an urban site Dharanikota (*IAR*, 1963-64: 2-4; 1964-65: 2-3). It is certain that during the Mauryan phase of contact, the Amravati-Dharanikota complex was emerging as a dynamic religious and commercial urban complex in the lower Krishna regions. This transition to an urban settlement in the 3rd century B.C. can be best explained in terms of the organising role of the Mauryan state and the intrusion of iron technology. The available evidence further suggests that Amravati emerged as a focal point of the diffusion of material culture in the area (Bagle, 1986: 304-305). The very process of state formation in the region, manifest in the form of the Sātavāhan polity, seems to have been influenced by the introduction of the new elements of material culture. The new rulers sought to legitimise their rule by adopting some of the dominant idioms of Mauryan administration and culture.

The process of the diffusion of the material culture of the Ganga valley in outlying regions was also aided by Buddhism. It is significant that the earliest surviving Buddhist monastic sites emerged in the 3rd century B.C. (Heitzman, 1984). The original zone of their distribution was located in the Yamuna-Ganga valley with stray examples in central India (Bharhut and Sanchi) and the Deccan (Amaravati). All the early Buddhist centres, lying either close to non-monastic sites or along routes connecting urban locations, lay within the distribution areas of either the Aśokan inscriptions or the NBP. Thus in the first phase of its evolution the Buddhist monastic settlements "were closely linked to the major forms of political and economic organization that were developing at the same time." Buddhism seemingly strengthened the hands of the Mauryan state in the former's endeavour to extend the material culture of the Ganga valley to outlying areas. The emergence of an all-India economic pattern by the 2nd century B.C. bears testimony to the positivity of this effort.

By *c.* 200 B.C. iron technology thus came to occupy a position of centrality in the mechanism of production. The improvement in its techniques and the wide area of its expanse helped large-scale growth of stable settlements. This was a prelude to economic unity of the whole of the sub-continent. The Mauryas played a decisive role in this process by formalising, standardising and diffusing the earlier trends. The trend, inaugurated in *c.* 1300 B.C. due to the beginning of the use of iron in India, had remained dormant till *c.* 1000 B.C. The succeeding centuries witnessed its gradual social acceptability through a process of mutual adaptations.

The real break-through was achieved after *c.* 500 B.C. The complimentary role played by the metal in the preceding period, however, created a congenial social climate helping its transition from a socially dependent to a socially dominant technology.

NOTES

1. *Taittiriya Samhitā* I.8.10.
2. *Jaiminiya Brāhmaṇa* I.43.
3. *Śatpatha Brāhmaṇa* V. 3.3.2.
4. *Ibid.*, V. 3.3.6.
5. *Ibid.*, V. 1.4.14.
6. *Taittiriya Samhitā* I.8.10.
7. The production of surplus is indubitably attested by the beginnings of trade, i.e. the commercialisation of a part of the available surplus and the diversification of arts and crafts. The *Atharvaveda* (III. 15.3) refers to three qualities of a successful trader—balam (Physical Strength) *Budhih* (wisdom) and *ruchi* (interest), a reference that indicates the emergence of a mercantile community. Reference to commercial transactions strengthen this hypothesis. Barley and rice were traded for lead (*Śatapatha Brāhmaṇa* XII 7.2.10) and some spices like Jambila i.e. a variety of large lemon (*Maitrayani Samhitā*, III. 15.3) haridra i.e. turmeric (*Atharvaveda* XII 24.2) and pipalli, (*Ibid* VI 10.9) were decidedly mercantile commodities. Sheep's wool was used as a medium of barter for obtaining grains and other commodities (*Śatapatha Brāhmaṇa* XII 7.2.10). Similarly the arts and crafts enumerated in the context of Purusmedha indicate considerable progress in differenciation of functions. There are references to smelters, smiths, carpenters, cloth manufacturers, leather workers, potters and jewellers (B.P. Roy). One is tempted to add that iron was aiding the two activities-iron objects may have been traded in and this metal did help certain crafts to develop. One may seriously debate the role of iron in craft specialization, but one can not doubt its relational dependence on surplus and its commercialisation.
8. *Taittiriya Samhitā*, I.8.9; *Maitrayani Samhitā* II. 6.5; IV 3; *Kathakā Samhitā* XV 4; *Taittiriya Brāhmaṇa* I 73 ff. *Śatapatha Brāhmaṇa* V 3.1.
9. *Taittiriya Samhitā* (I.8.9) the *Maitrayani Samhitā* (II 6.5; IV. 3) and the *Kathakā Samhitā* (XV. 4) gets the fifth spot, while the *Taittiriya Brāhmaṇa* (I.7.3. ff) accords him the sixth place.
10. The role of social climate for the use of technology is further illustrated if one contrasts the PGW developments with that of the pre 600 B.C., iron yielding cultures of eastern India. The evidence from Chirand, Taradih and Sonepur in Bihar alongwith the finds from Pandu Rajar Dhibi, Mahisdal and Tulsipur in W. Bengal suggest a date around 800-700 B.C. for the beginning of iron in this area. The point of interest, however, is the lack of any evidence for any consequential

role played by this metal in these cultures. Iron in the PGW culture did play a role in social stratification and class formation due to the social demands of the changing production mechanism. The absence of a similar situation i.e. a complete lack of a congenial social climate in the early iron cultures of eastern India may explain this non-acceptance of the new technology, situation that can compare favourably with the first phase of iron in Northern India. The situation in the area changes, however, with the superimposition of the material culture of the later Vedic period from 600 B.C.

11. The Mauryan state seems to be a very good example of secondary state. The secondary states, it has been argued, are formed by primary states conquering, non-states (R. Cohen & E.R. Service ed. *Origins of State* 1978, Philadelphia p. 6 ff). It is also economically integrated into the conquering state. Despite Romila Thapar's objections (*From Lineage to State*, Bombay 1984, pp. 159-60) the Mauryan state seems to have performed both the functions.
12. Seven (Rock Edicts - I, II, III, IV, VIII, IX, XI) out of the fourteen major rock edicts, to cite an instance of Aśoka's concern for non-violence to animals, repeat this request. Kauṭilya too is familiar with this issue. Almost echoing the same idea, he states, cattle such as a calf, a bull or a milch cow shall not be slaughtered (*Arthaśāstra*, II 26).
13. R.K. Mukherji & S.K. Maity, *Corpus of Bengal Inscriptions* Bearing on History & Civilization of Bengal, Calcutta 1967, pp. 38-39. This is incidentelly the first inscription forthcomming from Bengal suggesting the introduction of a script in the area during the Mauryan times. It refers to Pundranagar as a prosperous city where grains and coins were stored to meet the consequences of natural calamities. The establishment of village settlements and the availability of considerable amount of agrarian surplus in the area, therefore seems indubitable.
14. *Indian Archaeology, A Review* 1956-57: 29 ff.; 1957-58: 51; 1959-60: 56; 1960-61: 39; 1961-62: 62; 1962-63: 56; 1963-64: 63; 1964-65: 52; 1966-67: 48.
15. The Kalinga rock edicts I-II, Project this new policy formulation. It has been suggested, however, that Kalinga was a well populated geographical entity before the war implying thereby the marginal impact of Mauryan rule over the area (B.P. Sahu "The Archaeology of early Historic Orissa," *Proceedings of the Orissa History Congress*, 9th Session, Rourkela 1981, p. 2. Some aspects of Early Orissa Economy and Society, *Proceedings of the Indian History Congress*, 41st Session, Bombay, 1980, pp. 126-127). The figure of casualities and captives recorded in the Aśokan edicts are decidedly inflated and a hypothesis based on those numbers can hardly be reliable. Moreover, Kalinga association with Magadha from earlier period is very much admitted but the evidence at our disposal, both archaeological and literary, does not permit us to picture the pre - 300 B.C. Kalinga as an area of stable settlements.

    The Ganganagar minor irrigation project area and Hiratgarh have also yielded iron implements in post 200 B.C. context (R.N. Das, 1978 "Adya Lauha Yugair Orissara Sanskriti" *Paurus*, Cuttak, June pp. 69-72).

REFERENCES

Agrawal, D.P., 1971. *Copper Bronze Age in India*, New Delhi.

Agrawal, D.P. & Kusumgar, S., 1979. "Radiocarbon Chronology in Indian Protohistoric Cultures" in *Essays in Indian Protohistory*, ed. D.P. Agrawal & D.K. Chakrabarti, pp. 371-386, Delhi.

Banerjee, M.N., "Metals and Metallurgy in Ancient India", *Indian Historical Quarterly* Vol. III, pp. 793-802.

Banerjee, N.R., 1965. *Iron Age in India*, New Delhi, p. 179.

Barnal, J.D., 1965. *Science in History*, Vol. I, 3rd edition, Harmondsworth.

Bagle, Vimal, 1986. "From Iron Age to Early Historical in South Indian Archaeology" in J. Jacobson (ed.) *Studies in the Archaeology of India and Pakistan*, New Delhi, pp. 304-305.

Bharadwaj, H.C., 1979. *Aspects of Ancient Indian Technology*.

Bhattacharya, Sibesh. "Political Authority and Brahmana-Kshatriya Relationship in Early India - An Aspect of the Power Elite Configuration", *The Indian Historical Review*, Vol. X, Nos. 1-2, pp. 1-20.

Chakrabarti, D.K., 1973. "Beginning of Iron in India: Problems Reconsidered" in *Perspectives in Palaeo-anthropology*, A.K. Ghosh (ed.), pp. 345-356.

Chakrabarti, D.K., 1979. "Iron in Early Indian Literature", *Journal of Royal Asiatic Society of Bengal*, Great Britain and Ireland No. 1.

Chakrabarti, D.K., 1988. *Theoretical Issues in Indian Archaeology*, New Delhi, p. 119.

Champaklakshmi, R., "Archaeology and Tamil Literary Tradition" *Puratattva*, No. 8, p. 114.

Childe, V. Gordon, 1942. *What Happened in History*, Harmondsworth.

Gille, Bertrand 1990. "Technical Progress and Society: Historical Perspective" (tr. Sharon L. Romeo) in *Techniques to Technology: A Frech Historiography of Technology*, Delhi, p. 179.

Hegde, K.T.M., 1973. "Early Stages of Metallurgy in India", in D.P. Agrawal and A. Ghosh (eds.) *Radiocarbon and Indian Archaeology*, Bombay, pp. 401-405.

Heitzman, James, 1984. "Early Buddhism, Trade and Empire", in *Studies in Archaeology & Palaeo-anthropology of South Asia*, KAR Kennedy & G.L. Possehl (eds.), New Delhi.

King, Anna, 1984. "Some Archaeological Problems Regarding Gangetic Cultures in Early Historical India" in *Studies in the Archaeology and Palaeo-anthropology of South Asia*, KAR Kennedy and G.L. Possehl (eds.), New Delhi.

Kosambi, D.D., 1973. "The Beginning of Iron Age in India". *Journal of the Economic and Social History of the Orient*, Vol. VI, pp. 309-318.

Kosambi, D.D., 1975. *An Introduction to the Study of Indian History*, 2nd ed. Bombay.

Lal, B.B., 1948. "Sisupalgarh: An Early Historical Fort in Eastern India", *Ancient India*, No. 5, p. 79.

Lal, B.B., 1954. "Excavations at Hastinapur and other Explorations in the Upper Ganga & Sutlej Basin 1950-52", *Ancient India*, No. 10-11.

Macdonell & Keith, 1958. (Reprint), *Vedic Index of Names and Subjects* Vol. I. Especial reference to a metal share like that of a lance, Kathak Samhita XVI.

Macdonell & Keith Vol. I, p. 48; Vol. II, 275-276, 110, 309, 89, 87, 262.

Marshall, Sir John, 1951. Taxila Vols. I & II, Cambridge.

Marks and Engels, 1965. *Selected Correspondence* 2nd edition, Moscow.

Murray, 1983. Journal of the Royal Asiatic Society of Bengal, 3rd series Vol. VI, p. 101, cited in R.S. Sharma's *Perspectives in Social and Economic History of Early India*, New Delhi, p. 13.

Nagaraju, S. and Gururaja Rao, B.K., "Chronology of Iron Age in South India", in *Essays in Indian Protohistory*, (eds.) D.P. Agrawal and D.K. Chakrabarti, Delhi, pp. 321-329.

Narain, A.K. & Roy, T.N., 1976. Excavations at Rajghat, 1957-58, 1960-65, pt. I, Varanasi pp. 25-27.

Niyogi, P., 1914. *Iron in Ancient India*, Calcutta.

Parashar, Aloka. 1981. "Social Structure and Economy of Settlements in the Central Deccan (200 B.C. - 200 A.D.)", in Indu Banga ed. *The City in Indian History*, New Delhi, pp. 23-23.

Rau, W. 1976. *The Meaning of Pur in Vedic Literature*, Munchen, p. 750.

Raychaudhuri, H.C., 1958. *An Advanced History of India*, R.C. Majumdar, H.C. Raychaudhari & K.K. Datta eds., London.

Raychaudhuri, H.C., 1972. *Political History of Ancient India*, Calcutta, 7th Edition.

Rea, Alexander, 1912. Excavations at Amravati, *Annual Report, ASI*, 1908-1909, Delhi, pp. 88-89.

Rhys Davids, 1903. *Buddhist India*, London.

Roy, B.P., *The Later Vedic Economy*, Patna, pp. 142-150.

Roy, T.N., 1986. *A Study of Northern Black Polished Ware Culture: An Iron Age Culture of India*, New Delhi.

Sahi, M.D.N., 1974-75. *Indian Archaeology; A Review*.

Sahi, M.D.N., 1979. "Iron at Ahar" in *Essays in Indian Protohistory*, Agrawal, D.P. & Chakrabarty, D.K. (eds.), pp. 367-369.

Sahi, M.D.N., 1982. "Agricultural Production During the Early Iron Age in Northern India". *Proceedings of the Indian History Congress*, 43rd Session, Kurukshetra.

Sahi, M.D.N. "New Light on Life of PGW People as Revealed from the Excavations at Jakhera", *Man and Environment*, Vol. II, pp. 104-105.

Sarao, K.T.S., "Background to the Origin of Earliest Buddhism". *Indological Taurinensis*, vols. XV, pp. 305-318.

Sarkar, H., 1962. "Kesarpalli", *Ancient India*, No. 22, p. 95.

Sharma, G.R., 1962. Excavations at Kausambi, 1958-59, Allahabad.

Sharma, R.S., 1983. Cited from *Times of India*, New Delhi Edn. 26th June, 1981.

Singh, S.D., 1965. *Ancient Indian Warfare with Special Reference to the Vedic Period*, Leiden, pp. 94-95.

Sinha, K.K., 1967. *Excavations at Sravasti*. Varanasi.

Snodgrass, A.M., 1980. "Iron and Early Metallurgy in the Mediterranean" in *The Coming of the Age of Iron*, Wertime & Muhly (eds.). New Haven.

Thapar, Romila, 1966. *A History of India*, Baltimore, p. 81.

Thapar, Romila, 1973. *Asoka and the Decline of the Mauryas*, 2nd Edition, OUP, Delhi.

Thapar, Romila, 1986, *The Mauryas Revisited*, Calcutta.

Thakur, V.K., 1981. *Urbanization in Ancient India*, New Delhi.

Thakur, V.K., 1981. "Role of Iron in Second-Urbanization in India: A Resurvey of Evidence", *Proceedings of the Indian History Congress*, Bodh Gaya.

Thakur, V.K., 1984. "Origins of Buddhism: A Resurvey of Material Factors" in *Religion and Society in Ancient India*, P.N. Jash (ed.), Calcutta.

Thakur, V.K., "A Marxist View of Ancient Indian History", *The Journal of Bihar Research Society*, LXII, Part I to IV.

Tripathi, Vibha, 1973. "Introduction of Iron in India - A Chronological Perspective" in *Radiocarbon & Indian Archaeology*, D.P. Agrawal & A. Ghosh (eds.), Bombay, pp. 272-278.

Tripathi, Vibha, 1976. *Painted Grey Ware - An Iron Age Culture of Northern India*, New Delhi.

Tripathi, Vibha, 1985. "From Copper to Iron — A Transition", *Puratattva*, No. 15.

Tripathi, Vibha, 1986. Metal Technology vis-a-vis Economic Growth During the First Millennium B.C. Varanasi (Rajghat) in *Varanasi Through the Ages*, T.P. Verma (ed.), Varanasi.

Tripathi, Vibha, 1990. "Origin and Development of Iron Technology in India & Its Economic Implications" in *History of Science and Technology*, Vol. IV, G. Kuppuram and K. Kumudmani (eds.), pp. 14-21, New Delhi.

Varma, K.C., 1982. "The Iron Age, the Veda and the Historical Urbaniation" in *Indian Archaeology: New Perspectives*, R.K. Sharma (ed.), Delhi, pp. 155-157 & 172-174.

Wheeler, R.E.M. 1947. Brahmagiri and Chandravalli: Megalithic and other cultures in Chitaldurg, Dist. Mysore State. *Ancient India*, No. 4, p. 187 ff.

List of Illustrations

PRANAB KUMAR CHATTOPADHYAY AND BIRENDRA PRATAP SINGH

Pl. 1. (A-D) Senuwar (Illustrations of Metallographs)

HARI NARAIN, JAI PRAKASH, M.V. NAIR AND O.P. AGRAWAL

Pl. 2. A corrosion affected area near the periphery in a bangle Cu 96, from Manjhi 50 X.

Pl. 3. Fine equiaxed twinned crystals showing slip band in section from Copper bangle Cu 96 from Manjhi. Etched in ferric chloride, 500 X.

Pl. 4. Another corroded area (intergranular corrosion) in the same bangle, Cu 96, 50 X.

Pl 5. Fine grain equiaxed with twins and slip bands and inclusions in a section from Copper bangle Cu 94 from Manjhi. Etched in ferric chloride, 200 X.

Pl. 6. Corrosion affected area (intergranular corrosion) near the periphery and fine equiaxed twinned crystals with slip band in copper Cu 95 from Manjhi. Etched in ferric chloride, 500 X.

Pl. 7. Equiaxed twinned crystals of small grain size and inclusions of lead in a section from antimony rod Cu 97, site Manjhi. Etched in ferric chloride, 500 X.

Pl. 8. Equiaxed with twins and slip bands of moderate grain size and lead inclusions in a section from tool Cu 98, site Manjhi. Etched in ferric chloride, 500 X.

HARI NARAI, JAI PRAKASH, O. P. AGRAWAL AND M.V. NAIR

Pl. 9. Photomicrograph of the section from broken spoon lab no Fe 185, showing widmanstatten, pattern of ferrite at 200 X etchant 2% Nital.

Pl. 10. Another area of the same section from broken spoon lab no Fe 185 showing lammeller pearlite at 500 X, etchent 2% Nital

Pl. 11. Axe, Lab no Fe 178, section from near the cutting edge showing transition from low carbon area (core) to carburized region on outside (near the cutting edge). Ferritre (white), coars lamaller pearlite (dark) and high carbon area (dark), 50 X, etchant 2% Nital.

Pl. 12. Another area of the same section from the axe lab no Fe 178, showing widmanstatten structure in the interface of two regions of ferrite and pearlite and slag inclusions at 500 X, Etchant 2% Nital.

Pl. 13. Sickle, lab no Fe 182, Section from the cutting showing network of relic carbide structure in completely mineralized metal at 1000 X, Etchant 5% Nital.

Pl. 14. Photomicrograph of a section from bent knife lab no Fe 247 revealing alternate layers of wrought and low carburized iron in a section from the knife 125 X 2% Nital etchant.

B. PRAKASH

Pl. 15. Macrostructure of 'wootz' steel ingot.

Pl. 16. Microstructure of 1.33% C steel showing platelets of cementite in a matrix of ledeburite. (X 100, 2% Nital).

Pl. 17. 'Damascusizing' pattern the hallmark of wootz steel swords.

Pl. 18. Swords and dagger of 17th-18th C provided by Bharat Kala Bhawan (B.K.B), B.H.U. for surface study.

Pl. 19. The 'Watering Pattern' seen on the B.K.B. dagger (No. 3/9172).

Pl. 20-21. The 'Watering Pattern' scene on the B.K.B. Shamsheer (No. 11/1604)

Pl. 22. Microstructure of L.C. steel consisting of colonies of pearlite in a matrix of ferrite (X 200, 2% Nital).

Pl. 23. Microstructure of H.C. steel consisting of Ferrite grain boundry in a matrix of pearlite (X 200, 2% Nital).

Pl. 24. Photograph of the shape of the five forged blades.

Pl. 25. (a) Microstructure of the sample No. 3 showing Ferrite and Pearlite grains on the L.C. face (X 200, Picral), and (b) the fine dispersion of cementite globules in tempered Martensite on the reverse (H.C.) side (X 500, Picral), (c) shows the bond between the two layers.

Pl. 26. Microstructure of the cross-section of sample No. 4 showing the carbide pattern on the high carbon plate which has some similarity to the watering mark (X 50, 2% Picral).

Pl. 27. Microstructure of sample No. 5 near the hardened edge. It shows carbide network and tempered martensite (X 200, Picral).

Pl. 28. Microstructure of the cross-section of sample No. 5. It shows discontinuity in the joint of the L.C. plate and the H.C. plate shows carbide network similar to that shown in Pl. 26 (X 50, 2% Picral).

P. T. CRADDOCK ET AL.

Pl. 29. Entry to the underground mine from the surface on hill slop.

Pl. 30. Ancient underground stope, showing smooth wall and back, characteristic of fire setting.

Pl. 31. Iron tools recovered from ancient Mochia mine, Zawar.

Pl. 32. Vertical shaft on surface extending down underground stope.

Pl. 33. Wooden ladder, Zawarmala, dated 2150 ± 35 years before present.

Pl. 34. Wooden launder used to channelled mine water, Zawarmala, dated 2140 ± 110 years before present.
Pl. 35. Clay lamp recovered from Zawarmala.
Pl. 36. Part of timber revetment excavated at East Lode, Rajpura-Dariba, dated 2140 ± 100 years before present.
Pl. 37. Mortar hole carved into the hard calc silicate rock for ore crushing, RD mines.
Pl. 38. Small cupel with fragment of tuyere, Agucha, 3rd century BC
Pl. 39. Used retorts with broken condenser necks.
Pl. 40. Banks of zinc distillation furnace excavated at Zawarmala hill, Zawar, 15th century AD.
Pl. 41. Portion of perforated plate which separate the upper furnace chamber from cooler chamber.

VIBHA TRIPATHI

Pl. 42(a) Furnaces in a row as excavated, Khairadih, Distt. Ballia.
Pl. 42(b) An Agaria furnace still in vogue in Iron smelting.

B. B. AGRAWAL, K. K. PRASAD, G. I. S. CHAUHAN, P. K. CHAUDHURI AND S. M. AERON

Pl 43. An Excavated furnace (presumably operated over fifty year ago).
Pl. 44. Feet operated bellows used for air blast.
Pl. 45. Tribal furnace, Garha Harup Village, Netrahat Plateau.
Pl. 46. Tribal furnace, Garha Harup Village, Netrahat Pleteau.

N.R. SRINIVASAN

Pl. 47. Modified Rural Iron-making furnace, handblowing of air.
Pl. 48. Wrought Iron Mag. 100X.

SHARADA SRINIVASAN

Pl. 49. Cast ingot being heated, Payangadi, Palaghat district, Kerala.
Pl. 50. Ingot being beaten for making bowl.
Pl. 51. The ingot after it has been beaten out into a bowl is quenched in water.
Pl. 52. Inner surface of wrought bowl is scraped and polished with scribers.
Pl. 53. Contrast between polished and worked bowl which now has a golden inner surface, and the bowl after quenching which is blacked - is evident.

R. DASGUPTA

Pl. 54. Bānbāṭi or recptacle for holding curry, made of bell metal.
Pl. 55. Bāṭa or a vessel made of bell-metal for keeping *pan* and betelnuts (after Gait).
Pl. 56. Sarai or ritual vessel for placing offerings of food during religious services.

Pl. 57. Bronze lamp stand, Assam.
Pl. 58. Bronze lamp stand, Assam.
Pl. 59. Ṭemī, or a bell metal box used for keeping lime to be eaten with *pan*.
Pl. 60. An Ahom canon, Assam State Museum, Gauhati.

MADHURI SHARMA & D.P. SHARMA

Pl. 61. Stone anthropomorphic figure, 1000 B.C. Mathura.
Pl. 62. Bisauli anthropomorphic, Allahabad Museum.
Pl. 63. Anthropomorphic figure, Type II, Shiv-Raspur (Kanpur) Lucknow Museum.
Pl. 64. Anthropomorphic figure, Chandausi, Muradabad, National Museum.
Pl. 65. Anthropomorphic figure, Type I, Nurpur, Chisti Collection.
Pl. 66. Anthropomorphic figure, Type III, Patna Museum.
Pl. 67. Detail of Bead necklace of anthropomorphic shape, Yaksha from Pitalkhora.

D.P. SHARMA

Pls. 68-69. Lugged shouldered copper axe, Sitapur, National Museum.

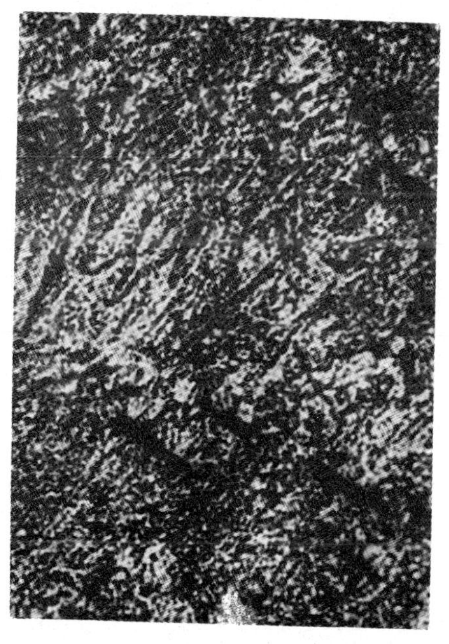

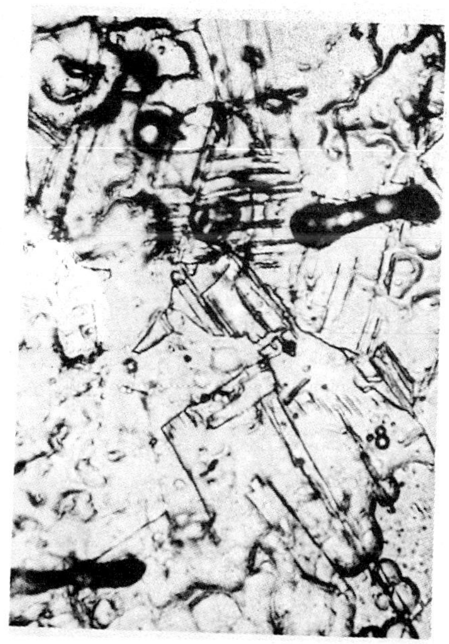

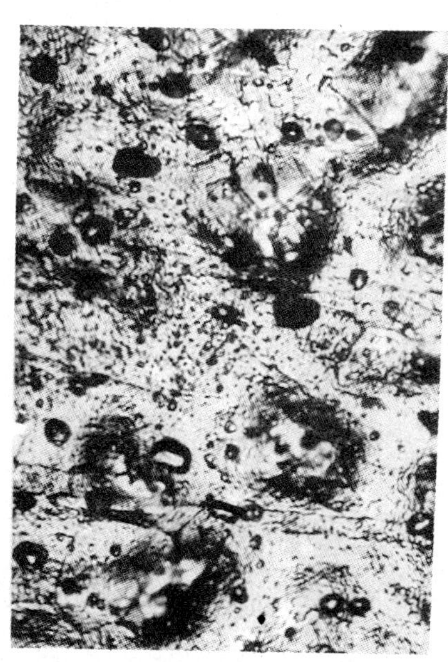

Pl. 1

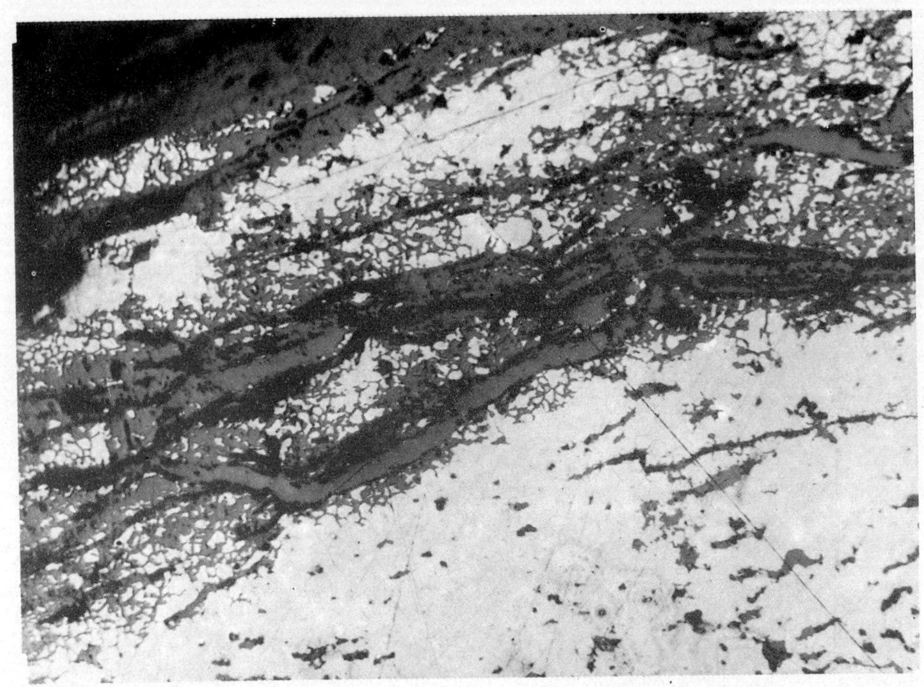

Pl. 2

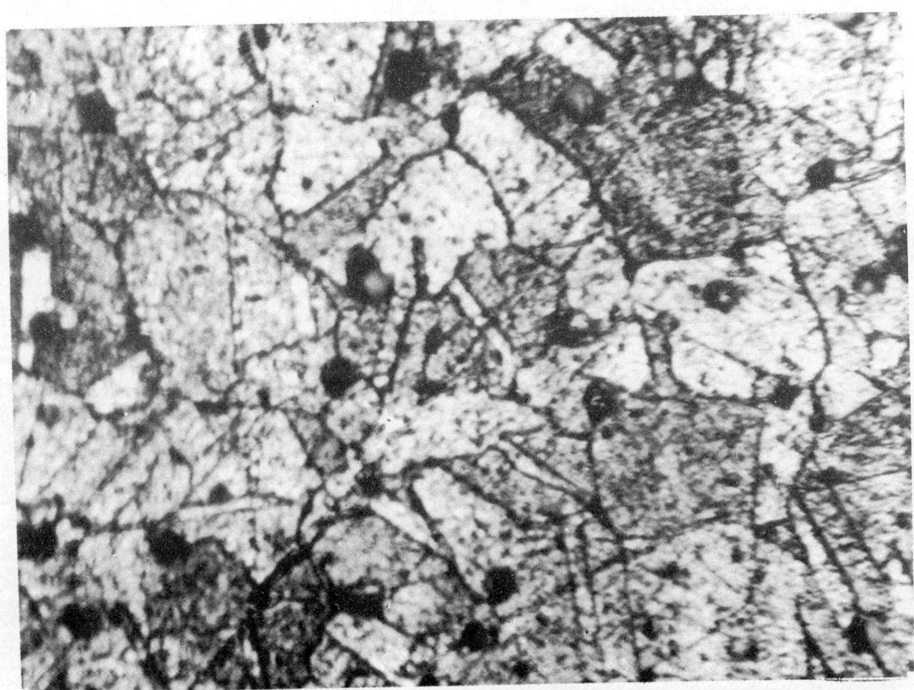

Pl. 3

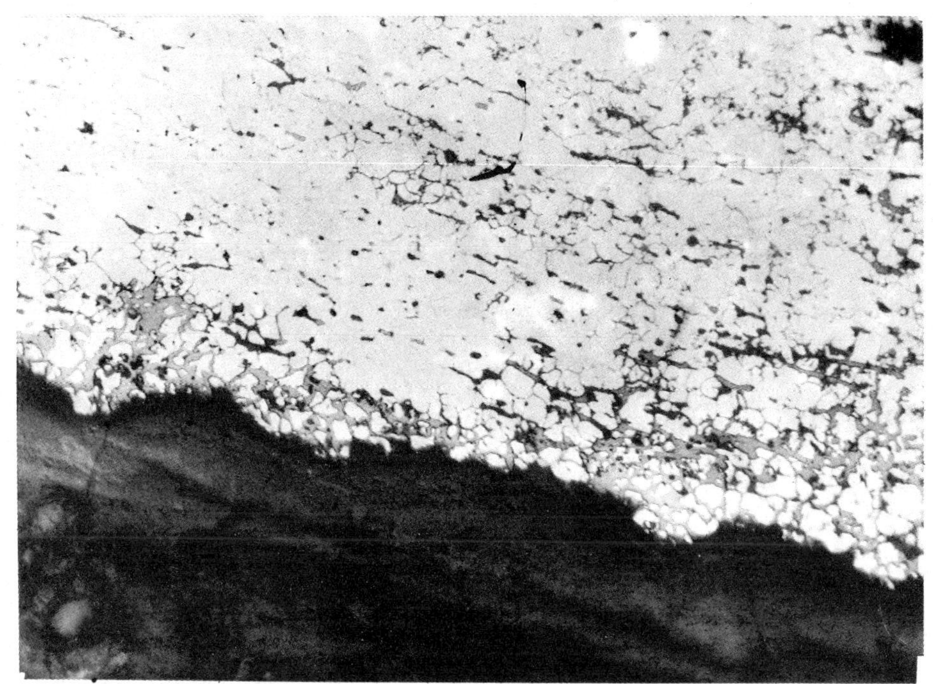

Pl. 4

Pl. 5

Pl. 6

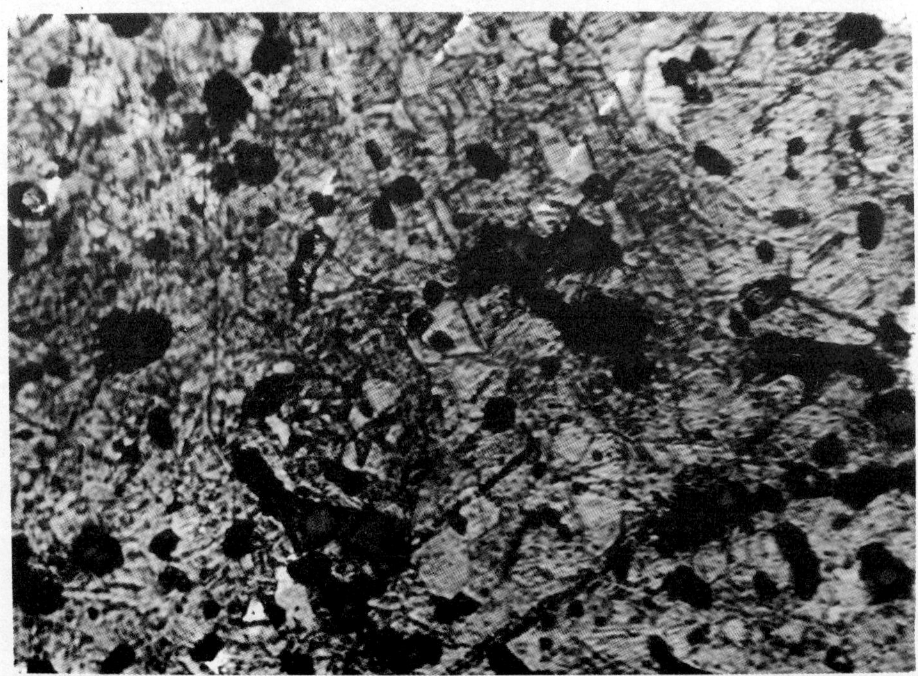

Pl. 7

Pl. 8

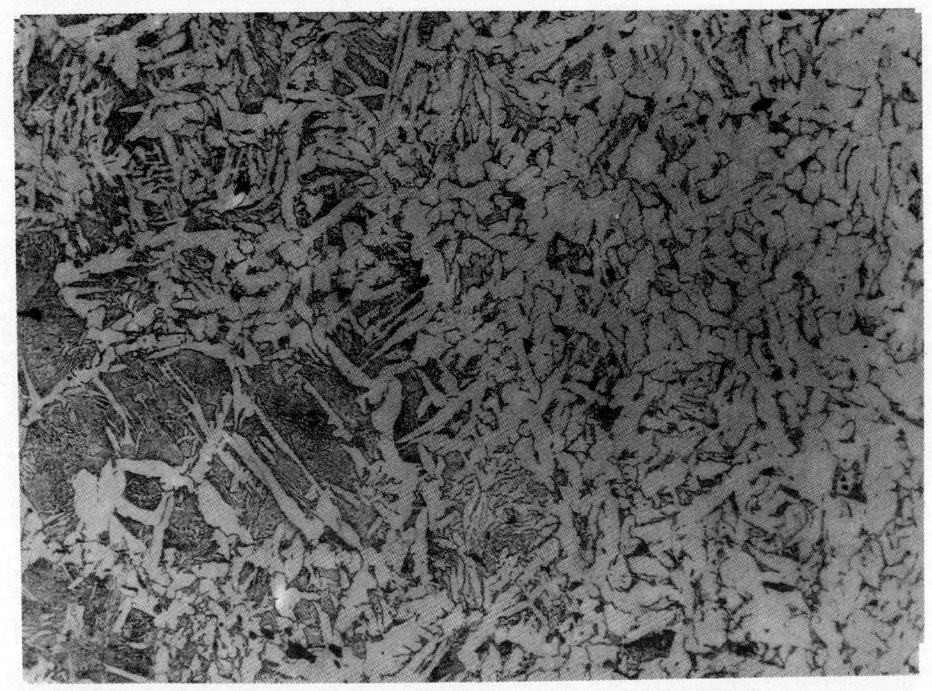

Pl. 9

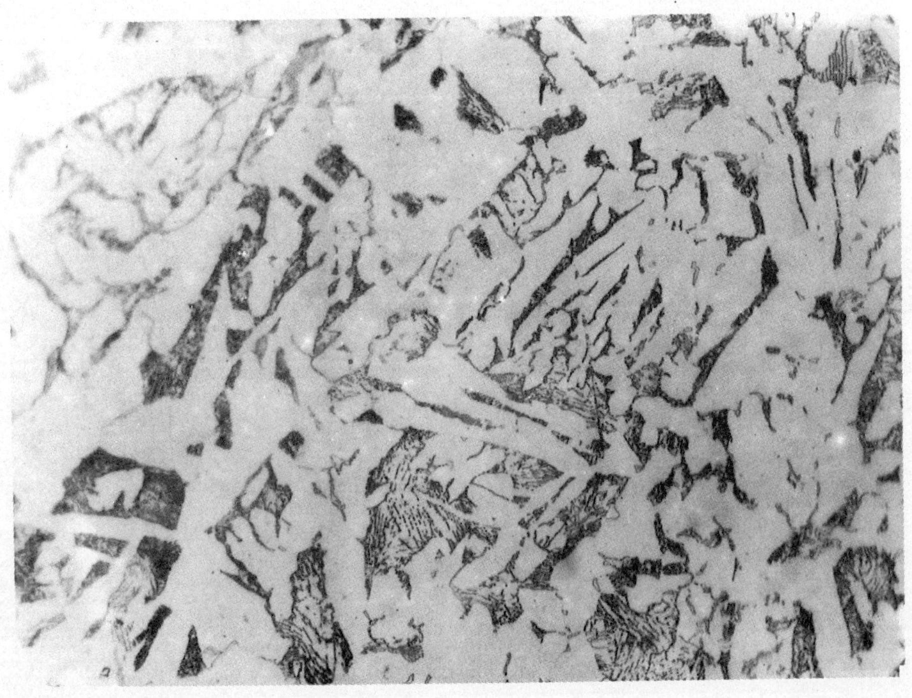

Pl. 10

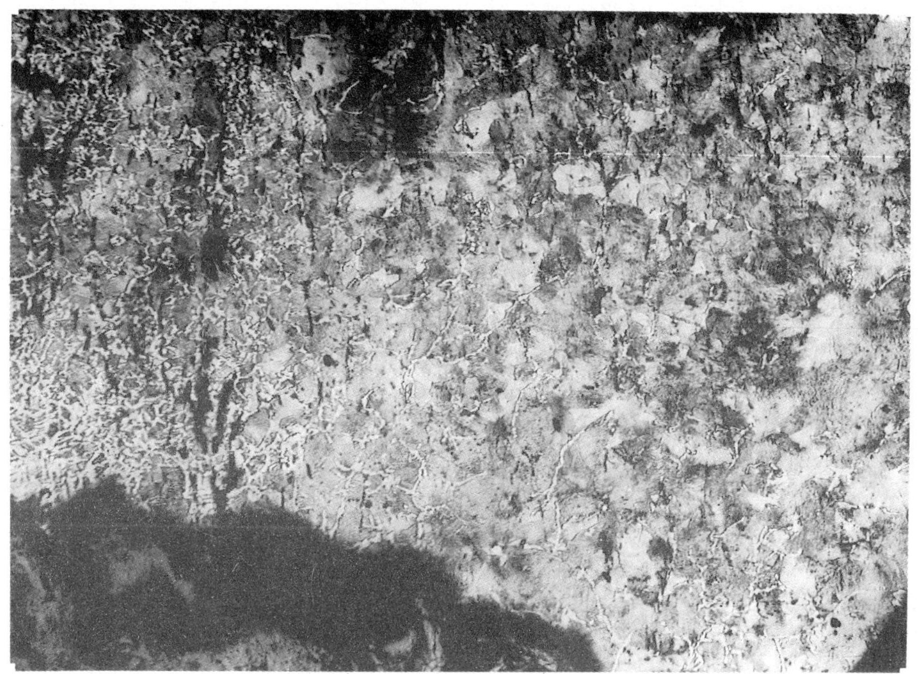

Pl. 11

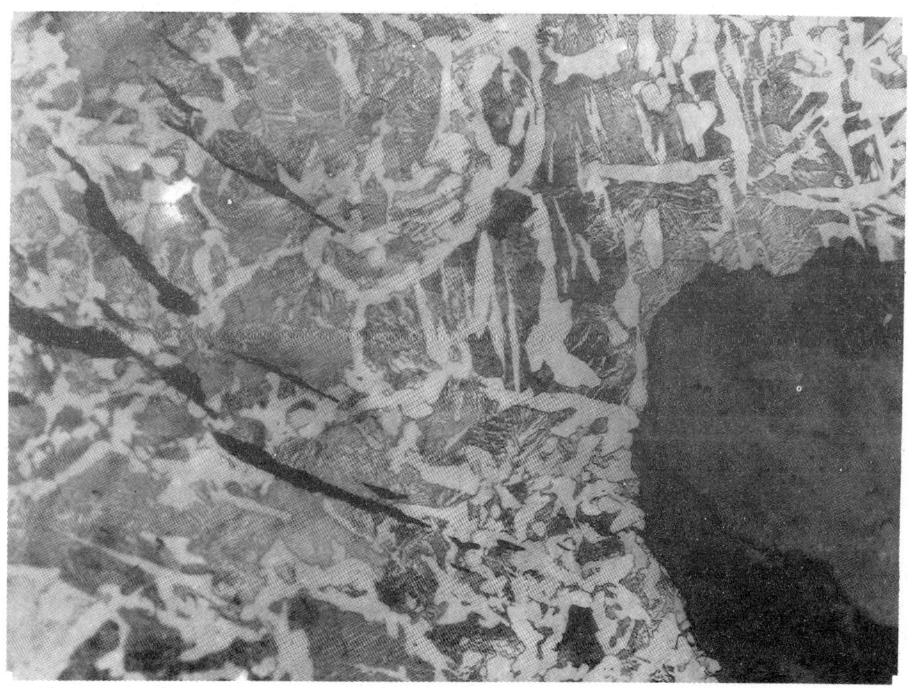

Pl. 12

Pl. 13

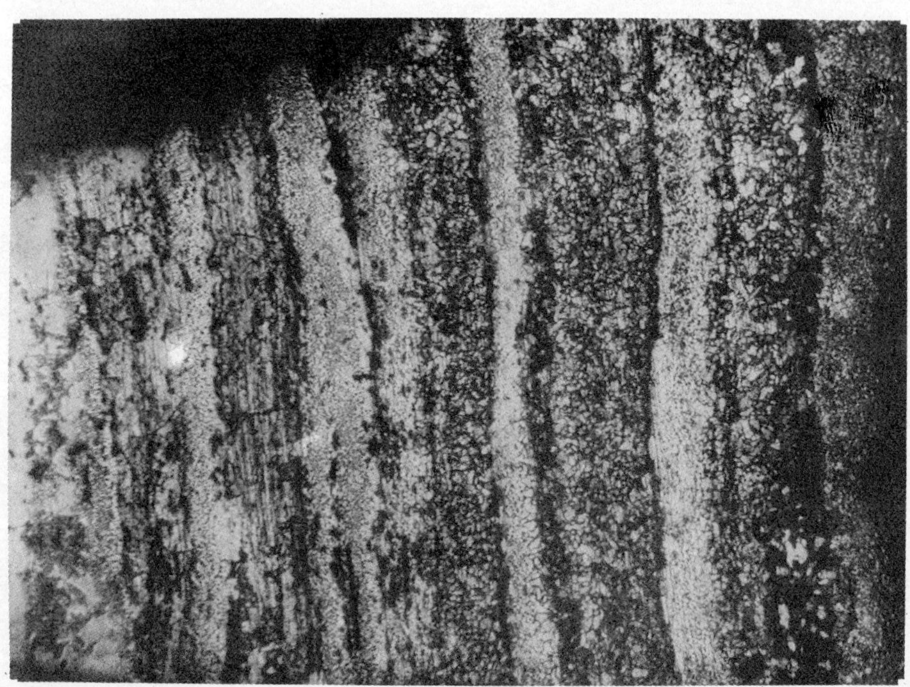

Pl. 14

Pl. 15

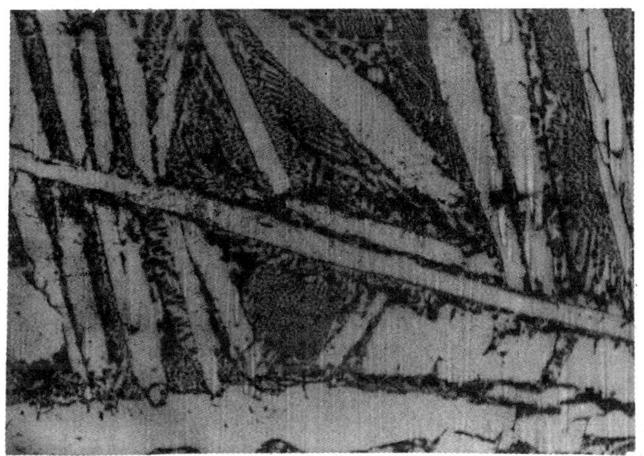

Pl. 16

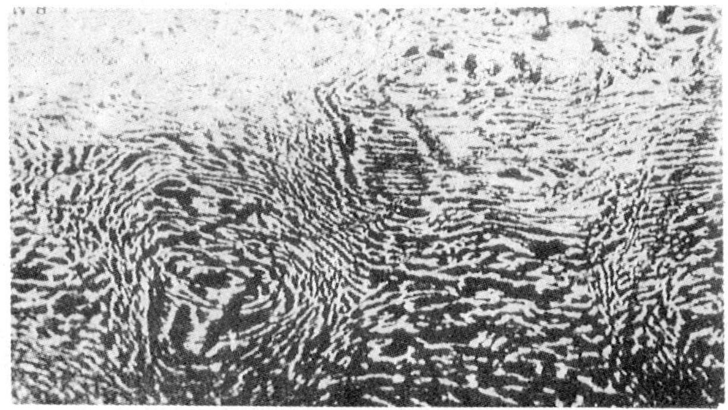

Pl. 17

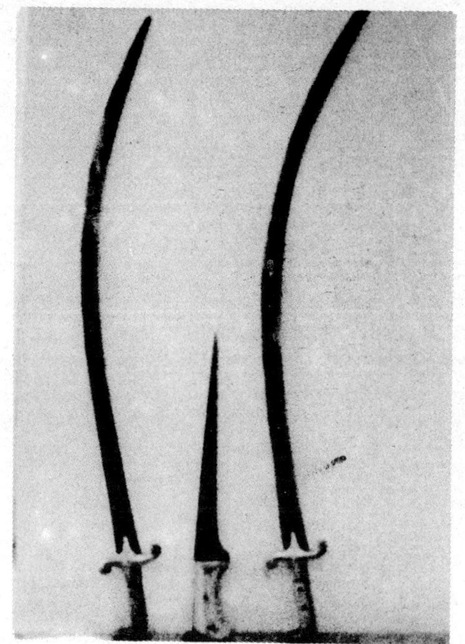

Pl. 18

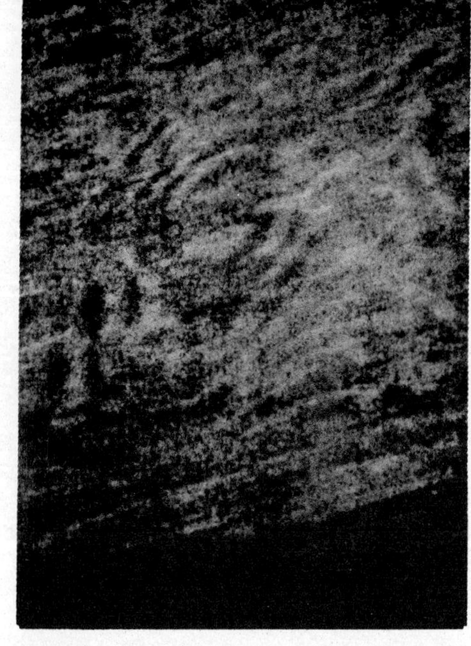

Pl. 19

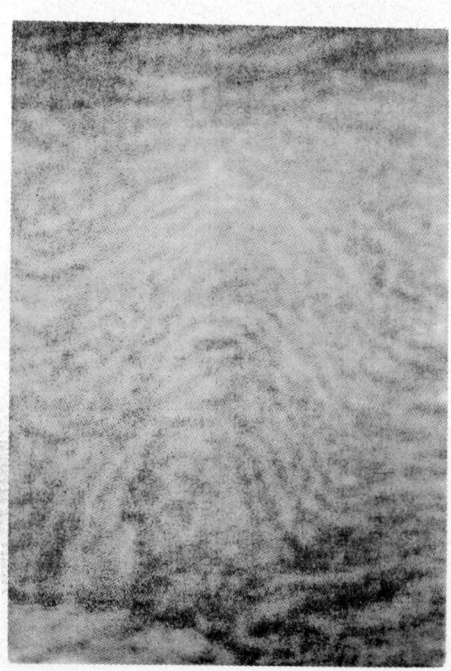

Pl. 20

Pl. 21

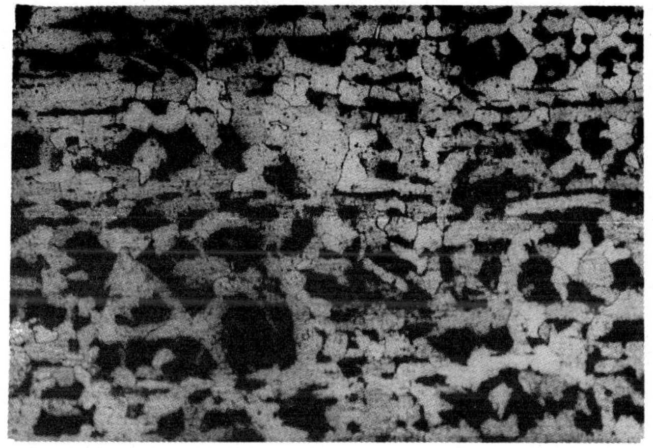

Pl. 22

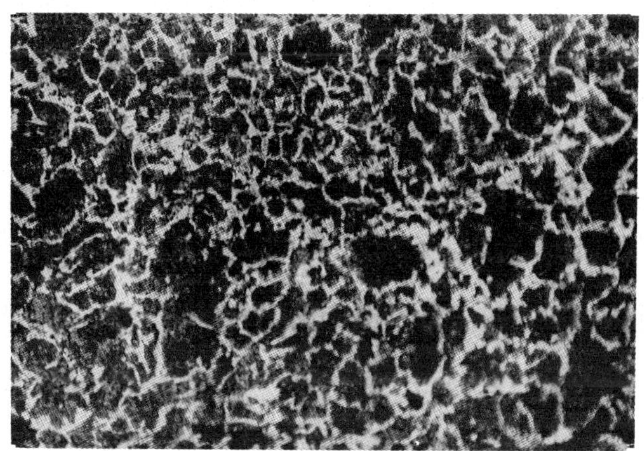

Pl. 23

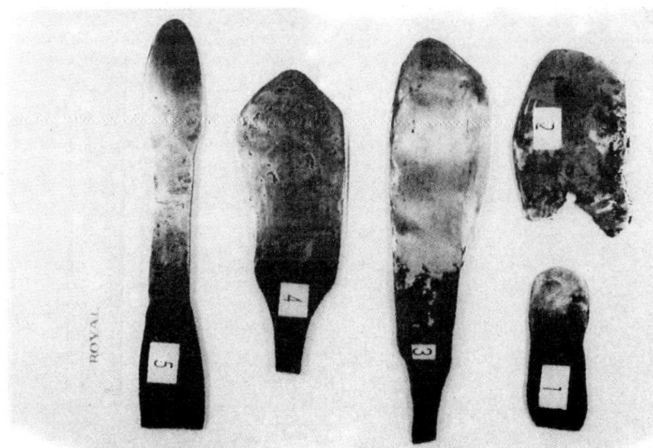

Pl. 24

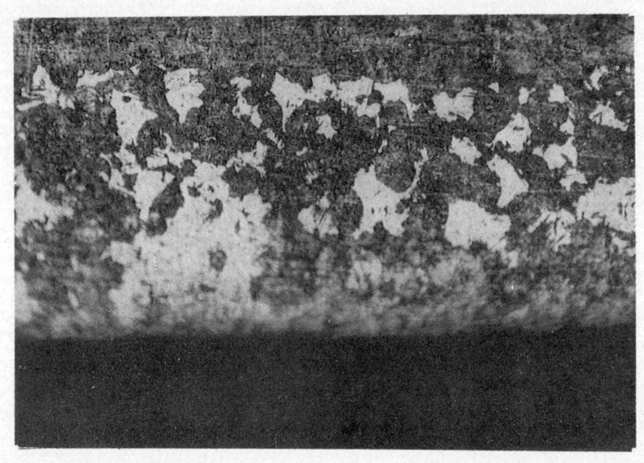

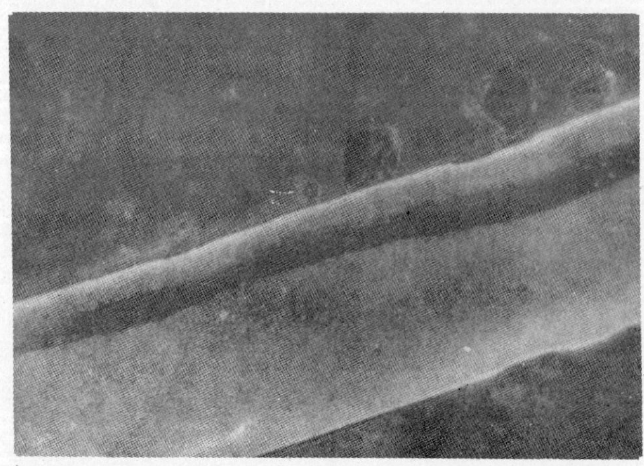

Pl. 25

Pl. 26

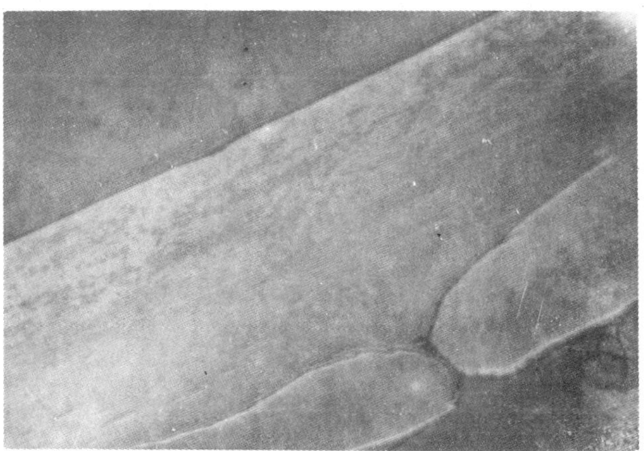

Pl. 27

Pl. 28

Pl. 29

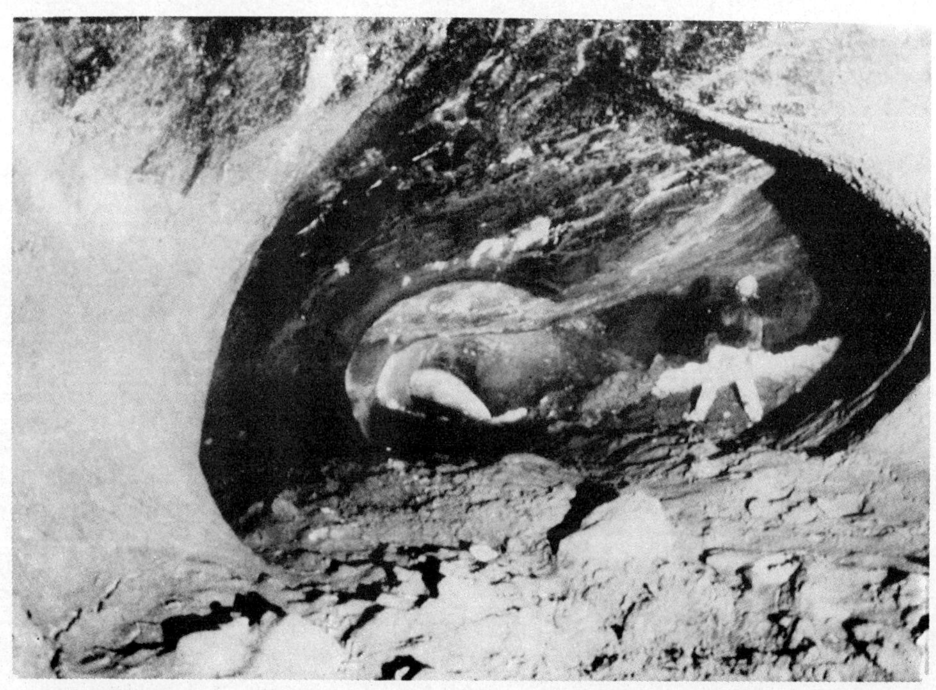

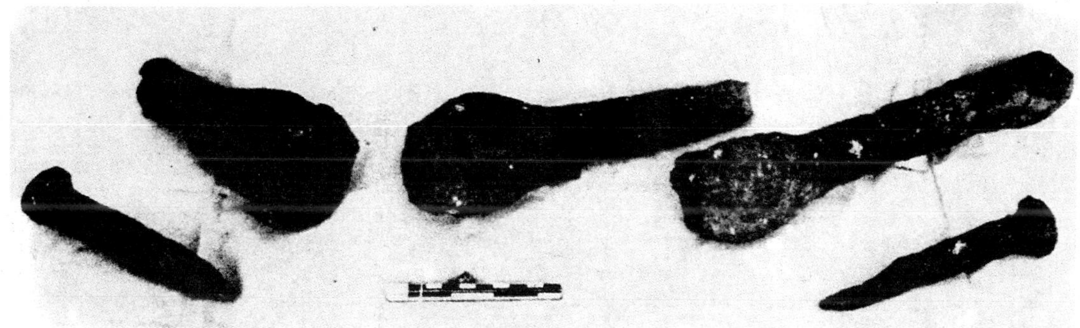

Pl. 31

Pl. 32

Pl. 33

Pl. 34

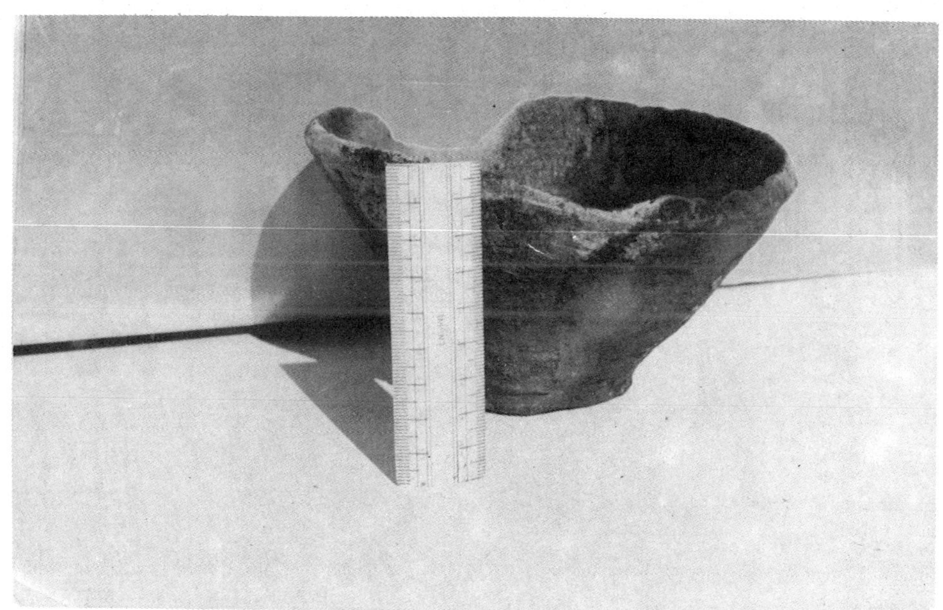

Pl. 35

Pl. 36

Pl. 37

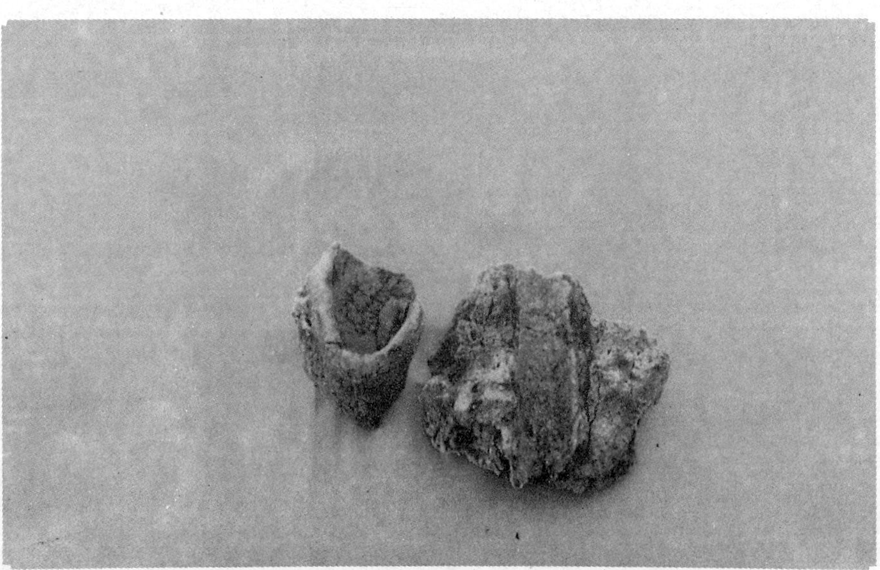

Pl. 38

l. 39

Pl. 40

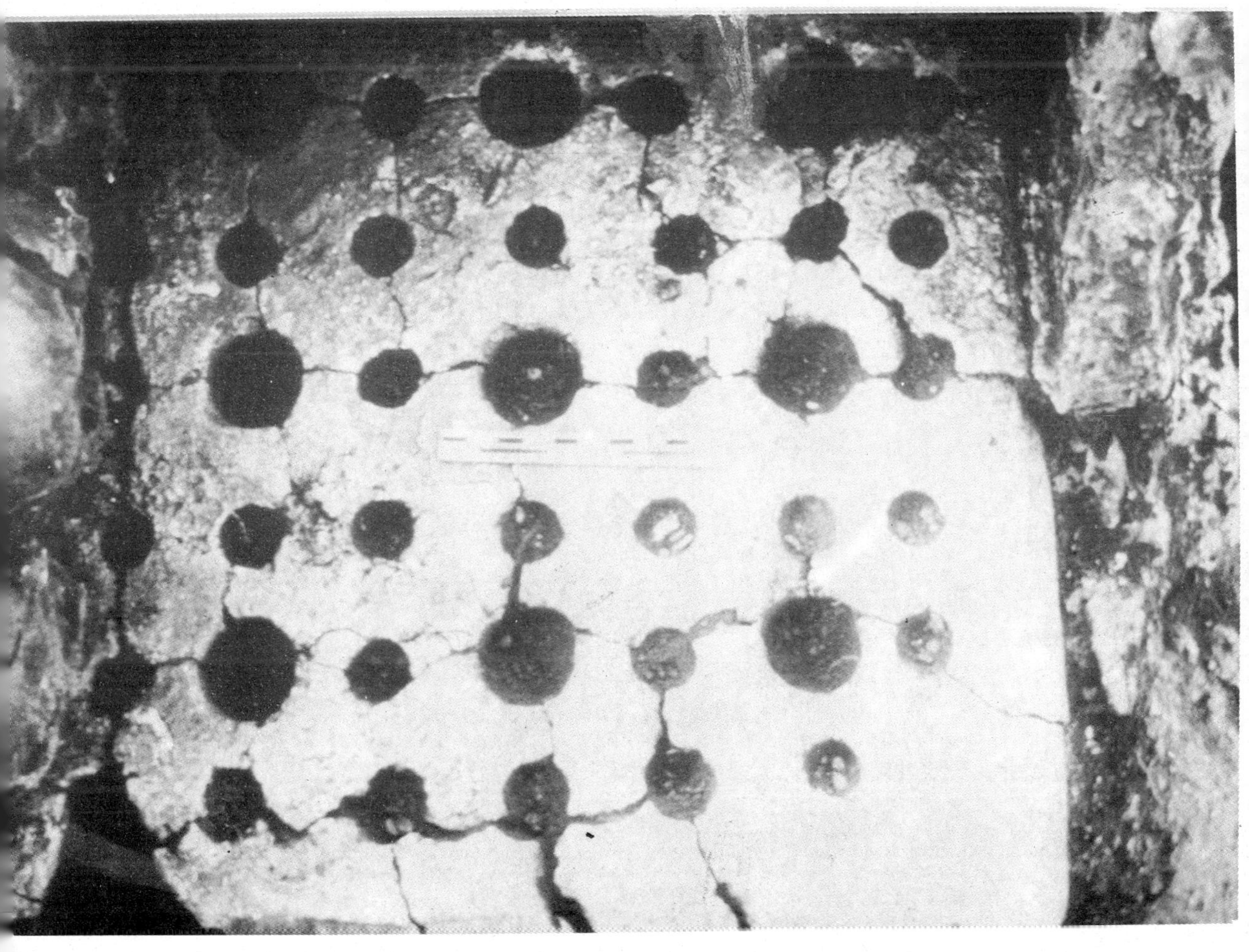

Pl. 41

Pl. 42a

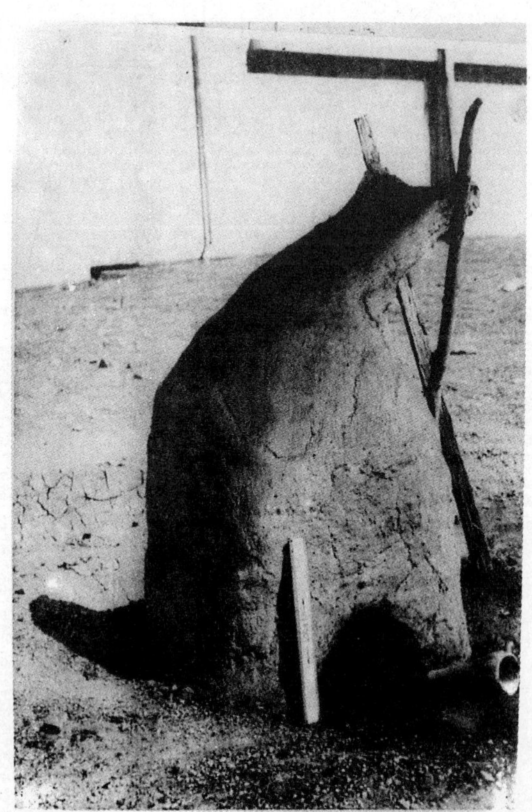

Pl. 42b

Pl. 43

Pl. 44

Pl. 45

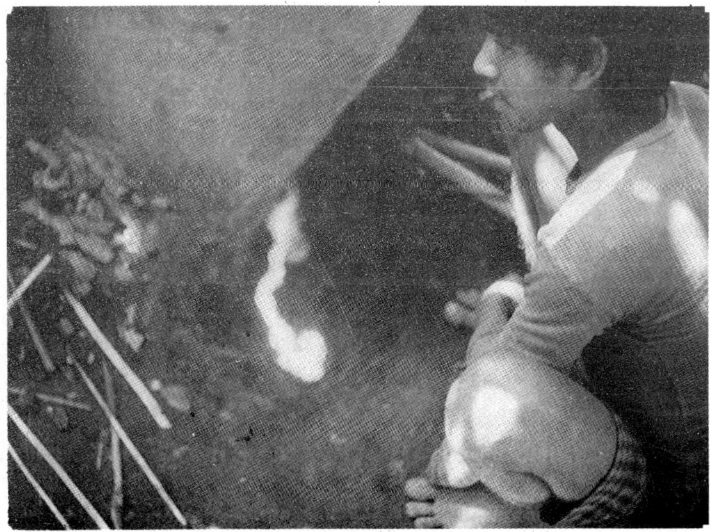

Pl. 46

Pl. 47

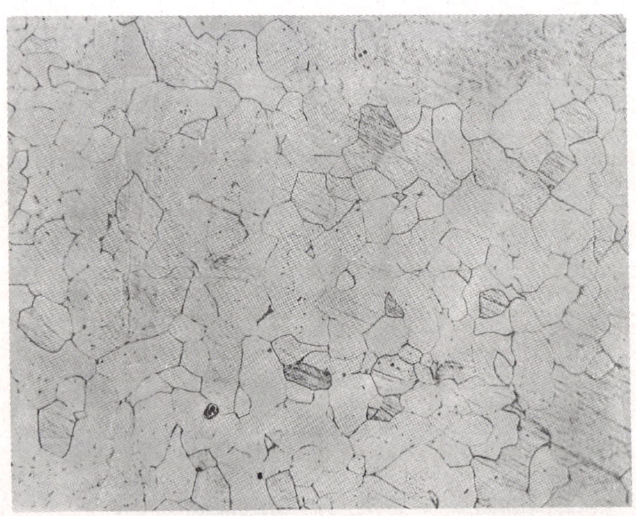

Pl. 48

Pl. 49

Pl. 50

Pl. 51

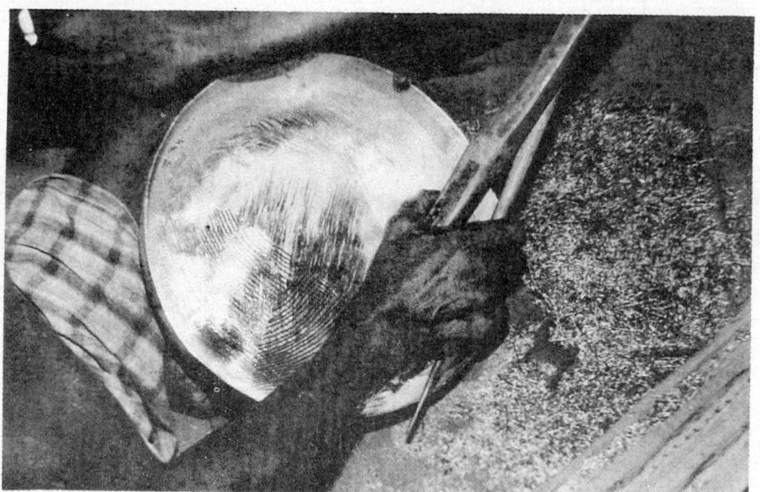

Pl. 52

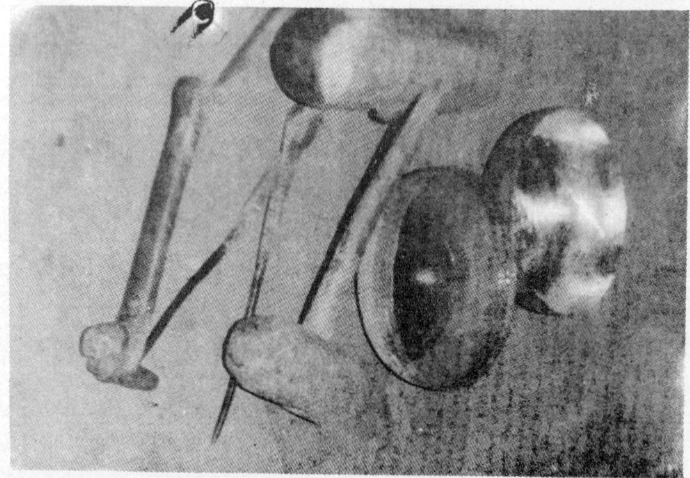

Pl. 53

Pl. 54

Pl. 55

Pl. 56

Pl. 58

Pl. 57

Pl. 59

Pl. 60

Pl. 61

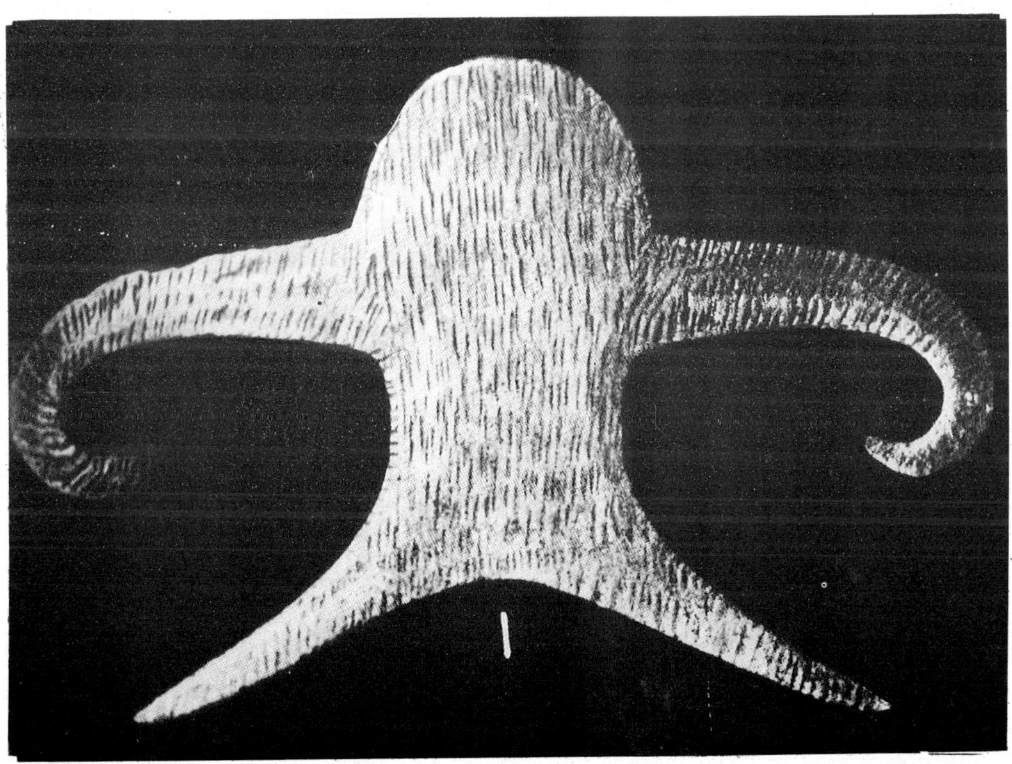

Pl. 62

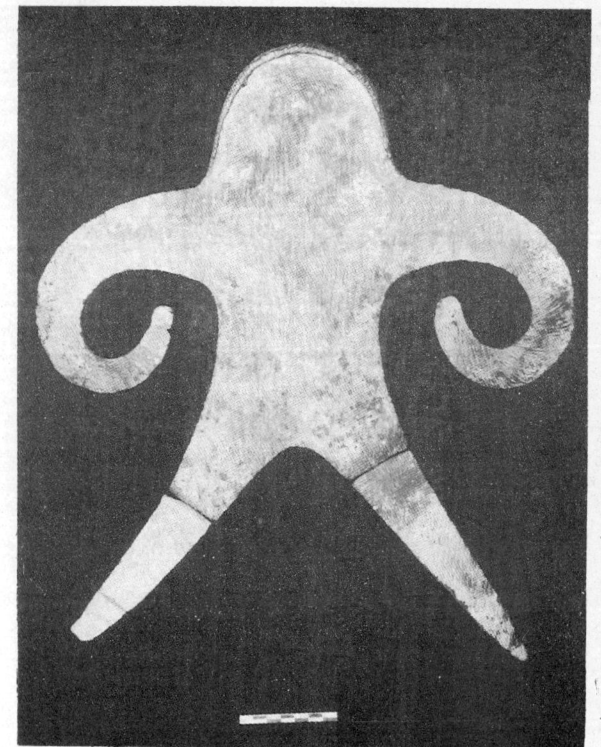

Pl. 63

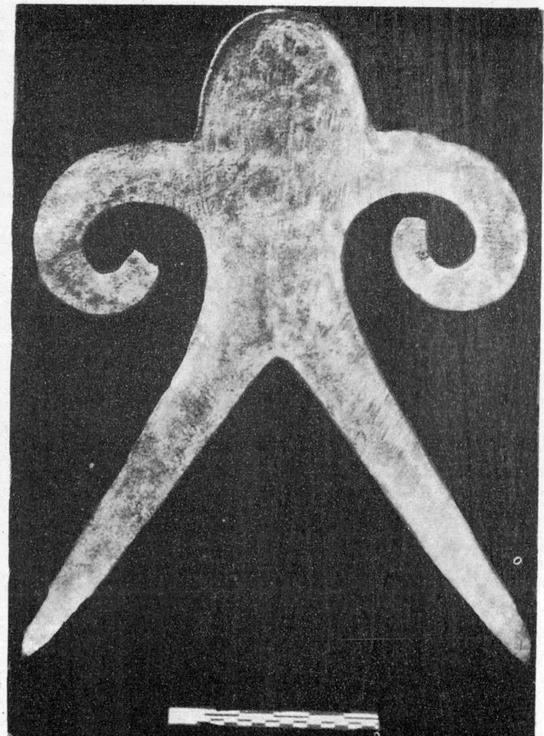

Pl. 64

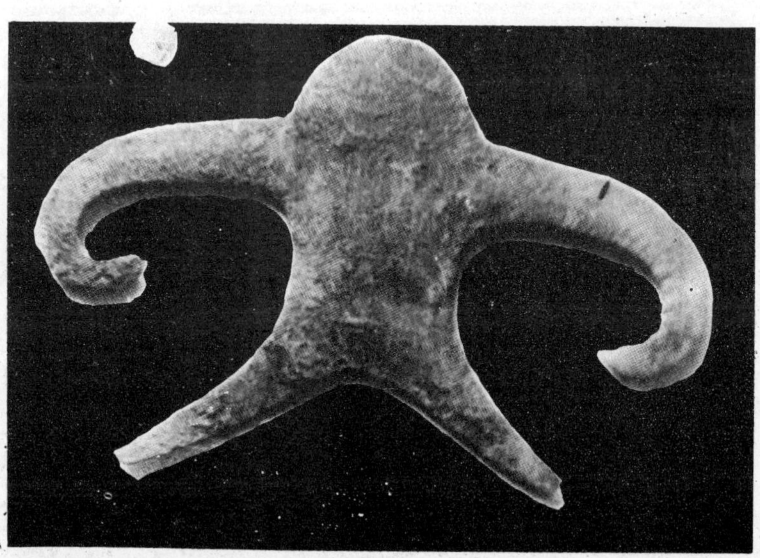

Pl. 65

Pl. 67

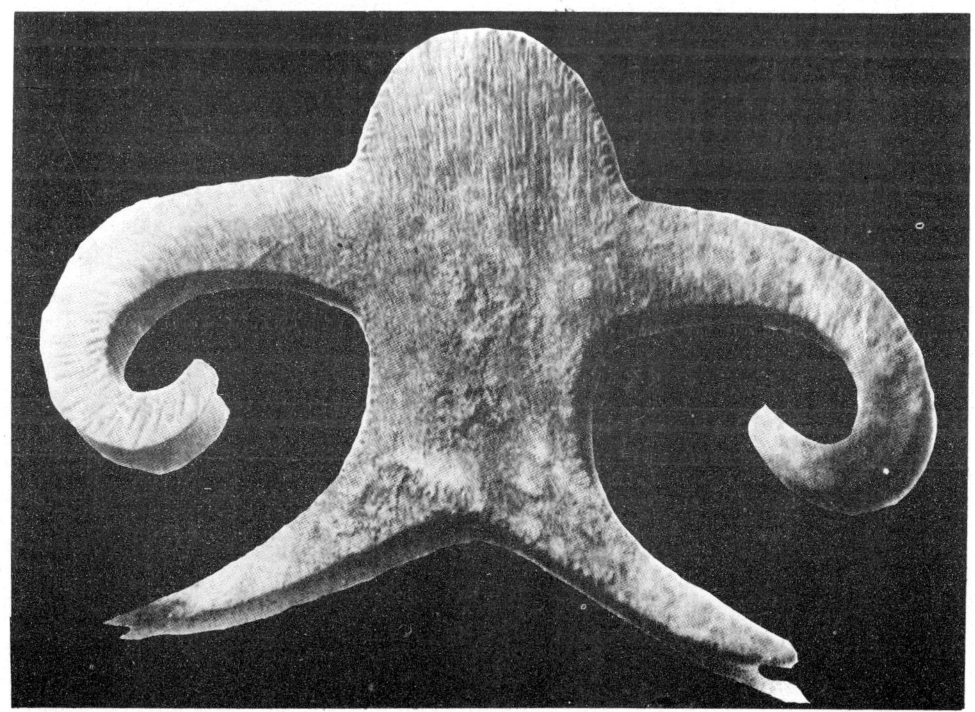

Pl. 66

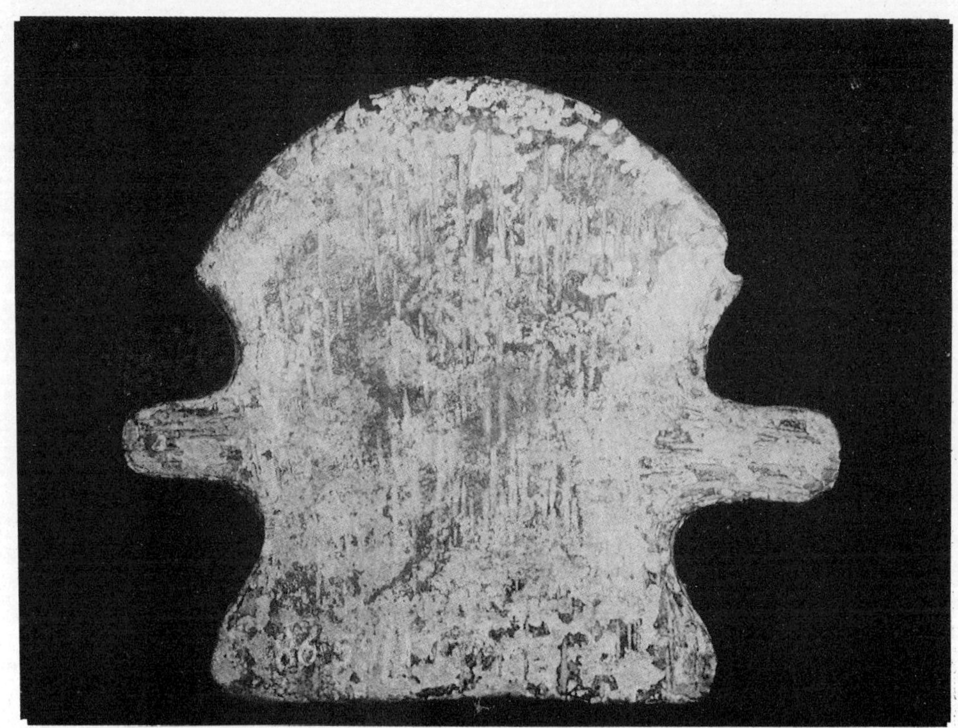
Pl. 68

Pl. 69